Lecture Notes in Computer Science

Edited by G. Goos, Karlsruhe and J. Hartmanis, Ithaca
Series: I.F.I.P. TC7 Optimization Conferences

4

5th Conference on Optimization Techniques

Part II

Edited by R. Conti and A. Ruberti

Springer-Verlag
Berlin · Heidelberg · New York 1973

Prof. Dr. R. Conti
Istituto di Matematica
"Ulisse Dini"
Università di Firenze
Viale Morgagni 67/A
I-50134 Firenze/Italia

Prof. Dr. Antonio Ruberti
Istituto di Automatica
Università di Roma
Via Eudossiana 18
I-00184 Roma/Italia

AMS Subject Classifications (1970): 68 A 20, 68 A 55, 90 A 15, 90 B 10, 90 B 20, 92-XX, 93 A XX

ISBN 3-540-06600-4 Springer-Verlag Berlin · Heidelberg · New York
ISBN 0-387-06600-4 Springer-Verlag New York · Heidelberg · Berlin

Offsetprinting and bookbinding: Julius Beltz, Hemsbach/Bergstr.

PREFACE

These Proceedings are based on the papers presented at the 5th IFIP Conference on Optimization Techniques held in Rome, May 7-11, 1973. The Conference was sponsored by the IFIP Technical Committee on Optimization (TC-7) and by the Consiglio Nazionale delle Ricerche (Italian National Research Council).

The Conference was devoted to recent advances in optimization techniques and their application to modelling, identification and control of large systems. Major emphasis of the Conference was on the most recent application areas, including: Environmental Systems, Socio-economic Systems, Biological Systems.

An interesting feature of the Conference was the participation of specialists both in control theory and in the field of application of systems engineering.

The Proceedings are divided into two volumes. In the first are collected the papers in which the methodological aspects are emphasized; in the second those dealing with various application areas.

The International Program Committee of the Conference consisted of: R. Conti, A. Ruberti (Italy) Chairmen, Fe de Veubeke (Belgium), E. Goto (Japan), W. J. Karplus (USA), J. L. Lions (France), G. Marchuk (USSR), C. Olech (Poland), L. S. Pontryagin (USSR), E. Rofman (Argentina), J. Stoer (FRG), J. H. Westcott (UK).

Previously published optimization conferences:

Colloquium on Methods of Optimization. Held in Novosibirsk/USSR, June 1968.
(Lecture Notes in Mathematics, Vol. 112)

Symposium on Optimization. Held in Nice, June 1969.
(Lecture Notes in Mathematics, Vol. 132)

Computing Methods in Optimization Problems. Held in San Remo, September 1968.
(Lecture Notes in Operation Research and Mathematical Economics, Vol. 14)

TABLE OF CONTENTS

URBAN AND SOCIETY SYSTEMS

COMPUTER AND COMMUNICATION NETWORKS

---

*paper not received

ENVIRONMENTAL SYSTEMS

ECONOMIC MODELS

*paper not received

Contents of Part I
(Lecture Notes in Computer Science, Vol. 3)

SYSTEM MODELLING AND IDENTIFICATION

DISTRIBUTED SYSTEMS

*paper not received

*paper not received

OPTIMAL CONTROL

STOCHASTIC CONTROL

MATHEMATICAL PROGRAMMING

NUMERICAL METHODS

*paper not received

SOME ASPECTS OF URBAN SYSTEMS OF RELEVANCE TO OPTIMISATION TECHNIQUES

- D. BAYLISS GREATER LONDON COUNCIL

1. Introduction

I would like to start by making clear what will shortly be obvious anyway - that I am not an expert in information science or optimisation techniques. I come here as an urban planner to try and explain the nature of some of the problems to be faced in practice, in a way which I hope is relevant to this important branch of technology and also to try and learn something of the possibilities for the application of optimisation techniques in tackling the difficult problems of planning and managing towns and cities.

My theme will be that the essential characteristics of urban systems are so complex that they raise conceptual measurement and computational problems of an order which preclude the possibility of building formal comprehensive optimisation models which are of value to the practising planner. There is, however, a role for sectoral optimisation models operated outside, but better, within a comprehensive analytical framework.

2. Types of System Parameters

As I have already said urban planning is distinguished from many other types of planning by its complexity but also it is characteristically confronted with large arrays of actual and potential conflicts and the resulting judicial role presents one of the most difficult challenges for formal optimisation techniques.

FIGURE 1

Policy Parameters in Urban Systems

Physical	Location
Social	Timing
Economic	Agencies?
Political?	

Structural change
Management regimes

Figure 1 shows some of the main parameters of interest to the urban policy maker. The physical, social and economic aspects of environmental change have to be considered, the way in which the structure of the city operates and might be developed under alternative management regimes, not simply, but differentiated in historic time and space.

Optimisation models typically operate within one of these sectors, some of the more adventurous linking two areas but rarely all three. Where the model deals with structural and management aspects or the dynamics of the system again they are usually confined to a single sector.

Another major problem is that of defining optimal conditions. Urban systems are typified by a multiplicity of client groups with conflicting and often undefined preferences. Thus simple objective functions are rarely meaningful except in single sector models with single client groups.

FIGURE 2

Objective Parameters in Urban Systems

Feasibility
Efficiency
Equity
Quality

Also it is rarely possible or proper to predetermine the weights to be attached to the criteria under these four heads (at least those which can be measured) or to the interests of different client groups as those weights are traditionally determined by political rather than technical processes.

3. An Analytical Framework

Having suggested that there are both fundamental and practical reasons why global optimisation models cannot be usefully developed for urban system the question must be faced as to how partial optimisation models should be most appropriately developed. A major danger in the use of partial optimisation models is that they attract too much weight to that part of the system with which they deal and therefore, unless their scope dominates the system, their exercise can be counter productive to good decision making. This danger can be lessended by establishing a comprehensive view of the structure of the system as a context for identifying the scope of relevance of sectoral optimisation.

Figure 3 outlines in simple diagrammatic form an analytical framework for an urban system. Paradigms of this kind can reveal the partiality of particular models and show linkages between individual sub-model. The value of basic frameworks of this kind can be increased by the development of behavioural models of the relationships they portray and a number of such models exist. Perhaps the best known is that for the transportation system - illustrated in Figures 4 and 5.

FIGURE 3

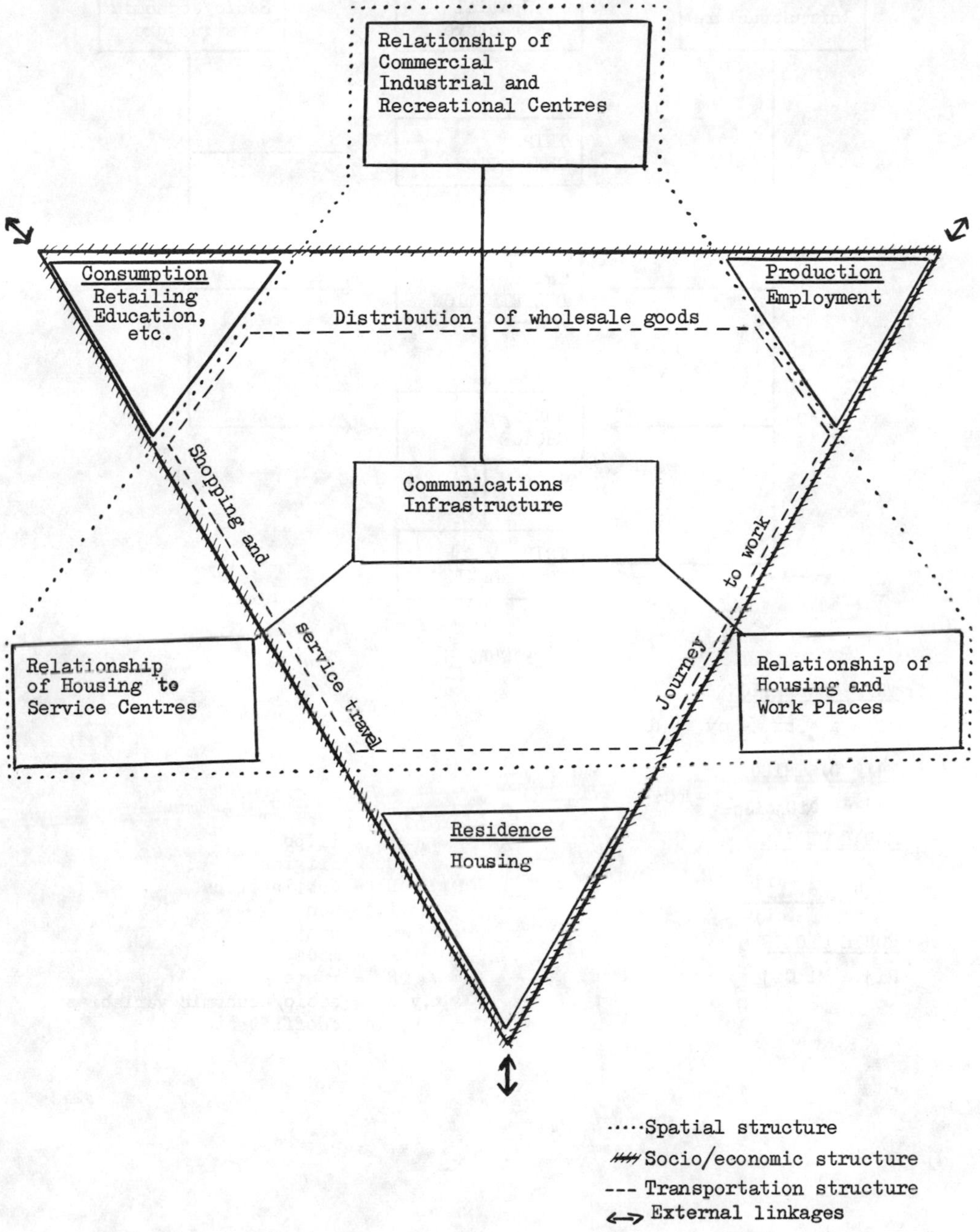
Relationship of Commercial Industrial and Recreational Centres
Consumption
Retailing Education, etc.
Production
Employment
Distribution of wholesale goods
Communications Infrastructure
Shopping and service travel
Journey to work
Relationship of Housing to Service Centres
Relationship of Housing and Work Places
Residence
Housing
·····Spatial structure
Socio/economic structure
--- Transportation structure
External linkages

FIGURE 4

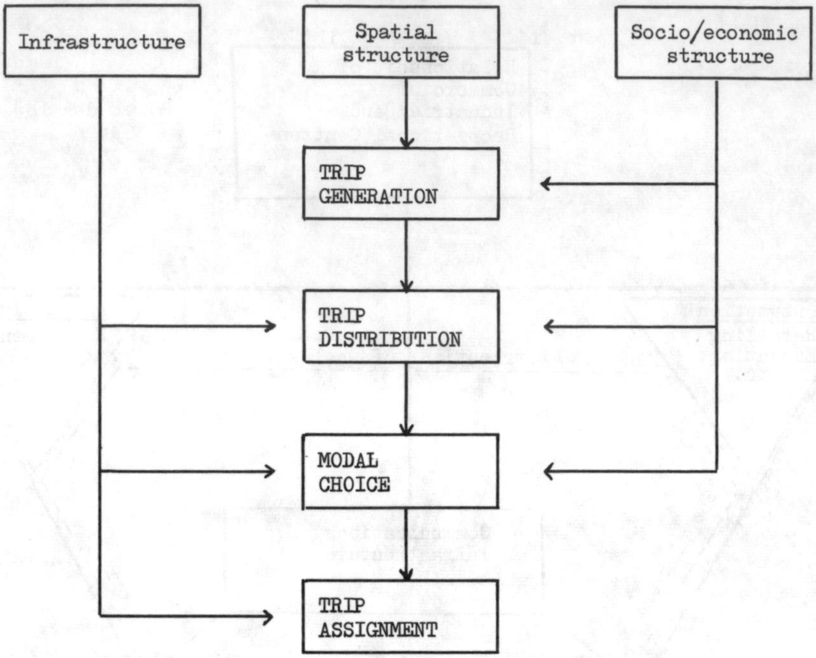

FIGURE 5

<u>TRIP GENERATION</u>

$Oi = a + bx_i + cy_i + dz_i$

<u>TRIP DISTRIBUTION</u>

$Tij = OiDjAiGje^{-\beta Cij}$

<u>MODAL CHOICE</u>

$$Tij^k = \frac{E^{-cij^k}}{E^{-cij^*}}$$

<u>ROUTE CHOICE</u>

$Rij = MinCij$

T = trips
O = origins
D = destinations
i,j = zones
C = cost
k = mode
R = route
x,y,z = socio/economic variables
a,b,c,d = coefficients

Models of this kind are often well developed and tested against extensive empirical evidence. In addition to specifying important interactions between other sub-models urban transport models can provide the basis for sectoral optimisation models. There are many examples of this but one of the most interesting was developed in the London Transportation Study in order to devise an optimum traffic restraint policy - Figure 6.

FIGURE 6

$$\alpha_1^a A'a + \alpha_1^b A'b + \alpha_1^c A'c + \dots\dots\dots \leq C_1$$

$$\dots\dots\dots + \alpha_x^n A'n \leq Cx$$

There being 'n' zones and 'x' links

Max $\sum_n A'_i$

Subject to $A'_i \leq Ai$

$A'i \geq kAi$ (Equity constraint)

A'_i = attractions allowed in zone 'i'

A_i = attractions desired in zone 'i'

C_j - capacity of link j

K = constant

α_j^i = number of trips generated on link 'j' by a unit attraction at zone

More extensive models exist which crudely embrace aspects of the transport, spatial and socio-economic characteristics of Urban Systems. Perhaps the best known is that developed by Lowry originally for the Pittsburg area. As can be seen from Figures 7 and 8 this is a combined spatial structure/transportation structure built around the journey to work/service journey elements of the transportation system. Simple optimisation can be based upon maximising choice or accessibility or minimising transport costs.

FIGURE 7

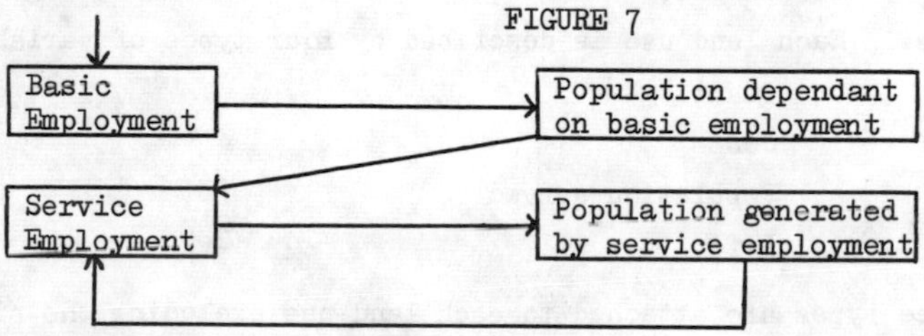

LOWRY MODEL - STRUCTURE

FIGURE 8

$$T_{ij} = A_i(E_b)_i P_j^a e^{-uc_{ij}}$$

$$\sum_j T_{ij} = (E_b)_i$$

$$(E_{bp})_j = \sum_i T_{ij}$$

Where

T_{ij} = Basic work trips between zones i and j

C_{ij} = Cost of travel between zones i and j

$(E_b)_i$ = No. of basic jobs in zone i

$(E_{bp})_j$ = No. of basic workers in zone j

a, U = constant (calibrated)

A_i = $1/\sum_j P_j^a e^{-uc_{ij}}$

P_j = 'population' of zone j

LOWRY MODEL - DISTRIBUTION OF EMPLOYEES IN BASIC EMPLOYMENT

4. An Alternative Approach

What I would like to do now is to go on to illustrate a slightly different approach which allows an interactive process between formal optimisation and informal optimisation which is more congruent with the nature of the urban decision making process. Thus non-controversial optimisation can be buried and treated exactly within the model and controversial optimisation is exposed and treated subjectively outside the model.

The model referred to is the General Urban Planning model currently being developed by Dr. Young in the Department of Planning and Transportation of the Greater London Council.

The number of variables and constraints makes it necessary to deal with the urban area by dividing into nine sections for which individual sub-models are operated a tenth sub-model reconciles the results of the sectoral sub-models.

5. The Sector Sub-models

The model deals in terms of land uses (see Figure 9) and their states through four consecutive time periods. Each land use is described by four types of variables:

Quantity
Cost
Population served
Life

Also constraints of three types are attached to each land use excluding the simple logical constraints necessary to programming models; these are

Compulsory   Arbitrary   Balancing

FIGURE 9

LAND USES

(a) housing
(b) hospitals
(c) schools
(d) factories
(e) offices
(f) shops
(g) public open space
(h) transport
(i) vacant
(j) others

6. Objectives

Objectives are set for housing density cost, land and resource utilisation, etc., and the model run to maximise these subject to the constraints, the nature and value of the arbitrary constraints and the objective function are varied to expose conflicts, illuminate the scales of trade between objective elements as well as discerning composite optimum conditions using a linear programme.

Although limited by data availability and the simplicity of assumptions employed this model is yielding valuable insights into the relationships between sectoral standards and objectives. What is more the treatment of variables in both formal and subjective modes creates a richness and sympathy which minimises the probability of alienating the practicing urban planner.

7. Conclusions

I am afraid that my comments have been too brief and patchy to do full justice to the nature of the problems of urban systems. However, I hope they will help a little to clarify my closing summary comments:-

(1) The problems of optimisation in urban systems are created by:-
   (a) The need, in real life, to treat a range of complex sub-systems with quite different theoretical bases at one and the same time.
   (b) The inherent lack of concensus between client groups and agencies as to what the objectives of urban planning, both in general and in particular, are.
   (c) The paucity and complexity of the models of system behaviour on which good optimisation models must be founded.
   (d) The need to treat the system in a high disaggregated form to get anywhere near to reality - this produces data and computational problems of a high order.

(2) Optimisation techniques have been relatively little used in urban planning in general, although important applications have been made in some narrowly defined areas. This is largely because of a lack of sympathy between system scientists and urban planners brought about by a failure by systems scientists on the one hand to set their work in a context seen to be relevant to urban planners and on the other an unreasonable mistrust by many urban planners of quantitative techniques.

(3) We are now at a stage where it is possible to construct general models of urban systems which have a reasonable degree of correspondence with the concepts of urban planners and one sufficiently formally defined to allow meaningful optimisation within and in some instances between individual sectors. However the value of sectoral optimisation models will be enhanced by being set in an overall analytical framework. Beware excessively behavioural assumptions as these have prevented much of the interesting work in regional science and urban econometrics from finding its way into urban planning practice.

(4) Finally both the challenge and opportunities (see Figures 10 and 11) are greater than ever before. The growing problems of cities are there for all to see and yet this is one of the aspects of modern society where modern problem solving methods have made least impact. Even if systems analysis and optimisation techniques do no more than provide a formal arbitrator in the plan development process they will have done urban planning a great service.

FIGURE 10

CHALLENGE

(1) Cities the cradle, haven and tomb? of civilisation.

(2) Urban problems of an unprecedented kind and degree.

(3) Modern problem solving techniques little applied.

FIGURE 11

OPPORTUNITIES

(1) Increased availability of the right kind of information.

(2) Better understanding and behavioural models.

(3) Continuous improvement in problem solving techniques.

(4) Developing sympathy amongst urban planners.

SELECTION OF OPTIMAL INDUSTRIAL CLUSTERS FOR REGIONAL DEVELOPMENT

Stan Czamanski[1]

The Research Problem

The importance of a new industry to a depressed region resides not only in the volume of new employment and income which it provides but very often primarily in its indirect or multiplier effects. A common feature of depressed regions is the general weakness of multiplier effects capable of being generated in their economies. It is due mainly to the patchiness of interindustry linkages. The absence of substantial indirect effects which ordinarily accompany new investments constitutes one of the greatest obstacles to efforts aimed at invigorating the economies of underdeveloped regions. The weakness of the multiplier effects makes remedial programs expensive and the resulting progress slow.[2]

The problem can also be viewed in a slightly different way. The study of industrial location patterns brings to the fore the phenomenon of significant progressive clustering of economic activities in a small number of urban-industrial agglomerations. This remarkable feature of locational preferences of industries is increasingly exploited for fostering regional progress by promoting the emergence of growth poles or more generally by furthering spatially imbalanced development. Explanations of the emergence of spatial concentrations of industries revolve around the extent to which geographical proximity between certain classes of plants confers significant advantages. The benefits attending the reduction in the friction of space may be due to savings on transportation costs, especially in the case of products which are weight-losing, transported in hot state, or capable of being transferred over a short distance without packaging, by pipes, belts, or conveyors. In other cases, especially those in which storage and related interest costs are substantial, and advanced planning difficult, the advantages associated with spatial proximity may be due to savings in transportation time rather than in transportation costs. Many industries are attracted to existing clusters because of the importance of human face-to-face contacts, of presence of external services, of an existing pool of trained labor, or more generally because of the possibility of realizing savings either in investments or production costs.

In a depressed, open regional economy a major breakthrough can result only from

---

[1]The author is Professor of City and Regional Planning at Cornell University, Ithaca, N. Y., U. S. A., and Research Associate at the Institute of Public Affairs, Dalhousie University, Halifax, N. S., Canada. The financial support of the Economic Development Administration, U. S. Department of Commerce, and of the National Science Foundation is gratefully acknowledged, as is the research assistance of Abby J. Cohen and Stephen B. Ellis.

[2]For some early studies bearing on this problem see [6, 5, 4]; and for later work. [3].

the introduction of new productive activities. Hence the empirical question frequently faced by regional planners or political decision makers is the selection of industries to be supported or attracted into the region. Recognition of the importance of the multiplier effects for both short-run and long-run development strategies adds a new criterion to those customarily used, such as output-labor, capital-labor, or capital-output ratios, relative rates of growth of industries considered, or the extent of required outlays on infrastructure.

Increased interest in the theory of imbalanced growth and in the growth poles hypothesis has led to the formulation more recently of the idea of promoting clusters of related industries rather than scattered individual plants.[3] The main reason seems to be a desire to realize some of the external economies which are internalized in a cluster of related activities. Moreover, it is assumed, so far without any empirical proof, that the growth of multiplier effects in a region experiencing an influx of new activities is more than proportional to the growth of its economy. The hypothesis is based on the plausible idea that the introduction of new industries progressively reduces leakages and reinforces the indirect impact of new activities until a point is reached when in a group of related industries linked by flows of goods and services the multiplier effects become significantly stronger, signalling a qualitative breakthrough.

It is, however, not immediately obvious which industries form groupings subject to external economies. Traditionally, industrial complexes have been identified on the basis of studies of either technological processes, or of spatial associations. Both methods present some distinct advantages but also some rather obvious difficulties.[4] A different approach was followed in the research reported here, with major emphasis focussed on subsystems of industries capable of generating external economies and identified on the basis of flows of goods and services connecting them either directly or often only indirectly.

2. Identification of Clusters of Industries

The main data basis for the study was the United States 478 x 478 input-output table for 1963 reduced to size 172 x 172.[5] The grouping of the matrix was carried out with the help of a program described elsewhere.[6]

---

[3]For an overview see [13]; and for some of the original points of view, [16] and [2].

[4]The various approaches are presented in [14; 8; 10; 11; 7; 15; 17; and 1].

[5]The reduction was necessary because such detailed data could not be reconciled with those available for the study of spatial patterns for which SMSA's were the basic units. In addition the operation of such a large table presented numerous computing difficulties involving very high costs for computer time which did not appear to be justifiable.

[6]For details see [12].

Clusters of industries were defined as groups of sectors with relatively stronger ties among themselves than with the rest of the economy. Three different but complementary methods were developed for the purpose of identifying clusters according to this definition. The first two methods were based on network analysis, while the third applied principal components analysis. The first method used as the criterion for including an industry into a cluster a single strongest link between any of the industries already in the cluster and any of the remaining industries. The criterion of the second method was the strength of the links with all the industries already in the cluster. The third method used as criterion for including an industry in a cluster similarity between its total profile of suppliers and customers (not only those belonging to a cluster). The research started with an examination of the relative importance of flows between pairs of industries. The following four coefficients derived from the input-output flow table described the relative importance of the links, either for the supplying or for the receiving sector:

$$a_{ij} = \frac{x_{ij}}{\sum_i x_{ij}} \; ; \quad a_{ji} = \frac{x_{ji}}{\sum_j x_{ji}} \; ;$$

$$b_{ij} = \frac{x_{ij}}{\sum_j x_{ij}} \; ; \quad b_{ji} = \frac{x_{ji}}{\sum_i x_{ji}} \; ;$$

where

x_{ij} = yearly flow between industry i and j.

The starting point of the first method was a triangular E matrix, the elements e_{ij} of which were formed by defining

$e_{ij} = \max(a_{ij}, a_{ji}, b_{ij}, b_{ji})$ for $i > j$, and

$e_{ij} = 0$ for $i \leq j$.

The model made use of an e_i column vector ($i = 1....172$) for any of the 172 industries (all were used in turn in the starting position). The entries in this column vector were ranked by interchanging rows and columns in the triangular matrix E. The second entry in the column vector was then the strongest link that the original industry (the first entry) had with any other sector in the economy. This second industry was represented by the adjoining column vector.

Next, the entries in the second vector were ordered and the top entries in both vectors were examined for the strongest link with any other sector in the economy. In the first variant the process was repeated until the mean value of the entries among the industries in the cluster, or $\sum_i \sum_j e_{ij}/n$ for $i, j = 1...n$, began to decline, where n is the number of industries in the cluster.

The second method also makes use of the triangular E matrix defined above, and follows a similar procedure of ranking industries by interchanging rows and columns. The ranking is done, however, on the basis of the sum, for all industries already in the cluster, of the links to the other sectors. This method employed the same stopping criterion that was used in the first method.

Not unexpectedly, the clusters identified by the second method were larger than those identified by the first method. The second method included industries which have links to several industries in the cluster but no strong links to any of the industries in the cluster.

Both methods had a tendency to become "side-tracked" whenever industries being added to the new cluster had stronger ties with the industries of another cluster. In such cases the program kept adding to the cluster being formed all members of the tight cluster already identified. The "side-tracking" would limit the number of clusters identified to a small number of the strongest groupings. This had been expected since the program, while not bent upon finding the maximum maximorum, clearly tries to find the strongest groupings by including industries with links represented by the highest e_{ij} values. Once industries belonging to a previously identified tight cluster were included in the cluster being formed, and the mean value of links began to decline, the program stopped without revealing the weaker grouping of possibly great interest.

In order to overcome this difficulty, the preliminary clusters defined by the first and second methods were supplemented by including all industries with links (e_{ij} coefficients) greater than .200. This additional procedure was carried out with the help of a computer program which created for each of the 172 industries lists of all significant e_{ij} coefficients in decreasing order. Hence, to be included in the cluster an industry had to meet at least one of the following criteria:

(a) More than two-tenths (.200) of its output was absorbed by one of the industries of the preliminary cluster,

(b) More than two-tenths (.200) of the output of one of the industries in the preliminary cluster was absorbed by the industry examined,

(c) More than two-tenths (.200) of inputs of the industry examined came from one of the industries in the preliminary cluster, or

(d) More than two-tenths (.200) of the inputs of one of the industries in the preliminary cluster came from the industry examined.

The analytically more interesting indirect links among industries were identified with the help of the third method, described elsewhere.[7] It also started with the 172 x 172 input-output flow matrix, from which an n x 4n matrix of zero order correlation coefficients was derived.

$$r = \left[r(a_{ik}.a_{il}) \;\middle|\; r(b_{ki}.b_{li}) \;\middle|\; r(a_{ik}.b_{li}) \;\middle|\; r(b_{ki}.a_{il}) \right] ;$$

Next an n x n covariance matrix was formed in terms of derivations from the mean values.

$$K = E \left[(r - \bar{r}) \quad (r - \bar{r})^T \right] ;$$

[7]For details see [9].

In order to identify, from the set of all industries, the subgroup belonging to a cluster, an iterative process was applied, eliminating all industries having a null column or a null row vector. The relative strength of the links binding the remaining industries together was assessed with the help of eigenvalues of the R matrix. The ratios of the characteristic roots to the trace of the R matrix define an Index of Association

$$C_n = \frac{\lambda_n}{\text{tr } R} \times 100$$

where

λ = eigenvalue, or characteristic root.

This provided an aggregate measure of the strength of the ties connecting the industries remaining in the R matrix - each C_i indicating the existence of an industrial cluster. The indirect links revealed by the third method were analytically highly significant since two industries k and l may be members of an industrial complex in the absence of direct flows between them. The three methods, using different criteria, yielded results fully consistent with one another when applied to the Washington State input-output matrix and to the United States 1963 matrix. The second method yielded larger but less closely linked groupings of industries than the first, while the application of the third method resulted in still larger and more diffuse clusters. All three were, however, consistent in their ranking of industries.

3. Characteristics of Industrial Groupings

Seventeen clusters were identified in the 1963 U. S. economy. Two groups of industries which were too small and too weakly linked to be defined as clusters are also of some interest. Characteristics of the seventeen clusters are listed in the summary tables.

The first and most obvious classification of the seventeen clusters is by manufacturing or service, according to the nature of industries of which they are composed. There are few service sectors in the predominently manufacturing clusters, and vice versa. The general weakness, or rather relative unimportance, of links between manufacturing and service sectors was unexpected. It runs counter to any notions of complementarity and may have significant implications for regional development theories and strategies.

The seventeen clusters differ greatly, in terms of the number of industries included, and in terms of the total size of the cluster as measured by value added. The two smallest clusters, Recreation and Government, are each composed of only eight sectors, while the largest cluster, Construction, contains forty-two sectors. More striking is the range in terms of value added for each cluster. The two largest clusters, in terms of output, are Real Estate and Construction, with 171.6 and 155.9 billions of dollars of value added respectively. The smallest is the Services cluster, which has a value added figure of 25.0 billion dollars, Table 1.

Examination of the identified clusters revealed that the component industries widely differed in their relative importance to the cluster as a whole. While the removal of some of the sectors would have little effect upon the structure of the cluster, in many clusters the exclusion of a single industry would lead to disintegration. These important sectors were called central industries and were defined as having at least four links to other industries in the cluster. The centrality of an industry may be measured not only by its degree (number of links), but also by the strength of the links.

TABLE 1

SIZE OF CLUSTERS

	Cluster	Number of Industries	Size in $000 of Value Added	Percentage of GNP*
1	Foods & Agricultural Products	21	54,823,590	9.39
2	Construction	42	155,886,704	26.69
3	Textiles	12	40,527,318	6.94
4	Wood & Wood Products	15	45,349,988	7.76
5	Paper & Printing	15	34,321,747	5.87
6	Petrochemicals	22	71,411,137	12.23
7	Petroleum	10	38,632,704	6.61
8	Leather Products	13	56,753,656	9.71
9	Iron & Steel	23	60,501,678	10.36
10	Nonferrous metals	10	28,503,745	4.88
11	Communications & Electronics	14	47,534,129	8.14
12	Automotive	18	55,240,961	9.46
13	Real Estate	23	171,612,897	29.39
14	Services	12	24,973,159	4.27
15	Recreation	8	50,034,347	8.56
16	Medical Services	11	81,237,289	13.91
17	Government	8	34,136,089	5.84

*Percentages of GNP are not additive because many sectors are members of more than one cluster.

Each of the seventeen clusters contained at least one central industry; some clusters had two or three. It may be noted that nearly all of the central industries appear to possess some special technological significance in their respective clusters. This is a reflection of their position in the sequence of processing operations. Interestingly, however, central industries are not always the largest ones in a cluster. In only ten of the seventeen clusters were the central industries the leaders in terms of output or value added, and even then they were not significantly larger than some other industries.

The importance of a central industry can be measured in a number of ways, as shown in Table 2. The measures used were size of the industry, the number of links to other industries (degree), the percentage of all links to the central industry, and the relative average strength of links.

A simple way of classifying the clusters would be to distinguish between single and multi-centered clusters. Of the seventeen clusters considered eight groups had

TABLE 2

CHARACTERISTICS OF CLUSTERS

Cluster Number	Name of Cluster	Central Industries	Size of C.I. in $1,000 Value Added	Degree	Percentage of Links to Central Industry	Average Strength of All Links	Average Strength of Links to Central Industry	Ratio 8 ÷ 7
(1)	(2)	(3)	(4)	(5)	(6)	(7)	(8)	(9)
1	Foods & Agricultural Products	Agriculture & Related Services (1)	22,104	12	54.5	.402	.451	1.121
2	Construction	Nonmetallic Mineral Mining (3)	1,633	5	10.6	.461	.288	0.625
		Contract Construction (10)	34,830	34	72.3	.461	.524	1.136
		Hydraulic Cement (84)	786	4	8.5	.461	.373	0.809
3	Textiles	Floor Covering Mills (13)	395	4	23.5	.383	.368	0.961
		Apparel (14)	6,679	4	23.5	.383	.559	1.460
		Fabric & Yarn Mills (30)	3,540	8	47.0	.383	.358	0.934
4	Wood & Wood Products	Logging (35)	520	6	35.3	.386	.370	0.959
		Sawmills (36)	1,573	6	35.3	.386	.385	0.997
5	Paper & Printing	Commercial Printing (46)	2,961	4	25.0	.431	.453	1.051
		Paper Mills (48)	1,857	6	37.5	.431	.461	1.070
6	Petrochemicals	Basic Chemicals (60)	6,171	7	29.2	.369	.345	0.934
		Fibers, Plastics, Rubbers (61)	2,865	8	33.3	.369	.348	0.943
		Tires & Inner Tubes (70)	1,322	4	16.7	.369	.393	1.065
7	Petroleum	Petroleum Refining (58)	3,358	4	40.0	.560	.561	1.001
8	Leather Products	Leather Tanning, Finishing (75)	273	6	37.5	.454	.455	1.002
		Boot, Shoe Cut Stock (77)	91	4	25.0	.454	.545	1.200
		Leather Goods, n.e.c. (82)	33	4	25.0	.454	.343	0.756
9	Iron & Steel	Steel Rolling, Finishing (90)	8,617	16	72.7	.417	.405	0.971
10	Nonferrous Metals	Primary Nonferrous Metals (92)	1,013	6	54.5	.371	.475	1.280
		Nonferrous Rolling & Drawing (94)	2,128	5	45.5	.371	.312	0.841
11	Communications & Electronics	Communication Equipment (120)	5,341	4	25.0	.391	.427	1.092
		Aircraft & Parts (124)	7,867	5	31.3	.391	.474	1.212
12	Automotive	Motor Vehicles & Equipment (123)	12,781	15	88.2	.280	.280	1.000
13	Real Estate	Real Estate (158)	29,759	17	68.0	.353	.344	0.974
14	Services	Miscellaneous Business, Personal & Repair Services (160)	13,710	8	72.7	.397	.425	1.070
15	Recreation	Amusement & Recreation Services, n.e.c. (166)	2,904	5	71.4	.452	.387	0.856
16	Medical Services	Physicians, Surgeons, Dentists, etc. (167)	7,708	4	28.6	.443	.429	0.968
		Medical & Health Services	1,819	4	28.6	.443	.380	0.858
17	Government	Government Enterprises (171)	2,053	5	62.5	.461	.569	1.234

one central industry, five had two, and four clusters had three central industries. However, the number of central industries is not nearly as important as the relative strength of these industries. Some of the clusters appeared to be dominated by a single industry, while the influence of central industries in other clusters was not nearly as great. The degree of dominance can be judged by the number and intensity of the links involving the central industry. The important indicators of dominance appear in columns 5, 6, and 9 of Table 2.

Another measure of the tightness of a cluster is the mean strength of interindustry links, or $\sum_i \sum_j e_{ij}/n$, where n is the number of industries in the cluster. Considering only those links for which $e_{ij} \geq .200$, the mean value of links ranged from .280 for the Automotive cluster to .560 for the Petroleum cluster, indicating that, on average, the industries of the latter were the more interdependent, in terms of interindustry flows.

In order to proceed with the analysis of the structure of the various clusters, a diagram was constructed of each identified grouping. The diagrams are two-dimensional representations of the multi-dimensional clusters. Each circle represents one sector; central industries are identified by a double circle. The area of each circle is proportional to the size of the industry it represents, using value of shipments as an index of industry size. Value added has been used in the cases where value of shipments cannot be obtained, or is meaningless, as for some service industries.

A line between two circles represents a link between two industries. The length of a line between a given pair of industries is inversely proportional to the respective e_{ij} coefficient. In cases where an industry is a member of another cluster, the industry appears within a dotted box indicating that there are a number of links to the other cluster that are not shown.

Among the seventeen clusters there were five yielding tree diagrams. In these clusters there is only one link, or set of links, connecting two industries. The clusters giving tree diagrams were Iron & Steel, Automotive, Petroleum, Recreation, and Services. Clusters which did not form trees, containing two or more circuits between industries, were called complex groupings. A small subgroup of clusters has been termed "snowflakes" because of the appearance of their diagrams. To be classified as a "snowflake", a cluster had to meet the following three conditions: (1) have only one central industry, (2) be classified as a tree, and (3) have at least two-thirds of all links in the cluster involve the central industry. Four such clusters were identified. These were Iron & Steel, Automotive, Recreation, and Services. Iron & Steel proved to be a typical representative of this group, of great interest for a number of reasons. The sequence of operations based upon engineering considerations is well preserved or only slightly distorted by the use of SIC classification. It is dominated by its central industry, Steel Rolling & Finishing. The strong revealed ties to the Construction and Automotive clusters are not unexpected.

U.S. ECONOMY IRON AND STEEL CLUSTER

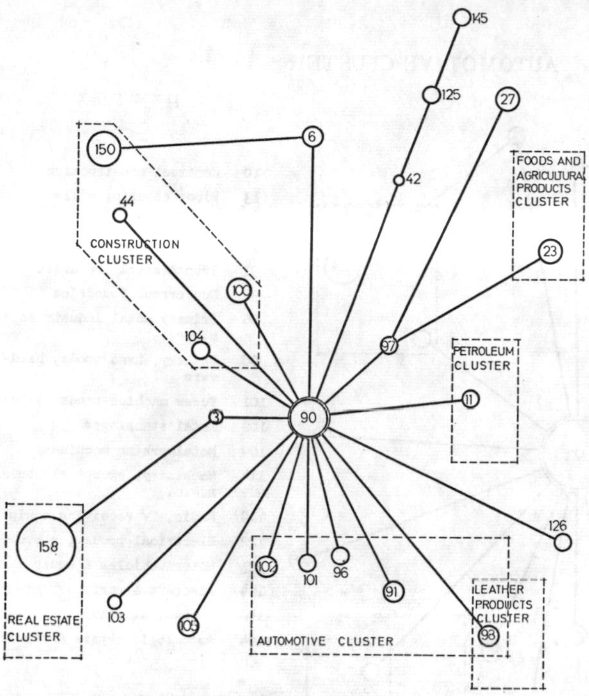

LEGEND

3	Iron & ferroalloy ores
6	Anthracite, lignite & bituminous coal mining
11	Oil & gas field services
23	Canned & frozen foods
27	Beverage industries
42	Office furniture
44	Partitions & fixtures
90	Steel rolling & finishing
91	Iron & steel foundries
96	Primary metal industries, n.e.c.
97	metal cans
98	Cutlery, hand tools, hardware
100	Fabricated structural metal products
101	Screw machine prods. & bolt
102	Metal stampings
103	Coating, plating, polishing engraving
104	Fabricated wire products, n.e.c.
105	Fabricated metal products, n.e.c.
125	Ships & boats*
126	Railroad equipment
145	Water transportation*
150	Electric, gas & sanitary service*
158	Real Estate*

*Industry size given by Value Added rather than Value of Shipments.

The Automotive cluster is itself highly significant, not so much in terms of its structure as because of its economic importance, its strong links to other groupings, and the presence of several nodal industries.

Of considerable significance for regional development strategies may prove to be a small group of industries, called nodal industries, which belong to more than one cluster or have strong links to another cluster. One industry, Contract Construction, was included in eleven of the seventeen clusters. Two others, Real Estate and Automotive, each belonged to six clusters. The following is an enumeration of the nodal industries and their number of occurrences:

Industry	No. of Clusters Appeared In
Contract Construction	11
Real Estate	6
Motor Vehicles	6
Agricultural Products	5
Advertising	4
Services	4
Floor Covering Mills	4
13 Others	3
58 Others	2

All of the nodal industries listed above are also central industries in their respective clusters, and most were highly correlated with population. The presence of these nodal industries is apparently highly significant for the formation of spatial industrial complexes.

U.S. ECONOMY AUTOMOTIVE CLUSTER

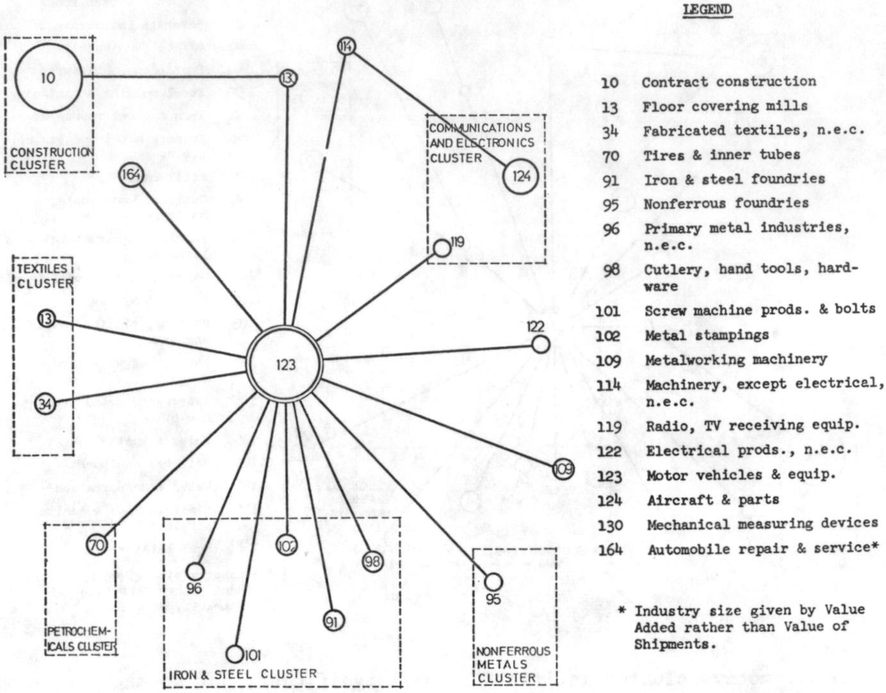

Several of the identified clusters are worthy of special consideration, either because of their structure or their relevance in any planning strategy. One of the most important groupings was the Construction cluster. This was due to its size and its ubiquitous nature. The central industry of the cluster, Contract Construction, repeatedly appeared in other clusters and made a substantial contribution to the value added totals of these clusters. Diagramatically, the Construction cluster was very nearly a perfect "snowflake", except for the presence of a small subgroup centered around the Nonmetallic Mineral Mining industry. The weakness of the ties within this subgroup did not justify defining it as a cluster, and it remained a part of the Construction group.

The Real Estate cluster was similar to Construction in two important respects. First, its central industry was extremely large and appeared in many clusters. Secondly, the diagram of the cluster had a "snowflake" appearance, but the presence of circuits prevented its classification as such.

Several clusters, particularly those primarily engaged in manufacturing rather than providing services, were composed almost exclusively of technically related industries. An example of this is the Textiles cluster, although the tree-like sequence of processes is obscured by the circuits present in the diagram. The lack of

U.S. ECONOMY CONSTRUCTION CLUSTER

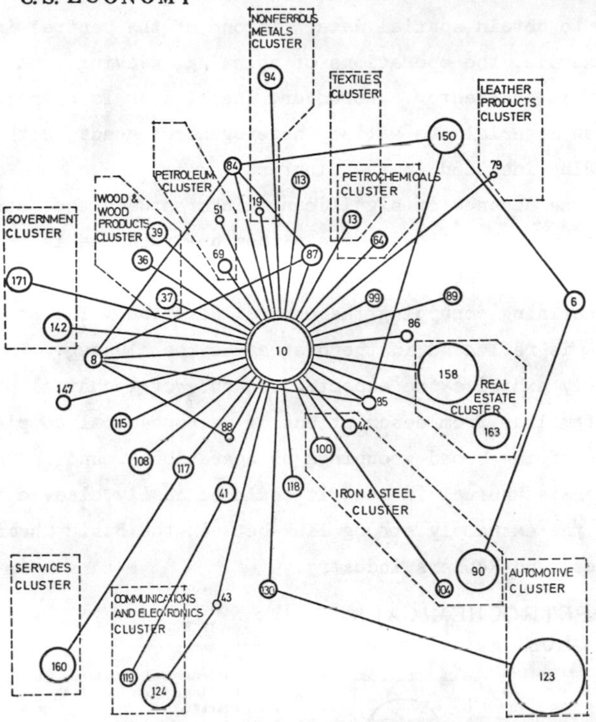

LEGEND

6 Anthracite, lignite & bituminous coal mining
8 Nonmetallic mineral mining
10 Contract construction
13 Floor covering mills
19 Small arms ammunition
36 Sawmills & planing mills
37 Millworks & related products
39 Misc. wood prods. & wood finishing
41 Household furniture
43 Public bldg. & related furniture
44 Partitions & fixtures
51 Bldg. paper & board mills
64 Paints, varnishes & allied products.
69 Paving & roofing materials.
79 Leather gloves & mittens
84 Hydraulic cement
85 Structural clay products
86 Pottery & related products
87 Concrete, gypsum & plaster products
88 Cut stone & stone products
89 Nonmetallic mineral prods., n.e.c.
90 Steel rolling & finishing
94 Nonferrous rolling & drawing
99 Plumbing & nonelectric heating equip.
100 Fabricated structural metal products
104 Fabricated wire prods., n.e.c.
108 Construction & like equip.
113 Service industry machines
115 Electric distribution prods.
117 Household appliances
118 Lighting & wiring devices
119 Radio, TV receiving equip.
123 Motor vehicles & equip.
124 Aircraft & parts
130 Mechanical measuring devices
142 Railroads & related services*
147 Pipe line transportation*
150 Electric, gas & sanitary ser.*
158 Real estate*
160 Services: misc. business, misc repair, personal*
163 Prof. services (except education), non-profit organizations*
171 Government enterprises*

*Industry size given by Value Added rather than Value of Shipments.

U.S. ECONOMY TEXTILES CLUSTER

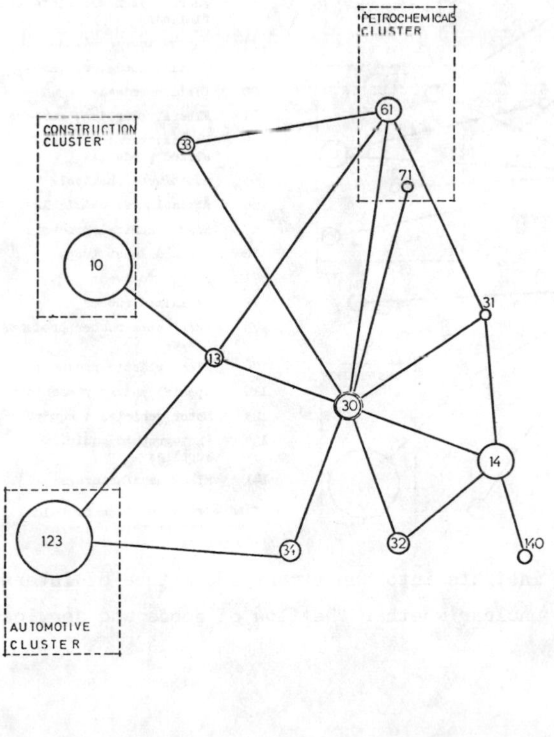

LEGEND

10 Contract construction
13 Floor covering mills
14 Apparel (except fur goods)
30 Fabric & yarn mills; textile finishing
31 Narrow fabric mills
32 Knitting mills
33 Textile goods, n.e.c.
34 Fabricated textiles, n.e.c.
61 Fibers, plastics, rubbers
71 Rubber footwear
123 Motor vehicles & equip.
140 Costume jewelry & notions

an orderly transition from one operation to the next is partly due to SIC classifications and the grouping required to obtain spatial data. In one of the central industries in this cluster, for example, the operations of spinning, weaving, and finishing have been lumped together in one sector. Moreover, the cluster is comprised of industries using different raw materials to satisfy heterogeneous needs, both natural fibers and synthetics being included in the cluster.

One surprising finding was the absence of significant links between the Petroleum and the Petrochemicals clusters. The links between them are technically crucial but not economically significant. The Petroleum cluster has an almost tree-like structure, with Petroleum Refining occupying the central position. The sequence of the various operations is reflected in the diagram of the cluster. The Petrochemicals cluster only weakly exhibits the expected sequence of technical process. The tree-like structure familiar from descriptions of petrochemical complexes is lost, at least partly because of the broad groupings of operations found in the SIC classifications. The cluster is unusual in that it could be easily cleaved into two separate groups, if not for the extremely strong link between the Basic Chemicals industry and the Fibers, Plastics, and Rubbers industry.

U.S. ECONOMY **PETROCHEMICALS CLUSTER**

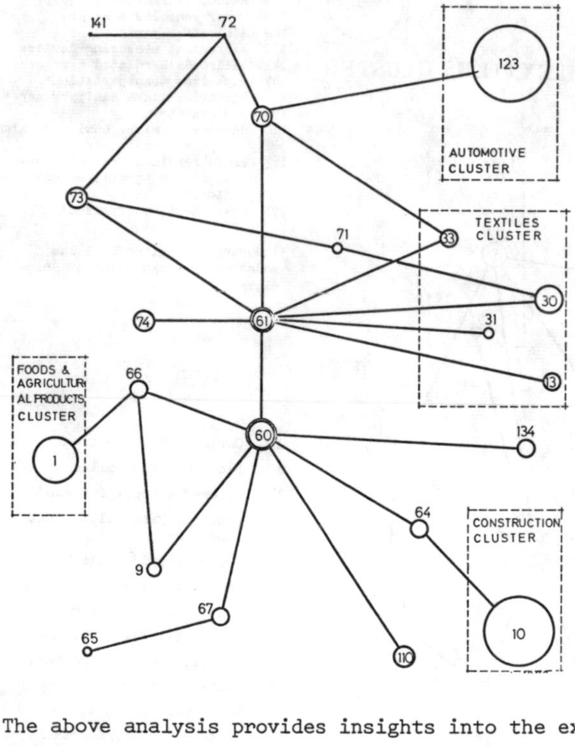

LEGEND

1 Agricultural services, hunting, trapping; fisheries*
9 Chemical & fertilizer mineral mining
10 Contract construction
13 Floor covering mills
30 Fabric, yarn mills; textile finishing
31 Narrow fabric mills
33 Textile goods, n.e.c.
60 Basic chemicals
61 Fibers, plastics, rubbers
64 Paints, varnishes, and allied products
65 Gum & wood chemicals
66 Agricultural chemicals
67 Misc. chemical products
70 Tires & inner tubes
71 Rubber footwear
72 Reclaimed rubber
73 Fabricated rubber products, n.e.c.
74 Misc. plastic products
110 Special industry machinery
123 Motor vehicles & equip.
134 Photographic equip. & supplies
141 Misc. manufacturers

*Industry size given by Value Added rather than Value of Shipments.

The above analysis provides insights into the extent and nature of intersectoral interdependence. However, it is unclear whether the flow of goods and services between sectors can help explain the phenomenon of spatial aggregation. Obviously, the relative value of flows between two industries need not be related to their mutual attraction in space. Work now under way is attempting to determine the correlation between flows and spatial association.

REFERENCES

[1] Bergsman, J., P. Greenston, and R. Healy, "The Agglomeration Process in Urban Growth", Urban Studies, IX, 3 (1972).

[2] Boudeville, J. R. Problems of Regional Economic Planning, 1966.

[3] Casetti, Emilio, "Optimal Interregional Investment Transfers", Journal of Regional Science, VIII, 1 (1968).

[4] Chenery, Hollis B., "Development Policies for Southern Italy", Quarterly Journal of Economics, LXXVI (November 1962).

[5] Chenery, Hollis B. and Paul G. Clark. Interindustry Economics, 1962.

[6] Chenery, Hollis B., Paul G. Clark, and V. Cao-Pinna. The Structure and Growth of the Italian Economy, 1953.

[7] Collida, A., P. L. Fano, and M. D'Ambrosio (F. Angeli, ed.). Sviluppo Economico e Crescita Urbana en Italia, 1968.

[8] Czamanski, Stan, "Industrial Location and Urban Growth", The Town Planning Review, Liverpool, XXXVI, 3 (October 1965).

[9] _____. "Linkages Between Industries in Urban-Regional Complexes", in G. G. Judge and T. Takayama, Studies in Economic Planning Over Space and Time, 1973.

[10] _____. "A Method of Forecasting Metropolitan Growth by Means of Distributed Lags Analysis", Journal of Regional Science, VI (1965).

[11] _____. "A Model of Urban Growth", Papers, Regional Science Association, XIII (1965).

[12] Czamanski, Stan, with the assistance of Emil E. Malizia, "Applicability and Limitations in the Use of National Input-Output Tables for Regional Studies", Papers, Regional Science Association, XXIII (1969).

[13] Hermansen, T. "Development Poles and Related Theories", in N. M. Hansen (ed.) Growth Centers in Regional Economic Development, 1972.

[14] Isard, Walter, Eugene W. Schooler, and Thomas Vietorisz. Industrial Complex Analysis and Regional Development, 1959.

[15] Klaassen, L. H. Methods of Selecting Industries for Depressed Areas, 1967.

[16] Perroux, F. "Economic Space: Theory and Applications", Quarterly Journal of Economics, LXIV (1950).

[17] van Wickeren, A. "An Attraction Analysis for the Asturian Economy", Regional and Urban Economics, II, 3 (1972).

OPTIMAL INVESTMENT POLICIES IN TRANSPORTATION NETWORKS

S. Giulianelli - A. La Bella (*)

1. INTRODUCTION

A relevant feature of land use planning is the design of transportation networks.

The problem which public administrators are often confronted with is that of apportioning a limited budget to the various branches of a network so that to achieve the goal put forward by the administration. Namely, given the demand for transportation among the centers connected by the network, one may wish to find the optimal investment policy, in order to minimize the total transportation time.

The main difficulty of this kind of problems is that the objective function turns out to be neither convex nor separable whereas the constraints turn out to be non linear. Some authors have tackled the problem using, in our opinion, too many simplifying assumptions. For example, the transit time on each branch of the network is assumed to be independent of the flow (1) (2) (3). Moreover, in case of several origins and destinations only zero-one investments in each branch are considered (1) (2), whereas discrete investments are analyzed for one origin only (3).

Although the hypothesis that the transit time is independent of the flow seems to be quite coarse, we keep it in this paper, where, as our first contribution to the problem, we propose to overcome all other restrictions. Namely, we consider the case of several origins and destinations with the investments in each branch taken as continuous variables. We also suggest a heuristic computational procedure which appears to be more efficient compared to those so far appeared in the literature.

2. DEFINITION OF THE PROBLEM

The basic hypotheses of this work can be summarized as follows:

Hypothesis 1: (transit time hypothesis): the transit time t_ℓ on each arc ℓ is independent of the flow.

Hypothesis 2 (behavioural hypothesis): the traffic flow distribution is such that the overall transportation time is minimized.

Hypothesis 3: the dependence of the transit time t_ℓ associated with arc ℓ on the apportionment c_ℓ is of the following linear type with saturation

$$t_\ell = \alpha_\ell - \beta_\ell c_\ell \qquad 0 \le c_\ell \le \bar{c}_\ell$$
$$t_\ell = \alpha_\ell - \beta_\ell \bar{c}_\ell \qquad c_\ell \ge \bar{c}_\ell$$

(*) Sandro Giulianelli - Agostino La Bella - Centro di Studio dei Sistemi di Controllo e Calcolo Automatici, C.N.R. - Istituto di Automatica, Università degli Studi di Roma - Via Eudossiana, 18 - 00184 Roma.

where

α_ℓ, β_ℓ, $\bar{c}_\ell$ are non negative constants.

Then the problem can be stated as follows:

Given:

1 - A transportation network G(N,A), where N is the set of nodes and A the set of arcs.

2 - The transportation demand matrix

$$R = \{r_{ij}\}$$

3 - A budget B

4 - An objective function T

$$T = \sum_{\substack{i\epsilon N\\ j\epsilon N}} \underline{t}^T(\underline{c})\underline{A}_{ij}\,\underline{x}_{ij}(\underline{c}) \tag{1}$$

(i.e. the overall transit time),

where

$\underline{x}_{ij}$: column vector, whose components x^k_{ij} represent the flow on the k-th path p^k_{ij} between i and j, i,jϵN due to the demand r_{ij}; $k=1,2,\ldots,q_{ij}$

q_{ij} : number of different paths between the nodes i and j, i,jϵN

P^k_{ij} : set of branches in the path p^k_{ij}

$\underline{A}_{ij}$: incidence matrix arcs-paths for the origin - destination pair i-j, i,jϵN, whose entries $a_{m,k}$, mϵA, $k=1,2,\ldots,q_{ij}$, are defined by

$$a_{m,k} = \begin{cases} 1 & \text{for } m \in P^k_{ij} \\ 0 & \text{for } m \notin P^k_{ij} \end{cases} \tag{2}$$

$\underline{t}(\underline{c})$: column vector with components $t_m(c_m)$, mϵA

$\underline{c}$: column vector with components c_m, mϵA

Minimize T, subject to the following constraints

$$\sum_{k=1}^{q_{ij}} x^k_{ij}(\underline{c}) = r_{ij} \qquad \forall i\epsilon N,\ \forall j\epsilon N \tag{3}$$

$$\sum_{m\epsilon A} c_m \leq B \tag{4}$$

$$\underline{c} \geq 0 \tag{5}$$

$$\underline{c} \leq \bar{\underline{c}} \tag{6}$$

where

$$\underline{\bar{c}} = \{\bar{c}_m\}, \qquad m \varepsilon A$$

3. PROPERTIES OF THE OPTIMAL SOLUTION

For each arc $m \varepsilon A$, it is possible to define a return as $\alpha_m y_m$, where y_m is the total flow on the arc.

Then one may think to obtain the optimal solution by putting the branches in order of decreasing return and investing in this order.

Actually, this approach would be fallacious, because by investing on each branch we change the relative transit time. Consequently a new distribution of flow may follow: this phenomenon modifies the problem due to the possible variation of the return associated with each branch.

It is evident that the investment on some branches albeit optimal relative to a certain amount of capital may fail to indicate the optimal solution for a larger budget.

Nevertheless the following properties and theorems hold:

Property 1 - The whole traffic demand for each origin-destination pair runs along one and only one path. In fact from the hypotheses 1) and 2) it results that each individual chooses the minimal time path from the origin to the destination (i.e. there is no interaction among individuals).

Remark 1 - For each investment vector $\underline{c}$ the problem of determining the traffic distribution in the network is reduced to that of finding the shortest route from each origin to all destinations.

This immediately follows from Property 1.

Property 2 - The flow originating at each origin branches out into a tree. Therefore, for each vector $\underline{c}$ the total traffic flow is distributed on a network determined by the union of as many trees as these are origins.

If now we put:

$$\underline{y}(\underline{c}) = \sum_{i \varepsilon N} \sum_{j \varepsilon N} \underline{A}_{ij} \, \underline{x}_{ij}(\underline{c})$$

where $\underline{y}(\underline{c})$ is a column vector with components $y_m(\underline{c})$, $m \varepsilon A$, the following theorem hold:

THEOREM 1 - Suppose that the optimal flow distribution $\underline{y}^*(B)$ is known. Consider the subnetwork $H^* = G(N_{H^*}, A_{H^*})$ obtained from $G(N,A)$ cancelling out all branches where the flow is zero (cfr. Property 2).

Then the optimal investment policy $\underline{c}^*(B)$ is obtained by investing the maximum amount of capital $\bar{c}_\ell$ in each branch in order of decreasing return until the budget is reached.

Proof - Under the hypotheses set forth, the problem (1), (3)-(6), is reduced to the following linear programming problem:

$$\min_{\underline{c}} \sum_{\ell \varepsilon A_H} y^*_\ell (\beta_\ell - \alpha_\ell c_\ell)$$

$$\sum_{\ell \varepsilon A_{H^*}} c_\ell \leq B$$

whose optimal solution is the investment policy stated in the theorem.

4. EXHAUSTIVE PROCEDURE

Based on the previous sections, the following procedure leads to the optimal investment policy.

Step 1 - Construct the set $\mathcal{H}$ of subnetworks $H = G(N_H, A_H)$ combining in all possible ways one tree for each origin to all destinations for which there is demand for transportation.

Assume that the traffic demand is satisfied using the tree chosen for the relative origin only. Let $\underline{y}(H)$ be the vector of traffic distribution.

Step 2 - For each $H \varepsilon \mathcal{H}$ solve the following linear programming problem:

$$\min_{\underline{c}} \sum_{m \varepsilon A_H} y_m(H) t_m(c_m) = \sum_{m \varepsilon A_H} y_m(H) t_m\left[c^*_m(H)\right]$$

$$\sum_{m \varepsilon A_H} c_m \leq B$$

(for the solution procedure cfr. theorem 1).

Step 3 - Compute

$$\min_{H \varepsilon \mathcal{H}} \sum_{m \varepsilon A_H} y_m(H) t_m\left[c^*_m(H)\right]$$

Comment - The exhaustive procedure singles out a finite number of points from the subset of the Euclidean space whose coordinates are c_m, $m \varepsilon A$, defined by

$$\sum_{m \varepsilon A} c_m \leq B$$

$$c_m \geq 0 \qquad \forall \, m \,\varepsilon\, A$$

5. HEURISTIC PROCEDURE

The procedure previously described examines a finite and discrete subset of solutions belonging to the infinite and continuous set of possible solutions, and chooses among them the optimal one.

Now this subset of admissible solutions, already defined, can be made up of a very large number of elements. It is therefore desirable to obtain an algorithm that makes it possible to find the optimal solution more rapidly. This algorithm can be set forth as follows:

Step 1 - Find the optimal traffic distribution, corresponding to the initial situation of no investment. Invest on the arcs in order of decreasing return until a new distribution of traffic is obtained.

Step i - Cancel the preceeding investment and, assuming traffic distribution obtained in the previous step, invest on the arcs in order of decreasing return until a distribution of traffic is obtained for an amount of capital greater than the previous investment.

If no redistribution is obtained, then stop.

The optimal solution is the one that corresponds to the investment made at the K-th iteration, such that:

$$T_k(B) = \min_i T_i(B)$$

where $T_i(B)$ is the overall transit time obtained in the i-th iteration for an investment equal to B.

Notice that some traffic distributions may occur in the euristic procedure which are not stable for the value B of the budget. In such cases, in order to compute $T_k(B)$, these distributions are replaced by those obtained starting with the unstable distributions with an investment equal to B.

6. NUMERICAL EXAMPLE

Let us consider the network of fig. 1, with the following traffic demand matrix:

$$R = \begin{bmatrix} 0 & 1 & 3 & 2 \\ 0 & 0 & 4 & 1 \\ 0 & 0 & 0 & 0 \\ 0 & 5 & 3 & 0 \end{bmatrix}$$

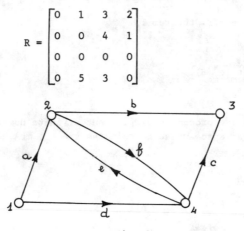

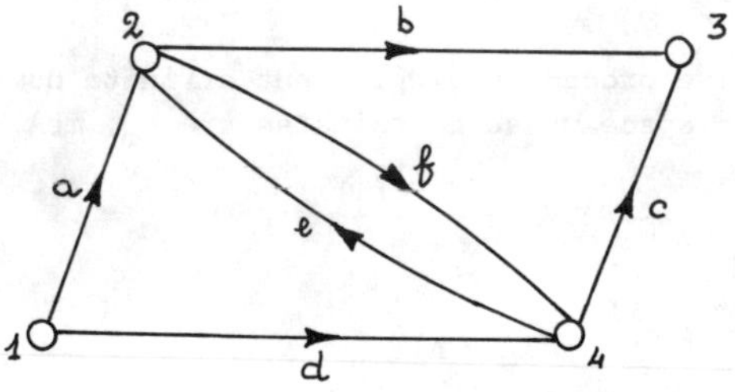

Fig. 1

Let the dependence of the transit time on each arc on the investment be as follows:

$$t_a = 10 - 2c_a \qquad 0 \le c_a \le 4$$
$$t_b = 6 - c_b \qquad 0 \le c_b \le 3$$
$$t_c = 6 - c_c \qquad 0 \le c_c \le 2$$
$$t_d = 9 - 2c_d \qquad 0 \le c_d \le 4$$
$$t_e = 7 - c_e \qquad 0 \le c_e \le 3$$
$$t_f = 20 - 3c_f \qquad 0 \le c_f \le 6$$

6.1 - Exhaustive procedure

Figures 2a, 2b and 2c represent all the trees associated with nodes 1, 2 and 4 respectively.

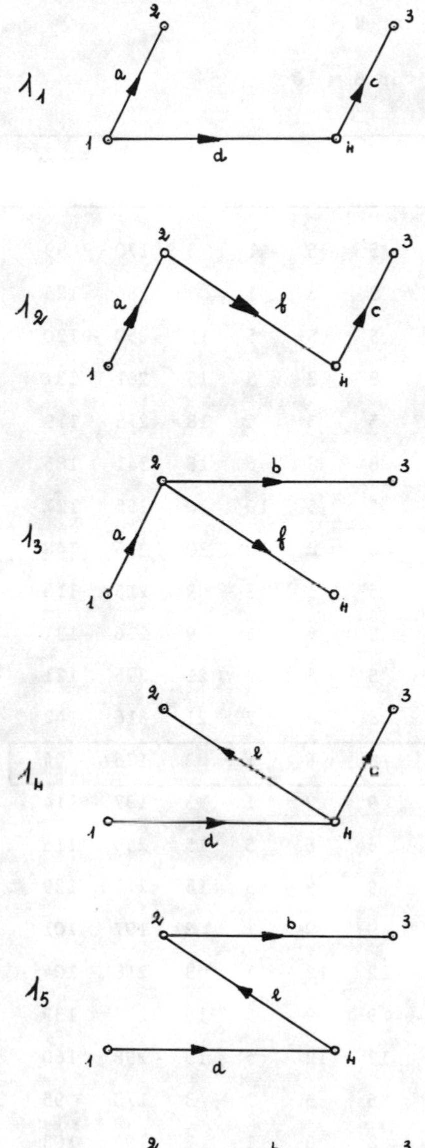

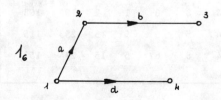

Fig. 2 a
Origin : node 1

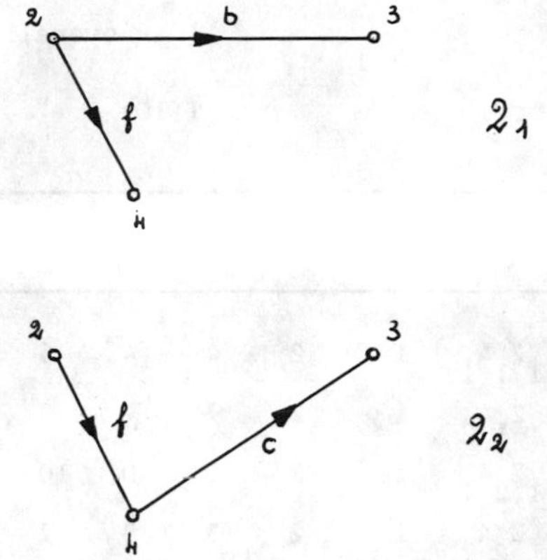

Fig. 2 b
Origin : node 2

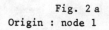

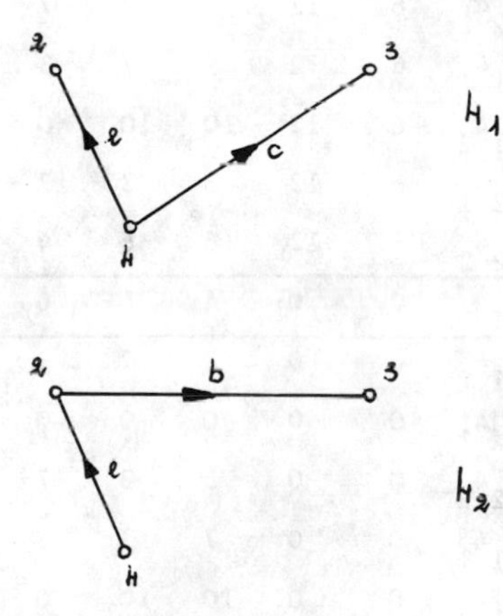

Fig. 2 c
Origin : node 4

Supposing that the budget is B = 10 and applying the procedure of section 4 we obtain the results of Table I, where

$$G_\ell = \alpha_\ell y_\ell$$

T_o = overall transit time for B = 0

T(10) = " " " for B = 10

H	y_1	G_1	y_2	G_2	y_3	G_3	y_4	G_4	y_5	G_5	y_6	G_6	T_0	T(10)
$1_1 2_1 4_1$	1	2	4	4	6	6	5	10	5	5	1	3	170	99
$1_1 2_1 4_2$	1	2	7	7	3	3	5	4	8	8	1	3	186	125
$1_1 2_2 4_1$	1	2	0	0	10	10	5	10	5	5	5	15	250	120
$1_1 2_2 4_2$	1	2	3	3	7	7	5	10	8	8	5	15	261	131
$1_2 2_1 4_1$	6	12	4	4	6	6	0	0	5	5	6	18	275	119
$1_2 2_1 4_2$	6	12	6	6	7	7	3	6	8	8	6	18	341	185
$1_2 2_2 4_1$	6	12	0	0	10	10	0	0	5	5	10	30	355	127
$1_2 2_2 4_2$	6	12	3	3	7	7	0	0	8	8	10	30	376	148
$1_3 2_1 4_1$	6	12	7	7	3	3	0	0	5	5	3	9	215	113
$1_3 2_1 4_2$	6	12	10	10	0	0	0	0	8	8	3	9	236	131
$1_3 2_2 4_1$	6	12	3	3	7	7	0	0	5	5	7	21	295	121
$1_3 2_2 4_2$	6	12	6	6	4	4	0	0	8	8	7	21	316	142
$1_4 2_1 4_1$	0	0	4	4	6	6	6	12	6	6	1	3	176	94
$1_4 2_1 4_2$	0	0	7	7	3	3	6	12	9	9	1	3	197	114
$1_4 2_2 4_1$	0	0	0	0	10	10	6	12	6	6	5	15	256	116
$1_4 2_2 4_2$	0	0	3	3	7	7	6	12	9	9	5	15	277	139
$1_5 2_1 4_1$	0	0	7	7	3	3	6	12	9	9	1	3	197	101
$1_5 2_1 4_2$	0	0	10	10	0	0	6	12	12	12	1	3	218	104
$1_5 2_2 4_1$	0	0	3	3	7	7	6	12	9	9	5	15	277	139
$1_5 2_2 4_2$	0	0	6	6	4	4	6	12	12	12	5	15	298	160
$1_6 2_1 4_1$	4	8	7	7	3	3	2	4	5	5	1	3	173	95
$1_6 2_1 4_2$	4	8	10	10	0	0	2	4	8	8	1	3	194	108
$1_6 2_2 4_1$	4	8	3	3	7	7	2	4	5	5	5	15	253	115
$1_6 2_2 4_2$	4	8	6	6	4	4	2	4	8	8	5	15	274	152

TABLE I

From Table I we infer that the optimal investment policy is

$$\underline{c}^{*T} = [0 \;\; 1 \;\; 2 \;\; 4 \;\; 3 \;\; 0]$$

and correspondingly we have T*(10) = 94.

6.2 - *Euristic procedure*

Applying the euristic procedure we obtain the results of Tables 2 and 3 where $\underline{y}_i$ represents the traffic distribution at the i-th step and B_{r_i} the value of the investment for which we have a redistribution.

Arcs	$\underline{y}_0$ $B_{r0}=3$	$\underline{y}_1$ $B_{r1}=20,5$	$\underline{y}_2=\underline{y}_0$ $B_{r2}=22$	$\underline{y}_3$ $B_{r3}=-$
a	1	0	1	4
b	4	4	4	7
c	6	6	6	3
d	5	6	5	2
e	5	6	5	5
f	1	1	1	1

TABLE 2

Distrib.	Investment Policy $\underline{c}^T$	T(B) B = 10
$\underline{y}_0$	$\underline{c}^T = [0\ 0\ 0\ 3\ 0\ 0]$	-
$\underline{y}_1$	$\underline{c}^{*T} = [0\ 1\ 2\ 4\ 3\ 0]$	94
$\underline{y}_2=\underline{y}_0$	$\underline{c}^T = [0\ 1\ 2\ 4\ 3\ 0]$	94(1)
$\underline{y}_3$	$\underline{c}^T = [4\ 3\ 0\ 0\ 3\ 0]$	105

TABLE 3

(1) The distribution $\underline{y}_2 = \underline{y}_0$ is clearly unstable for B = 10

It is worth to note that with the euristic procedure the optimal solution is obtained in 3 steps only.

REFERENCES

(1) Ridley T.M.: *An Investment Policy to Reduce the Travel Time in a Trasportation Network*. Transportation Research, vol. 2, pp. 409-424, 1968.

(2) Ridley T.M.: *Reducing the Travel Time in a Transportation Network*. Studies in Regional Science, Scott. A.J. Ed., Pion Limited, London, 1969, pp.73-87.

(3) Goldman A.J., Nemhauser G.L.: *A Transport Improvement Problem Transformable to a Best-Path Problem*. Transportation Science, vol. 1, n. 4, pp. ,295-307, 1967.

AN ON-LINE OPTIMIZATION PROCEDURE FOR AN URBAN TRAFFIC SYSTEM

C. J. Macleod and A. J. Al-Khalili

Department of Electronic Science,
University of Strathclyde, Glasgow.

INTRODUCTION

This paper describes a method for the on-line optimizing closed loop control of urban traffic systems. The method of approach is to use the theory of hierarchial systems as developed by Mesarovic[4] and the decomposition of the system follows the pattern used in Madrid by Fuehrer[2].

SYSTEM DECOMPOSITION

Basically the approach used is to decompose the system into smaller units and by optimizing the performance of these units and coordinating them, an overall optimum or near optimum is achieved. The control system consists of sub-systems which are located on levels of authority which, in this case, leads to two discrete levels as shown in Figure 1. The supreme authority lies with the computer or the supremal unit at level one. A subordinate level of authority is vested in the local controllers or the infimal units at level two. The two levels are interactive in that the local controllers are influenced directly and explicitly from the computer such that the parameter settings in the local controllers are decided by the computer whose decisions are based on information fed from the local controller and the process.

DESCRIPTION OF THE OVERALL SYSTEM

The Process

The process consists of the total traffic area being controlled sub-divided into sub-areas of common cycle length. Each sub-area consists of a small group of junctions and interconnecting roads. There are two types of signals in the process: (a) a control signal which relates to light settings, (b) inputs and outputs consisting of traffic flowing across the process boundary.

Local Controller

Each local controller controls a sub-process consisting of a junction of two or more interconnecting roads. The two inputs to the controller are: (a) a coordination input from the computer consisting of cycle length, optimum

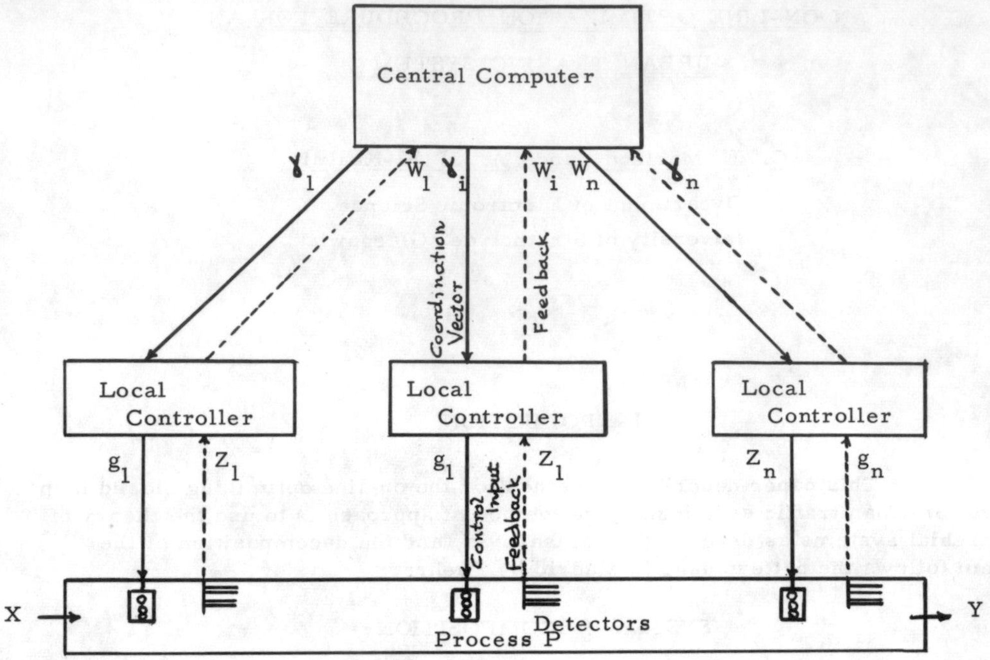

Fig. 1 The Overall System

offset, predicted inputs and a decision set to be specified in a later section, and (b) an input consisting of feedback information from the process. The output from the controller is the local control signal consisting of the green time (or split) allocated to one direction.

The Computer

The computer calculates the coordination vectors to all the Local Controllers based on the feedback information supplied to it from the process via the Local Controllers.

Feedback Information

1. Feedback from the process to the various local controllers, contains information about junction inputs and outputs.

2. Feedback received by the computer contains information concerning the behaviour of the local controller and the process.

COORDINATION

Introduction

This concept involves setting up conditions whereby the overall system is optimized or near optimized as a result of optimizing local problems.

The overall objective function to be optimized involves the minimization of the total vehicle hours spent in the network.

Since the delay in interconnecting roads is essentially constant, the overall system objective function used is expressed in terms of delays in junctions, i.e.

$$f(g) = \sum_{\text{all } j} f_j(g,\mu) = \sum_{\text{all } j} \sum_{i=1}^{m} (K'_{ji} D_i + K_{ji} S_i) \tag{1}$$

i.e. f(g) is a combination of weighted delays, D_i, and stops, S_i, and relates implicitly to the total flow through the process. D_i are the delays on the arms of the ith junction, S_i are the stops, K_{ji} and K'_{ji} are weighting functions and m is the number of approaches to the junction. The objective function for the jth junction is,

$$f_j(g,\mu) = \sum_{i=1}^{m} (K'_{ji} D_i + K_{ji} S_i) \tag{2}$$

The function $f_j(g,\mu)$ is the delay incurred on all arms of the junction. Minimization of this function with respect to g yields the optimum split $\hat{g}$ while minimization of delays on the priority arm only gives the optimum offset $\hat{\mu}$. The total delay incurred on the priority arm may also be expressed as

$$f_{\ell}(g,\mu) = \sum_{i=1}^{M} (D_i + K S_i) \tag{3}$$

and is actually used to find the optimum offset $\hat{\mu}$. In equation (3), D_i is the delay incurred by traffic for the ith value of offset, μ_i is the offset, S_i the number of stops on the priority link, K is a weighting factor and M covers a range corresponding to twice the journey time. Simulation studies have shown equations (2) and (3) to be concave functions with distinct minima as shown in Figures 2 and 3.

Coordinability with Least Delay as an overall goal

The system proposed is coordinable if essentially

1. f(g) and $f_j(g,\mu)$ are monotonically related
2. there is infimal harmony, i.e. the control decision results in achievement of minimum $f_j(g,\mu)$ for all j, between coordination instants.

Since $f(g) = \sum_{\text{all } j} f_j(g,\mu)$, a linear sum of local objective function, it is strictly order preserving and is therefore a monotonic function and provided the junctions are optimally synchronised by using an optimum offset $\hat{\mu}$, coordination is possible if other conditions are satisfied.

APPLICATION OF THE OPTIMIZATION PRINCIPLES

From the foregoing sections, the successful application of the on-line optimization procedure described hinges on the evaluation of the functions described in the following sections.

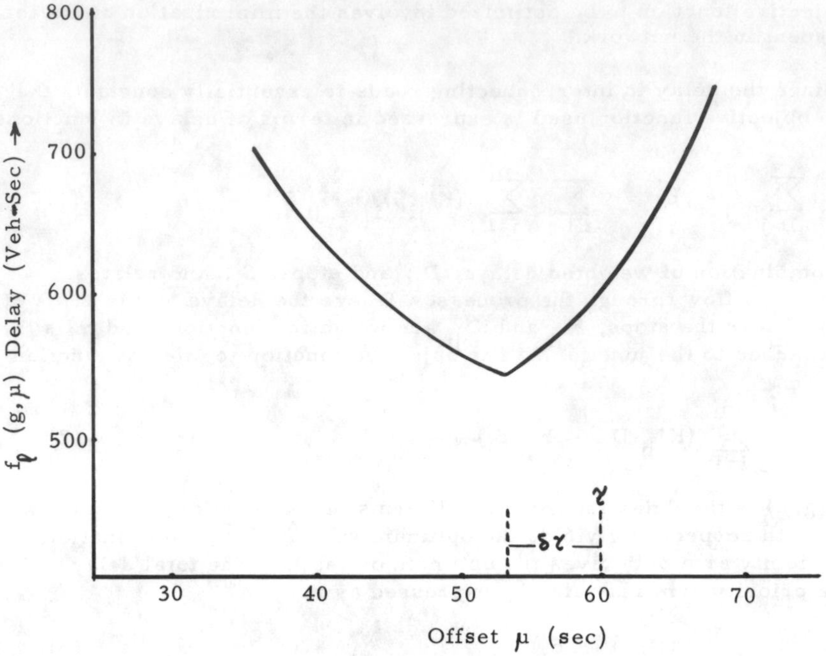

Fig. 2 Variation of f_ℓ (g, μ) with offset μ

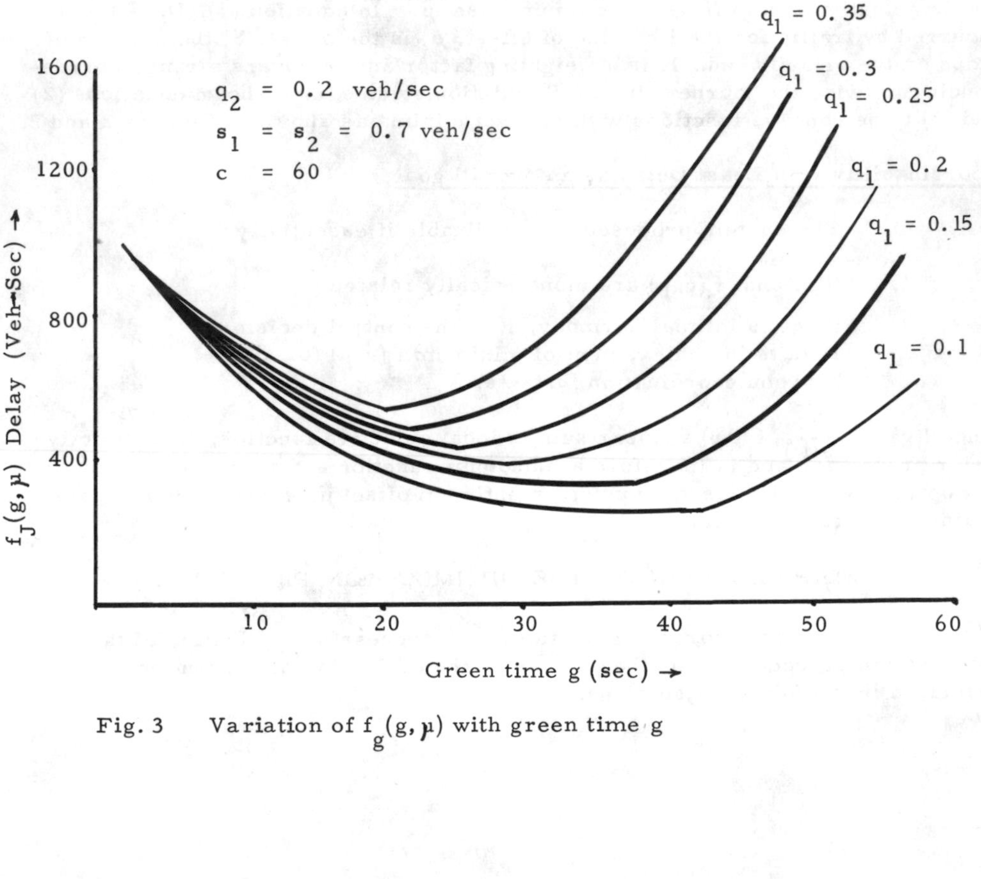

Fig. 3 Variation of $f_g(g, \mu)$ with green time g

The Coordination Vector γ.

This consists of the four components fed to the local controller.

(a) Optimum Cycle length, $\hat{c}$

Each sub-area will have a common optimal or near optimal cycle length. Three kinds of sub-areas are to be used:

1. Fixed sub-areas which consist of junctions in succession on one-way or two-way main streets.
2. Variable sub-areas constituting one major intersection and its satellites.
3. Input control junctions which control traffic entering the network.

The selection of variable sub-areas is formed based on geography and cycle lengths such that all the optimal cycle lengths within the area are within $\frac{3}{4}$ to $1\frac{1}{2}$ times the optimum cycle length of the major intersection. This is based on Webster's[6] suggestion that such lengths cause delays which are within 10-20% above that of the optimal cycle.

The expressions for optimum cycle length proposed by Webster[6] and Gazis and Potts[3] for undersaturated and oversaturated conditions respectively are to be used.

(b) Optimum Offsets

These enable the optimum synchronisation of junction control signals which ultimately allow the coordination principle to be applied in terms of an overall objective function expressed as a sum of junction objective functions.

In general, the optimum offset $\hat{\mu}_i$ is given by

$$\hat{\mu} = \tau - \delta\tau \qquad (4)$$

where τ is the average journey time of the link and $\delta\tau$ is the amount by which $\hat{\mu}$ differs from τ and is a function of degree of saturation, dispersion factor and τ. A plot of $f_{\ell}(g, \mu)$ is shown in Figure 3 and it shows that $\hat{\mu}$ does not coincide with τ but is displaced from it by an amount equivalent to $\delta\tau$. The applicability of aspects of Robertson's approach as used in Transyt[5] is also being investigated.

(c) Decision Set, B_o

B_o is a decision set which contains the decisions taken in the computer, based on traffic conditions, regarding formulae to be used to solve cycle length, optimum offset and the weighting factors used in the criteria functions.

(d) Predicted Inputs

These inputs u are predicted using the model described elsewhere by Al-Khalili and Macleod[1]. This constitutes feedforward of information and is an essential aspect of the technique described in that it is used to evaluate $\hat{g}_j$.

The Decision Set, B_j

B_j are decisions taken in the jth local controller, based on traffic conditions, regarding formulae to be used to solve for optimum split and weighting factors, if any, given to the junction arms to be used when evaluating the optimum split. The allocation of these tasks to the local controller ensures the independence of the local controller and helps to keep the central computer down to a reasonable size.

The Optimum Split, $\hat{g}_j$

This is the value of the split, g_j, for the jth junction which minimizes the junction criterion function

$$f_j(g, \mu) = \sum_{i=1}^{m} (K'_{ji} D_i + K_{ji} S_i) \qquad (2)$$

The procedure for achieving this is:

(i) (a) Expressing stops as delay functions which enables K to be evaluated.
(b) Evaluating K'.

(ii) Development of the Self-Reflective cycle leading to expressing delay in flow parameters.

(iii) Solution of $f_j(g, \mu)$ for the optimum split, $\hat{g}_j$.

(a) Evaluation of the weighting factors, K and K'

Assuming that the stop penalty can be expressed as delay and neglecting economic constraints the expression for K has been derived as

$$K = \frac{2}{a} \lambda_o \log_e (C_J/C) \qquad (5)$$

where
C_J = concentration at jammed state in veh/km
C = concentration in veh/km
λ_o = the speed at maximum flow in m/sec
a = average acceleration of vehicles in m/sec^2

Figure 4 shows the variation of K with concentration for different λ_o.

K' has been chosen to be constant multipliers of arrival rates on one arm relative to the others and is used when an arm is given a higher priority or a link leading to an approach of an intersection is of a low capacity.

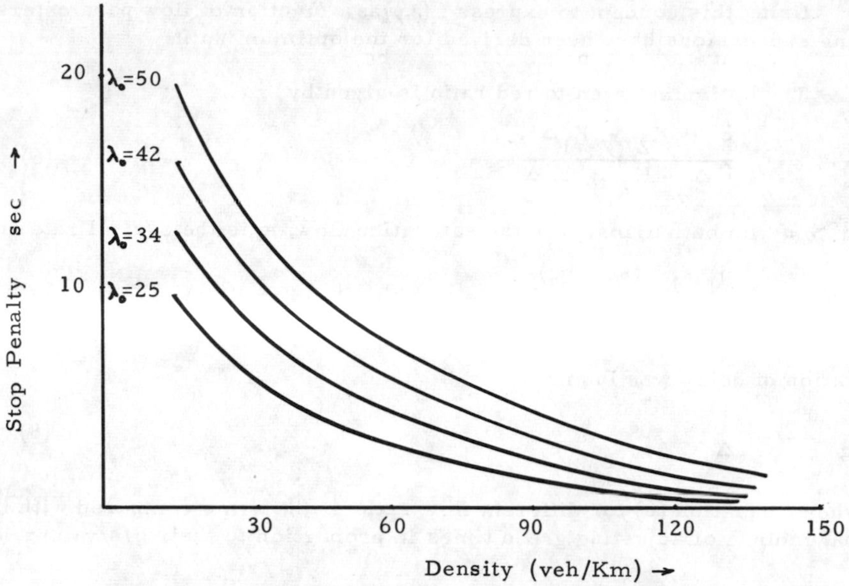

Fig. 4 Stop Penalty/Density Relationship

(b) Determination of Optimum Split, $\hat{g}$

The self-reflective cycle is a cycle where the queues developed at its ends are fed back to its input. Although based on mathematical convenience it has the practical effect due to its feedback property of stabilizing the queues developed at its ends during local optimization. It is shown in Figure 5.

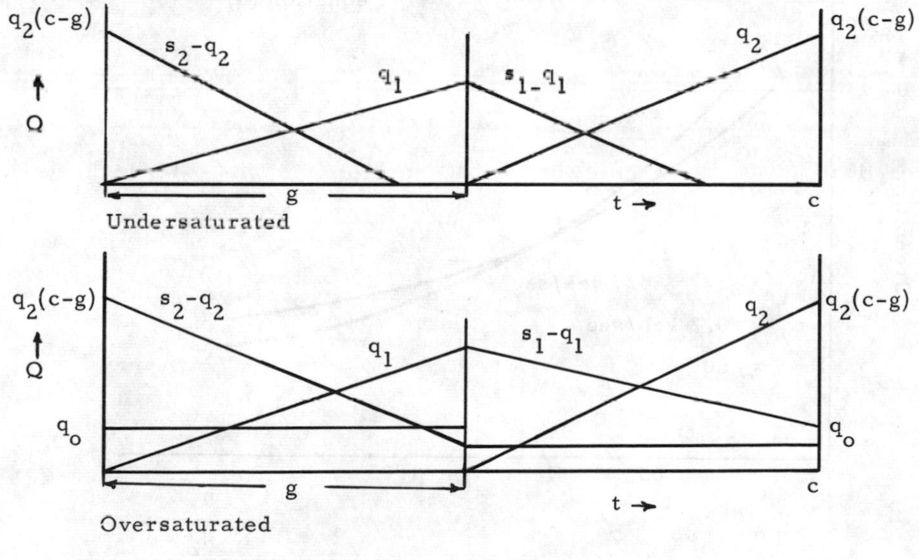

Fig. 5 The Self Reflective Cycle

Using this concept to express $f_j(g,\mu)$ as a function of flow parameters, the following expressions have been derived for the optimum split.

The optimum green to red ratio is given by

$$\frac{\hat{g}}{\hat{c}-\hat{g}} = \frac{\hat{c} + K_2 - K_1\Delta}{\hat{c}\Delta - K_2 + K_1\Delta} \qquad (6)$$

where $s-q \geqslant q$ on both arms, s is the saturation flow, q is the arrival rate and

$$\Delta = \frac{q_1\, s_1\, (s_2 - q_2)}{q_2\, s_2\, (s_1 - q_1)} \qquad (7)$$

If minimization of delay was intended

$$\frac{\hat{g}}{\hat{c}-\hat{g}} = \frac{1}{\Delta} \qquad (8)$$

Figure 6 shows equation (6) for different flow ratio of the arms compared with the conventional method of adjusting green times in proportion to their q/s values.

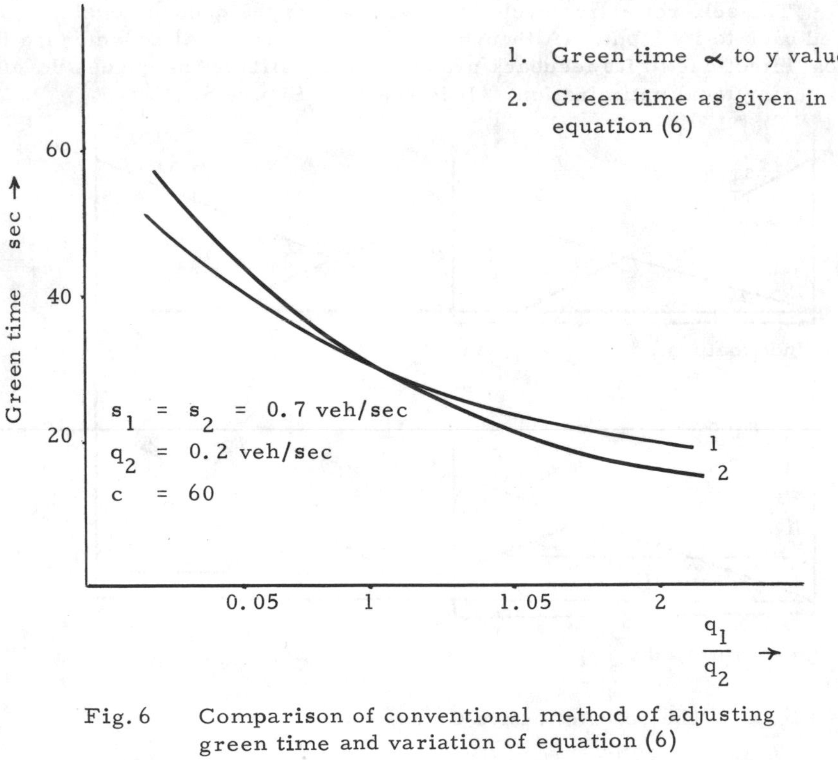

Fig. 6 Comparison of conventional method of adjusting green time and variation of equation (6)

if s-q < q on both arms,

$$\frac{\hat{g}}{\hat{c}-\hat{g}} = \frac{s_2 + q_2 - q_1}{s_1 + q_1 - q_2} \tag{9}$$

If s-q < q on any one arm while the other arm has light traffic, then

$$\hat{g} = \frac{q}{s} \cdot \hat{c} \quad \text{for the light arm} \tag{10}$$

COORDINATION PROCEDURE

In general terms, this becomes

1. $\hat{c}$ is evaluated on-line for each junction within the network.
2. Based on $\hat{c}$, the network is divided into sub-areas. These do not include the fixed sub-areas or input control junctions.
3. The value of cycle length used is based on a running average of $\hat{c}$ over J cycles, where J is an integer.
4. $\hat{\mu}$ is evaluated on-line for each priority arm in the network.
5. The value of offset used is based on a running average of $\hat{\mu}$ over J cycles.
6. The coordination vector ($\hat{c}$, $\hat{u}$, etc) is applied sequentially to the junctions within a sub-area.
7. Each junction having received its coordination vector starts its green. At the end of G_{min}, inputs to each junction is measured, the future inputs predicted. Based on this, the optimum split is calculated and applied.
8. At the end of J cycles, the above procedure is repeated.

The flow diagram in Figure 7 shows the line of control through the system.

COMMENT

Because of the short journey times involved in urban networks, i.e. journey times of links are in most cases shorter than cycle times, it is proposed to do a sensitivity analysis of the system to find for example, the effect on control variables of using a predictive model and the effect on the overall objective function of estimating certain control variables.

This paper is contributed by permission of the Director of Transport and Road Research, Department of the Environment.

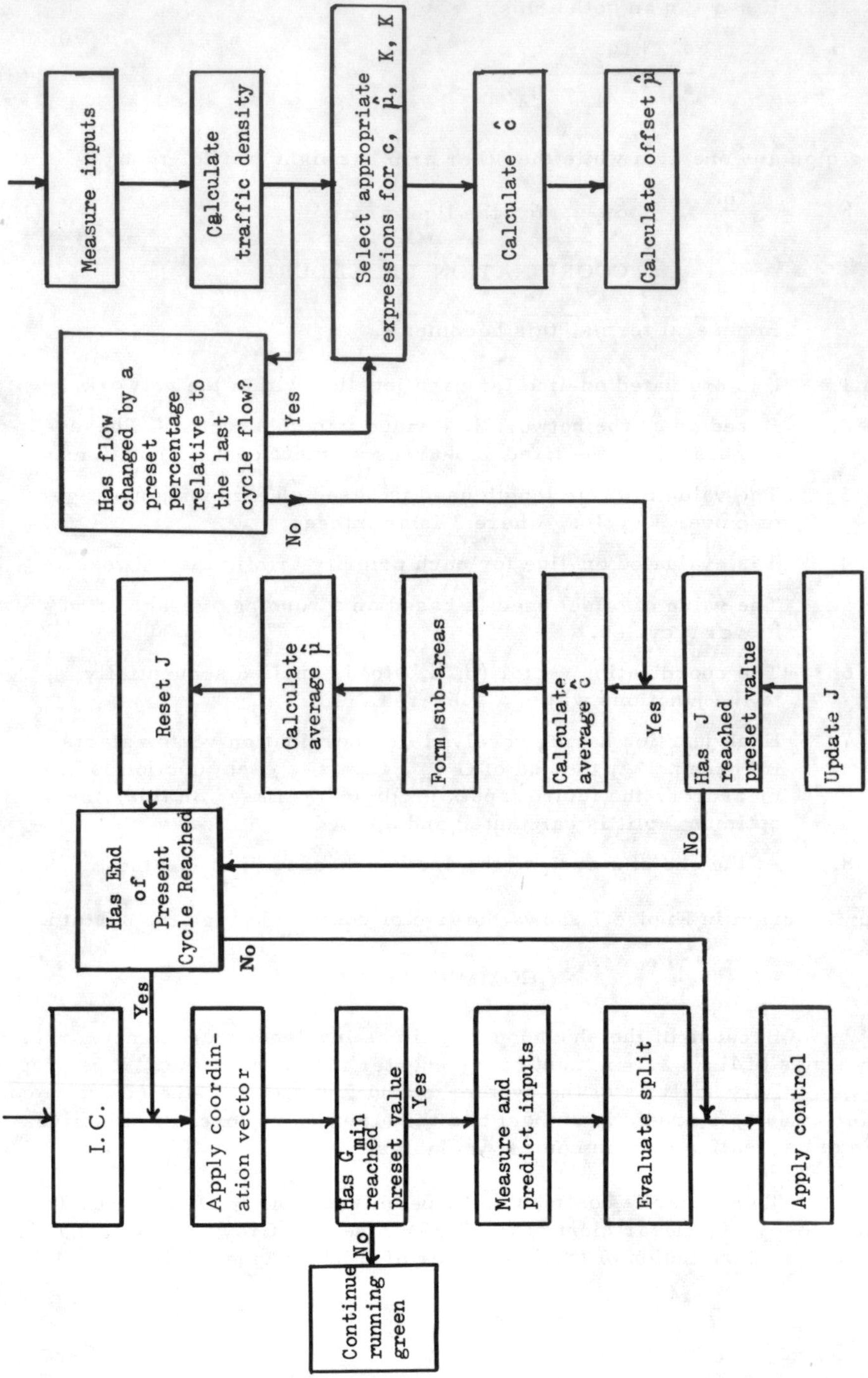
Measure inputs
Calculate traffic density
Select appropriate expressions for c, $\hat{\mu}$, K, K
Calculate $\hat{c}$
Calculate offset $\hat{\mu}$
Has flow changed by a preset percentage relative to the last cycle flow?
Yes
No
Reset J
Calculate average $\hat{\mu}$
Form sub-areas
Calculate average $\hat{c}$
Has J reached preset value
Yes
No
Update J
Has End of Present Cycle Reached
Yes
No
I. C.
Apply coordination vector
Has G_{min} reached preset value
Yes
No
Continue running green
Measure and predict inputs
Evaluate split
Apply control

REFERENCES

1. Al-Khalili, A. J. and Macleod, C. J.; "Development of a Model of an Urban Traffic System for Closed Loop Optimization Strategies", UTSG Conference, January 1972.

2. Fuehrer, H. H.; "Area Traffic Control - Madrid", IFAC/IFIP 1st Int Symposium, Versailles, June 1970.

3. Gazis, D. C. and Potts, R. B.; "The Oversaturated Intersection", Proc. 2nd Int Symposium on Traffic Theory, London, 1963.

4. Mesarovic, M. D., Macko, D. and Tahahara, Y.; "Theory of Hierarchial Systems", Academic Press, 1970.

5. Robertson, D. I.; "Transyt: A Traffic Network Study Tool", RRL Report LR. 253, UK Ministry of Transport, 1969.

6. Webster, F. V.; "Traffic Signal Settings, RRL Technical Paper No. 39, HMSO, London, 1958.

Hierarchical Strategies for the On-line Control of Urban Road Traffic Signals

by
M.G. Singh
Control Engineering Group
Department of Engineering
Cambridge University

Abstract

This paper examines the problem of dynamic optimisation of urban road traffic networks with a view to applying on-line control. From this point of view, the problem of the rush hour is quite distinct from that of the off peak period due to the different time scales involved. In the present paper, hierarchical strategies are developed for both the peak and the off peak periods. Due to their modest computational requirements, the strategies could enable optimal on-line control to be achieved. Simulation studies for both undersaturated and oversaturated systems illustrate the approach.

1. Introduction

The automobile is one of the largest sources of environmental pollution in cities. However, in the short term at least, cars are unlikely to be banned from urban areas and therefore it is worth investigating possible ways of reducing their impact upon the environment. One way of doing this is to attempt to improve the flow rates of vehicles within the city, since this would lead both to decreased wear and tear of vehicles and, more important, a reduction in the atmospheric pollution at present caused by vehicles continually stopping and starting, and moving for extended periods in low gears. Indeed, in Britain alone, the average annual delay to vehicles at traffic signals amounts to one hundred million vehicle hours, and this in itself represents a colossal social loss to the community.

With these considerations in view, a lot of work has been done by traffic engineers to reduce delays in urban road traffic networks culminating in various area traffic control schemes using digital computers, Hillier (1966). Substantial improvements in the flows, as measured by decreased delays, have been achieved. However, all the schemes used at present are off-line and use historical data to compute the signal control sequences, as a result of which they are unable to cope with disturbances. Facilities for on-line monitoring of most of the traffic variables do exist, however, and it is intuitively reasonable to believe that on-line control should make further improvements in delays.

This paper considers the dynamic optimisation problem for urban road traffic networks. From the point of view of control there are essentially two different problems, that of the rush hour and that of the off-peak period. The former is much more critical since this is the period when the service facilities are overloaded and the aggregate delays are very large. The off-peak problem is, in that sense, much less important. In the present paper, some preliminary ideas for the control of both undersaturated and oversaturated systems are formulated. A much more detailed study of realistic oversaturated systems is currently in progress and will be described elsewhere, Singh and Tamura (1973).

2. An Oversaturated intersection

An oversaturated intersection is one where queues remain at the end of the green and during peak periods most intersections in the average network would be oversaturated. Kulikowski,(Burhardt, K.K. and Kulikowski, R. (1970)) has proposed a simple model for such intersections and a discrete dynamical version of this model will be explained here. For simplicity of exposition, a simple one way, no turn intersection, as shown in Fig. 1 will be examined, although it is easy to extend the analysis to cover more complex junctions.

Fig. 1

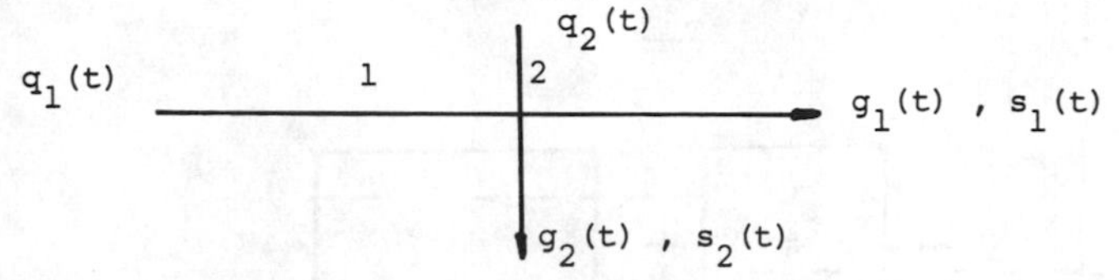

$q_i(t)$ denotes the arrival rates of vehicles (number of vehicles per minute) in the direction i. $i=1$ denotes the horizontal traffic direction and $i=2$ the vertical traffic direction. $s_i(t)$ denotes the saturation flow through the intersection in the direction i. This is the maximum number of vehicles per cycle which can pass through the intersection in the direction i if this direction had all the available green.

Let C be the duration of the cycle time and L the loss time in the cycle due to the amber phases then

$$C = G_1 + G_2 + L$$

where G_i $i=1,2$ is the duration of the green in the direction i. Let $g_i(t)$ be the average (per cycle) departure rate of vehicles in the direction i (in number of vehicles per C). Figures 2(a) and (b) show the essential relationships between these variables for a typical sequence of departures. Note that because of the oversaturation, queues never vanish so that the outflows are always approximately square wave. This enables one to choose the control variables $u_i(t)$ to be such that they are related very simply to the outflow rates. Let $u_i(t)$ be defined as the percentage of the green in the direction i.

i.e. $u_i(t) = g_i(t)/s_i(t) = G_i/C \qquad i=1,2$

In this formulation, the cycle time C is assumed to be a known constant. The justification of this is the classical experiment of Webster, Webster, F.V. and Cobb, B.M. (1966),which showed that for the cycle time to vary 50% on either side of the minimum delay point, the increase in delay is only 10%.

For reasons of safety, it is necessary to have minimum and maximum greens. This imposes inequality constraints on the control variables

i.e. $U_i \leq u_i(t) \leq V_i \qquad i=1,2 \qquad (1)$

where U_i , V_i are positive constants. Also, $u_1 + u_2 = 1-L/C = EG$ where EG is the effective green. From this it is possible to have only one control variable per intersection. If u_1 is denoted by u , then $u_2 = EG - u$.

Figs. 2(a) and (b)

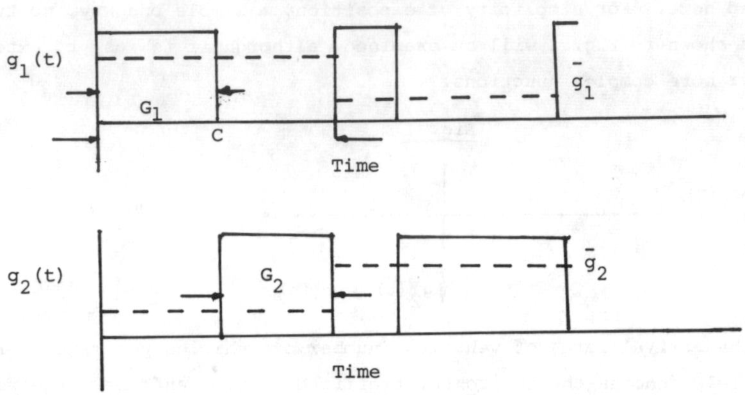

Since the delays at an intersection are given by the time spent queuing, a suitable state vector is the vector of queues on the two arms. If average values are used over one cycle then the dynamical evolution of the queues can be described by the linear difference equation

$$x_i(k+1) = x_i(k) + q_i(k) - g_i(k) , \quad i=1,2 .$$

and since $g_i(k) = s_i u_i(k)$ the dynamic equation becomes

$$x_i(k+1) = x_i(k) + q_i(k) - s_i u_i(k) \quad .... \tag{2}$$

Due to the physical limitations on the road network, the queues must be subject to inequality constraints

$$0 \leq x_i(t) \leq X_i \qquad i=1,2 .... \tag{3}$$

3. Dynamical optimisation of a single oversaturated intersection

Having discussed the elementary dynamical behaviour of an oversaturated intersection, it is now possible to consider its control. In order to apply "optimal" control, it is necessary to have a suitable criterion function which measures the performance of any given feasible control signal sequence. The essential objective of any control scheme during the rush hour should be to reduce the delays. Since the delays are measured by the time spent queuing, the summation of the queues over the rush hour should give an appropriate measure. A secondary objective is the degree of utilisation of the service facility. Thus a suitable cost function is:

$$I = \min_{u_1, u_2} \sum_{k=1}^{N} \sum_{i=1}^{2} K_1 x_i^2(k) + K_2 s_i (u_i(k) - V_i)^2 \quad \ldots . \qquad (4)$$

where N is the period of optimisation, K_1, K_2 are weighting factors and V_i is the maximum green in the direction i. The quadratic forms have been chosen primarily for analytical tractability although they do have a physical significance, this being that bigger queues and lower utilisation of the facility are penalised more heavily. In practice, since delays are the most significant factor, K_2 will be small.

The optimisation problem for a single intersection can be formulated as follows:

> Minimise the functional I given by equation (4) subject to the dynamic equality constraints given by equation (2) and the inequality constraints on the states and controls given by equations (3) and (1).

4. Dynamical optimisation of oversaturated networks

Networks comprise junctions and interconnecting roads arranged in a two dimensional cascade. Since for oversaturated junctions, only the macrobehaviour is important, it is adequate to model interconnecting roads as pure delay elements. Then the output $g_{ij}(k) = s_{ij} u_{ij}(k)$ of the jth intersection becomes the delayed input of the $j+1^{th}$ intersection

i.e. $q_{i\ j+1}(k+1) = s_{ij} u_{ij}(k)$ where l is the number of delay periods. Then the overall network can be represented by a linear vector matrix difference equation with pure delays in the controls:

$$\underline{x}(k+1) = A\underline{x}(k) + B_o\underline{u}(k) + B_1\underline{u}(k-1) + \ldots + B_l\underline{u}(k-l) + \underline{C} \quad \ldots . \qquad (5)$$

$$k=0,1,\ldots N-1$$

where $\underline{x}$ is the vector of queues on all the arms of all the junctions. A is an identity matrix, B_j, $j=0,1,\ldots l$ are control weighting matrices, $\underline{u}(k)$ is the control vector and $\underline{u}(k-j)$, $j=0,1,2,\ldots l$ are the delayed controls used to simulate the pure delays on the interconnecting roads. $\underline{C}$ is the vector of inputs which come from outside the network boundary.

Similarly, the cost function could be written as

$$J = \min_{\underline{u}(0),\ldots\underline{u}(N-1)} \sum_{k=0}^{N-1} \| \underline{x}(k) \|^2_{Q(k)} + \| \underline{u}(k) - V \|^2_{R(k)} \quad \ldots . \quad (6)$$

where Q and R are diagonal weighting matrices.

The states and controls are subject to inequality constraints:

$$\underline{O} \leq \underline{x}(k) \leq \underline{X} \quad \ldots . \qquad (7)$$

$$\underline{U} \leq \underline{u}(k) \leq \underline{V} \quad \ldots . \qquad (8)$$

The oversaturated network optimisation problem is therefore to minimise equation (6) subject to the constraints given by equations (5), (7) and (8). In general this is a fairly difficult problem to solve due to the presence of the delays in the control variables and inequality constraints on both the states and controls even though the actual system dynamics are linear and the cost function is quadratic. In addition, the dimensionality problems could easily become overwhelming even for fairly small networks. However, Tamura, Tamura, H. (1973), has recently developed a powerful hierarchical optimisation algorithm for this class of problems. The basis of the technique is that it is possible to formulate the delay problem as a goal coordination problem without increasing the dimensionality of the system due to the presence of the delays.

Define a dual function $\phi(\underline{p}) = \phi(\underline{p}(0), \underline{p}(1), \ldots, \underline{p}(N-1))$ .... (9)

Then the maximisation of the dual function $\phi(\underline{p})$ with respect to $\underline{p}$ yields the overall optimum. The dual function $\phi(\underline{p})$ can be written in terms of the Lagrangian function $L(\underline{x},\underline{u},\underline{p})$ as

$$\phi(\underline{p}) = \min_{\underline{x},\underline{u}} \left[L(\underline{x},\underline{u},\underline{p}) \ : \ \text{equations (7) and (8)} \right] \quad \ldots. \qquad (10)$$

The Lagrangian $L(\underline{x},\underline{u},\underline{p})$ is defined as

$$L(\underline{x},\underline{u},\underline{p}) = \sum_{k=0}^{N-1} \| \underline{x}(k)\|_Q^2 + \|\underline{u}(k)-\underline{V}\|_R^2 - \underline{p}(k)\left[\underline{x}(k+1) - A\underline{x}(k) - B_0\underline{u}(k) - B_1\underline{u}(k-1) - \ldots B_1\underline{u}(k-1) - \underline{C}(k)\right] \quad \ldots. \qquad (11)$$

The computational savings arise because the Lagrange multiplier $\underline{p}(k)$ is of the same dimension as the state vector $\underline{x}(k)$ despite the presence of the delays. The actual solution of the max min problem can be performed using a two level goal coordination algorithm as discussed below. It should be noted that the convergence to the optimum is always guaranteed due to the concavity of the dual function $\phi(\underline{p})$.

4.1 The two level goal coordination algorithm of Tamura

At some value of the Lagrange multiplier $p=p^*$, the gradient of the dual function is given by

$$\left.\nabla \phi(p)\right|_{\underline{p}(k)=p^*(k)} = -\underline{x}^*(k+1) + A\underline{x}^*(k) + B_0\underline{u}^*(k) + B_1\underline{u}^*(k-1) + \ldots B_1\underline{u}^*(k-1) + \underline{C}(k)$$

$$= \underline{g}(k) \qquad (12)$$

where $\underline{x}^*$, $\underline{u}^*$ are the state and controls at $p=p^*$. Thus the gradient is simply the error in the system equation. The simple two level algorithm is:

Step 1

Choose initial guessed trajectory of the Lagrange multiplier $\underline{p}^1 = (\underline{p}^1(0), \underline{p}^1(1), \ldots, \underline{p}^1(N-1))$ and send to level 1.

Step 2

Set iteration index $i=1$. Using the given values of $\underline{p}$, solve N independent minimisation problems (one for each time point). It is possible to decompose the problem in time because of the additive separability of the Lagrangian $L(\underline{x},\underline{u},\underline{p})$ with respect to $\underline{x}$ and $\underline{u}$ for fixed $\underline{p}$. For this linear quadratic case, it is possible to use an analytical solution at this level. Also since Q and R are assumed to be diagonal matrices, each of the N independent minimisation problems can be written as a set of $(n+r)$ independent one dimensional minimisation problems where n is the order of the overall state vector and r is the order of the overall control vector. For a one dimensional minimisation problem, it is of course very easy to include hard inequality constraints.

Step 3

Calculate the gradient g^i at p^i using equation (12). This gradient is then sent to level two.

Step 4

At level 2, the p^i trajectory is improved using say the conjugate gradient algorithm so that

$$\underline{p}^{i+1} = \underline{p}^i + \alpha^i \underline{d}^i$$

where $\underline{d}^i$ is the conjugate direction and α^i is that step length which maximises $\phi(\underline{p}^i + \alpha^i \underline{d}^i)$. The search direction $\underline{d}^i$ can be calculated using the relationahip

$$\underline{d}^i = \underline{g}^i + \beta^{i-1} \underline{d}^{i-1}$$

where $\beta^i = \underline{g}^{i^T} \underline{g}^i / \underline{g}^{i-1^T} \underline{g}^{i-1}$

Initially, $\underline{d}^0 = \underline{g}^0$ i.e. the steepest ascent direction.

The iterations are terminated whenever the gradient $\underline{g}^i$ reaches an acceptably low value and the corresponding control trajectory is recorded as the optimal one.

5. A network optimisation simulation study

As a simple example, a small network of two one-way junctions was simulated on the IBM 370 digital computer at Cambridge University using a program developed by Professor Tamura for general linear dynamical systems. Fig. 3 shows the network configuration.

Fig. 3

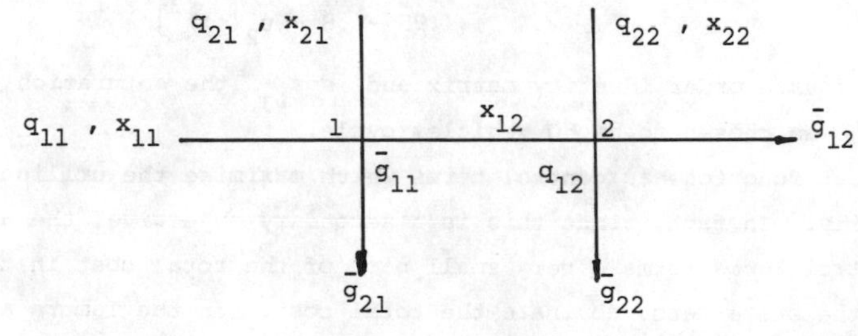

In this study it was assumed that the cycle time of each of the two junctions is 100 seconds and a pure time delay of one cycle exists between the two junctions, i.e.

$$q_{12}(k) = g_{11}(k-1)$$

The loss time L for each junction was assumed to be 5 seconds. The control variables were chosen to be the horizontal direction green to cycle ratio u_1 and u_2 for junction 1 and 2. The vertical direction controls are then $0.95-u_1$ and $0.95-u_2$ for junctions one and two.

Then, the state equations can be written as:

$$\begin{bmatrix} x_{11}(k+1) \\ x_{21}(k+1) \\ x_{12}(k+1) \\ x_{22}(k+1) \end{bmatrix} = \begin{bmatrix} 1 & & & 0 \\ & 1 & & \\ & & 1 & \\ 0 & & & 1 \end{bmatrix} \begin{bmatrix} x_{11}(k) \\ x_{21}(k) \\ x_{12}(k) \\ x_{22}(k) \end{bmatrix} + \begin{bmatrix} -s_{11} & 0 \\ s_{21} & 0 \\ 0 & -s_{12} \\ 0 & s_{22} \end{bmatrix} \begin{bmatrix} u_1(k) \\ u_2(k) \end{bmatrix}$$

$$+ \begin{bmatrix} 0 & 0 \\ 0 & 0 \\ s_{11} & 0 \\ 0 & 0 \end{bmatrix} \begin{bmatrix} u_1(k-1) \\ u_2(k-1) \end{bmatrix} + \begin{bmatrix} q_{11} \\ q_{21} - s_{21} \times 0.95 \\ 0 \\ q_{22} - s_{22} \times 0.95 \end{bmatrix}$$

the state constraints were chosen to be:

$$0 \leq x_{11} \leq 200 \;,\; 0 \leq x_{21} \leq 200 \;,\; 0 \leq x_{12} \leq 40 \;,\; 0 \leq x_{22} \leq 200$$

This implies that the interconnecting road is assumed not to be able to hold more than forty vehicles during any cycle whereas the other roads can hold much more. The control variables were constrained to lie in the region

$$0.2 \leq u_i \leq 0.8$$

these being reasonable values for minimum and maximum green. The cost function was chosen to be

$$J = \tfrac{1}{2} \sum_{k=0}^{N-1} \|\underline{x}(k)\|^2_{I_4} + s\big[(u_1(k) - 0.8^2 + (.95 - .8 - u_1(k))^2 + (u_2(k) - .8)^2 + (.95 - .8 - u_2(k))^2\big]$$

where I_4 is the fourth order identity matrix and $s=s_{ij}$ the saturation flow rate. In this study, s was chosen to be 60 vehicles/cycle.

The above cost function has control terms which maximise the utilisation of the service facility. In fact, since this is a secondary objective, the actual magnitude of the control terms forms a very small part of the total cost in the examples studied below as the state terms dominate the total cost. In the Tamura algorithm, it is necessary to ensure that $u_i(k) = 0$ for k less than zero. This can be achieved by choosing the control variable to be deviations around zero.

Let $\Delta\underline{u} = \underline{v}-\underline{u}^*$ where $\underline{u}^*$ is the steady state value and in this case $u^* = 114/2\times120 = 0.475$. Then $\underline{v} = \underline{u}^* + \Delta\underline{u} = 0.475 + \Delta\underline{u}$. Then the inequality constraints on the controls become

$$0.2 \leq 0.475 + \Delta\underline{u} \leq 0.8$$

or

$$-0.175 \leq \Delta\underline{u} \leq 0.225$$

where $\Delta\underline{u}$ is now the deviation about the $\underline{u}^*$ vector of $\begin{bmatrix}0.475\\0.475\end{bmatrix}$

The state equations therefore become

$$\underline{x}(k+1) = \underline{x}(k) + \begin{bmatrix}-60 & 0\\ 60 & 0\\ 0 & -60\\ 0 & 60\end{bmatrix}\Delta\underline{u}(k) + \begin{bmatrix}0 & 0\\ 0 & 0\\ 60 & 0\\ 0 & 0\end{bmatrix}\Delta\underline{u}(k-1) + \begin{bmatrix}q_{11}\\ q_{21} & -57\\ 0\\ q_{22} & -57\end{bmatrix}$$

$$+ \begin{bmatrix}-60 & 0\\ 60 & 0\\ 0 & -60\\ 0 & 60\end{bmatrix}\begin{bmatrix}0.475\\0.475\end{bmatrix} + \begin{bmatrix}0 & 0\\ 0 & 0\\ 60 & 0\\ 0 & 0\end{bmatrix}\begin{bmatrix}0.475\\0.475\end{bmatrix}$$

and the cost function is modified to:

$$J = \tfrac{1}{2}\sum_{k=0}^{N-1} \|\underline{x}(k)\|^2_{I_4} + \|\Delta\underline{u}(k)\|^2_{I_2}$$

5.1 Example 1

In an initial study, the inflow rates q_{11}, q_{21} and q_{22} were chosen to be 20, 30 and 30 vehicles per cycle and initial queues were assumed to be:

$$x_{11} = 100\ ,\ x_{21} = 100\ ,\ x_{12} = 40\ ,\ x_{22} = 100\ .$$

such a situation could arise during the peak of the rush hour when inflow rates are still quite large and big queues have built up. The object is to dissipate the queues optimally. In general, the rush hour in most cities is a fairly repeatable phenomenon and fairly accurate values of the average inflow rates coming from outside the overall system boundary are often available. In this study, N , the period of optimisation, was chosen to be 10 cycles. Figure (4) shows the simulated trajectories of the queues with the optimal signal sequence as calculated from the algorithm of section 4.1. The optimal control sequence for the 10 periods = 16.6 minutes is given in Table 1. Note that even though the inflow rate into the queue x_{11} is 33% less than the inflow rates q_{21} and q_{22}, minimum cost is achieved by allowing x_{11} to build up to 120 vehicles in the first three periods before discharging it monotonically. This allows x_{21} and x_{22} to be discharged more rapidly. In this example, the queue on the interconnecting road, x_{12} remains at its constraint level of 40 vehicles over the whole of the period of optimisation period. The reason for this is that no special weighting was assigned to this queue in the cost function and since none of the other queues ever get below 40 vehicles, its contribution to the overall cost is the smallest in any case. In practice, it may often be desirable to dissipate x_{12} more rapidly and this can be done by increasing its weighting in the cost function. The signal control sequence in table 1 rarely hits the constraints. The optimal and dual costs were computed to be identical at 102500. Convergence took place in 228 iterations and the total computation time for this problem with 40 dual variables was 1 minute 17 seconds.

TABLE 1

k	1	2	3	4	5	6	7	8	9	10
u_1%	20	20	26.9	41	41.5	41.3	40.6	40	41	77.2
u_2%	25.4	20	20	27.7	41.5	41.1	41.7	41.8	42.2	42.1

5.2 Example 2

As another example, the case when the inflows are much smaller is considered. Here $q_{11} = q_{21} = q_{22} = 10$ vehicles per cycle. Such a situation could arise near the end of the rush hour when the inflows are relatively low although large queues exist. Using the same constraints and initial conditions as for example 1 but for $N=5$ cycles, the optimal signal sequence was calculated. The optimal control sequence is given in Table 2 and the state trajectories corresponding to this control sequence are given in Fig. 5. In this case all the queues dissipate quite rapidly due to the relatively smaller inflow rates. Even x_{12} is reduced to less than half its initial value in 5 cycles. Again, the constraints on the control variables are rarely hit. For this problem with 20 dual variables, convergence took place in 91 iterations and the computation time was 14.4 seconds. The optimal cost was 22370 and the dual cost was 22360.

Comment: Where a single processor is used, as in the examples above, the computation time increases rapidly with the increase in the number of dual variables. Since the number of dual variables is the product of the number of queues and the number of periods of optimisation, large networks can be handled if the number of periods of optimisation is small. At the present time, offline optimisation is performed, Huddart (1972), based on average flows of two hours. It is estimated that the present algorithm could tackle the problem of 200 queues for one optimisation period in roughly 5 minutes of computation time on the IBM 370. For on-line control of large systems, it may be necessary to use parallel processors and it is for such systems that the hierarchical strategy is best suited.

TABLE 2

k	1	2	3	4	5
u_1%	20	47	53.5	53.7	63.2
u_2%	47.7	20	57.4	65.8	65.8

6. Undersaturated systems

The problem of optimal control of undersaturated intersections is not a particularly critical one since the aggregate delays are fairly small in magnitude. An undersaturated junction is one where no queues exist on an arm at the end of the green period. It is not valid therefore to use the coarse averaging period of one cycle as for oversaturated intersections since it is the time spent queuing during the cycle which causes the delays. Thus for real time control of undersaturated systems, it may be necessary to sample as often as once a second. Although this does not pose any technical problems for the actual monitoring of the variables since this is presently done at a much faster rate than once a second, the computational difficulties for the dynamic optimisation become acute. The reason for this is that as the period of optimisation increases, the computational burden becomes much larger. Also, the microbehaviour of the junctions invlove a number of Boolean functions which are very difficult to handle analytically in the control optimisation. However, the most significant difficulty that arises due to the increased sampling rate is due to the fact that it is no longer valid to use values of inflows and outflows averaged over a cycle; rather, it is necessary to use the detailed second by second trajectories of the inflows. In practice, this information may be available only up to one cycle ahead by placing suitable sensors outside the overall system's boundary. The inflow trajectories are not available over the whole of the off peak period. It is therefore not even theoretically possible to perform the optimisation over a long time horizon. In fact, it is only meaningful to perform the optimisation over the next cycle. This fact, however, simplifies the on-line control problem considerably as discussed in the following sections.

6.1 A model for undersaturated junctions

The general microbehaviour of single junctions is well known. When an input arm sees a red signal, queues build up. When the lights are green, discharge of the queue takes place at the saturation flow rate followed by free flow of traffic from input to output. Fig. 6 shows a single one way no turn junction. Let q_1 and q_2 be the inflow rates and g_1 and g_2 the outflow rates. Let x_1 and x_2 be the

Fig. 4

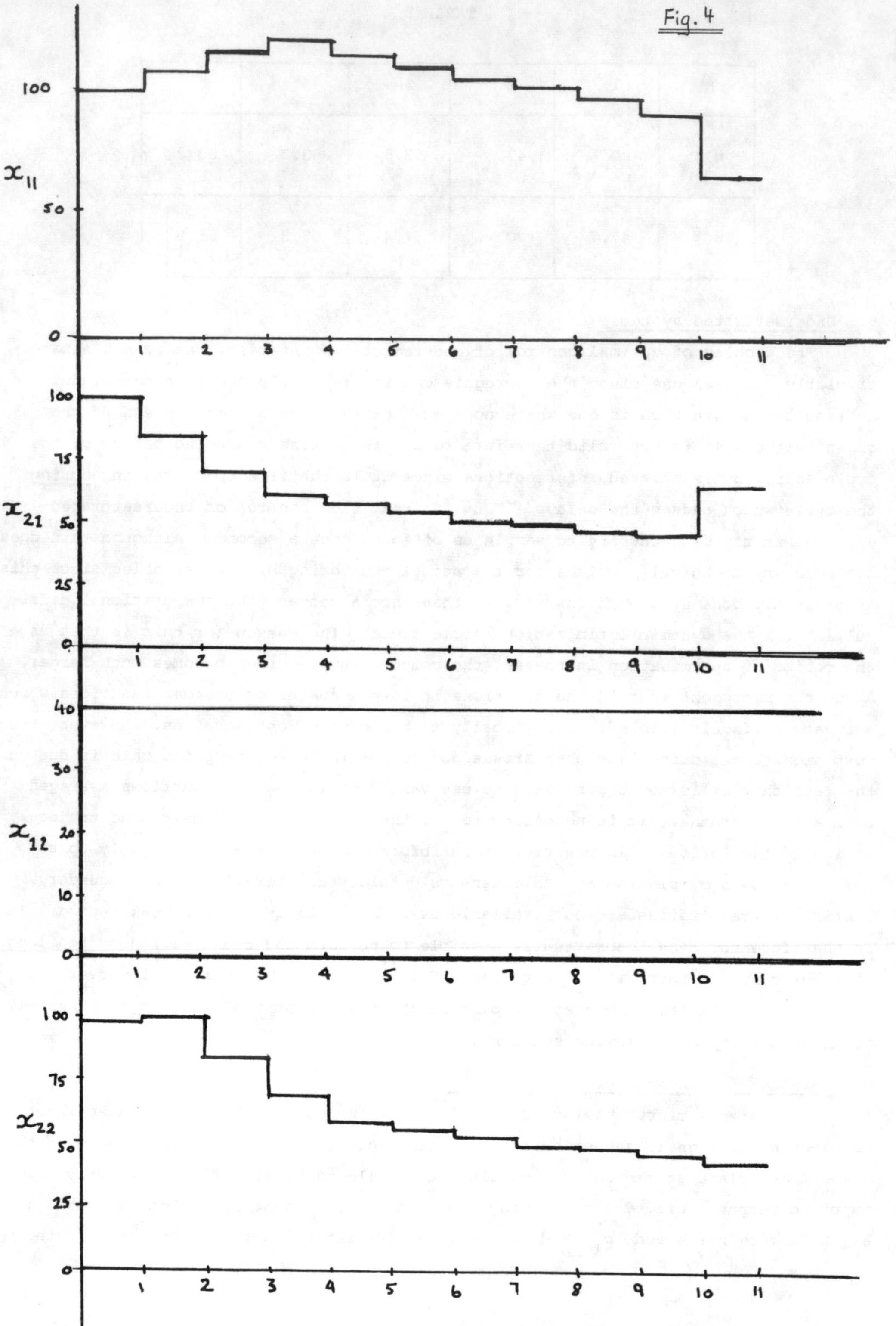

100
50
0
x_{11}
1 2 3 4 5 6 7 8 9 10 11
100
75
50
25
0
x_{21}
1 2 3 4 5 6 7 8 9 10 11
40
30
20
10
0
x_{12}
1 2 3 4 5 6 7 8 9 10 11
100
75
50
25
0
x_{22}
1 2 3 4 5 6 7 8 9 10 11

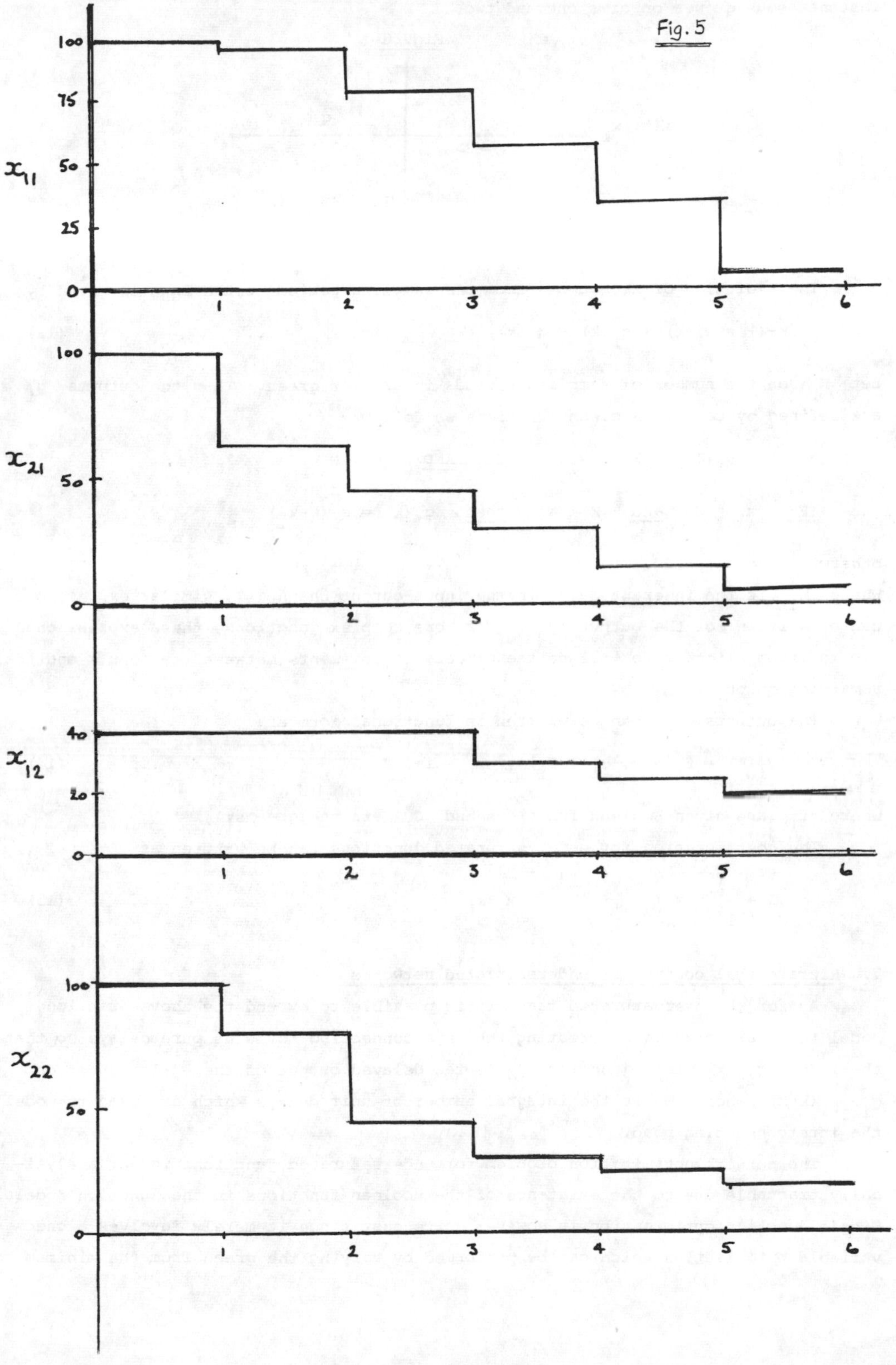

Fig. 5

instantaneous queues on arms one and two.

Fig. 6

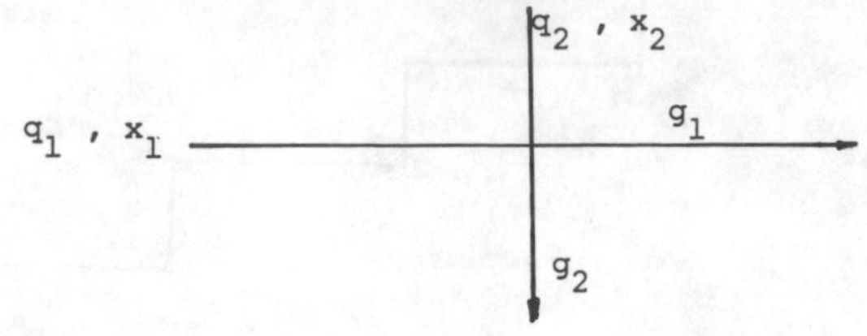

The evolution of the queues from one sampling period to the next is given by

$$x_i(k+1) = x_i(k) + q_i(k) - g_i(k) \qquad i=1,2 \qquad \ldots. \qquad (12)$$

Let N_1 be the number of sampling periods in arm 1's green. Then the outputs g_i are defined by certain Boolean functions as follows:

$$g_1(k) = s_1(k) \quad \underline{\text{IFF}} \quad x_1(k) > 0 \quad \underline{\text{AND}} \quad k < N_1$$
$$\underline{\text{IFF}} \quad x_1 = 0 \quad \underline{\text{and}} \quad k < N_1 \quad \underline{\text{then}} \quad g_1(k) = q_1(k-k_1) \qquad \ldots. \qquad (13)$$

otherwise, $g_1(k) = 0$.
Where k_1 is the inherent delay in the input-output channel 1, similar expressions can be written for the output g_2 . For more complex junctions, these expressions are modified slightly to include the effects of movements between the inputs and the remaining outputs.

The outputs g_i can be written in functional form as:

$$g_i(k) = f_i(q_i\ ,\ s_i\ ,\ G_i)\ , \qquad i=1,2 \qquad \ldots. \qquad (14)$$

where f_i are given Boolean functions and G_i are the greens.

The cost function for under saturated junctions can be written as

$$I = \sum_{i=1}^{2} \sum_{k=1}^{N} x_i(k) \qquad \ldots. \qquad (15)$$

7. Hierarchical control of undersaturated networks

As for the oversaturated case, it is possible to extend the above junction model to cover networks by treating the interconnecting roads as pure delays so that the input $q_{ij}(k)$ into junction j is the delayed output of the j-1th junction, $g_{i\ j-1}(k-M)$ where M is the integral number of unit delays which are used to model the interconnecting road.

The network optimisation problem for undersaturated junctions is not analytically tractable due to the existence of the Boolean functions in the junction models. Single junction optimisation is however quite easy since it merely involves a one variable optimisation which can be performed by varying the green from the minimum to the maximum and finding the value which gives minimum I in equation (16). Now, since the meaningful period of optimisation is <u>one cycle</u>, and because of the pure delays between the junctions, all the junctions become decoupled for the duration of the cycle provided all the inputs into the junctions are available. The reason for this decoupling is that the inputs q into the junctions are independent of each other during any cycle, depending as they do on outputs from previous cycles as opposed to current cycle outputs. For this reason, it is possible to set up a simple hierarchical control structure for undersaturated systems as shown in Fig. 7.

<u>Fig. 7</u>

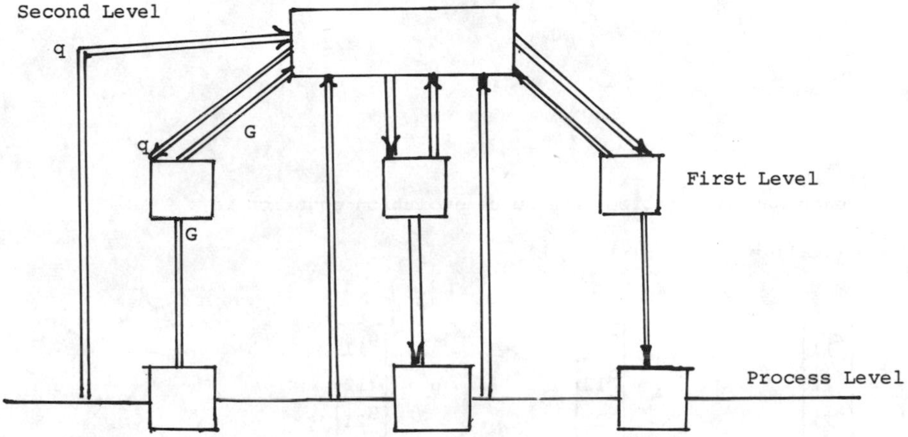

On level 1, the simple one dimensional minimisation of equation 15 is performed subject to the constraints of equations 12 and 14. For this minimisation, it is necessary for each junction controller to have knowledge of inputs q_i into the junction. This information is provided by the second level which calculates, in fast time, the inputs using the controls used in the previous cycle as calculated by the first level.

Such a scheme can be used for on-line control since its computational requirements are quite minimal.

8. <u>Example</u>

As an example of the approach to the control optimisation of undersaturated networks, consider the simple two junction, one way, no turn system shown in Fig. 8. This network has the same configuration as for the oversaturated example in Section 5. Let the delay between the two junctions be again one cycle and let the saturation flow rate for all the outputs be 70 vehicles per cycle. Let the sampling rate be 10% of the cycle time and let the inflow rates be uniform during the cycle at $q_{11} = 10$ vehicles/cycle, $q_{21} = q_{22} = 20$ vehicles/cycle and during the first cycle, let $q_{12} = 10$ vehicles/cycle. Let the control constraints be

$$0.3 \leq u_i \leq 0.7 \qquad i=1,2$$

and the state constraints

$$0 \leqslant \underline{x} \leqslant 20$$

Fig. 8

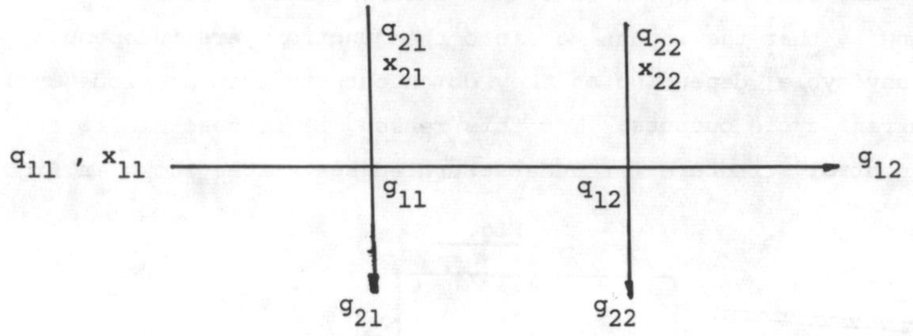

For each sampling period, the queue evolution equation is

$$\underline{x}(k+1) = \underline{x}(k) + \underline{q}(k) - \underline{g}(k)$$

where

$$\underline{x} = \begin{bmatrix} x_{11} \\ x_{12} \\ x_{21} \\ x_{22} \end{bmatrix} \qquad \underline{q} = \begin{bmatrix} q_{11} \\ q_{12} \\ q_{21} \\ q_{22} \end{bmatrix} \qquad \underline{g} = \begin{bmatrix} g_{11} \\ g_{12} \\ g_{21} \\ g_{22} \end{bmatrix}$$

and the outputs $\underline{g}$ are given by Boolean relationships of the type described in Section 6. But, $q_{12}(k) = g_{11}(k-10)$

The cost function for this example is

$$J = \min_{u_1, u_2} \sum_{k=1}^{10} \underline{x}(k)$$

The simple hierarchical strategy was simulated for 20 sampling periods = 2 cycles . For the first cycle, the one dimensional minimisations for each junction performed on the first level yielded the optimal controls:

$$u_1 = G_1/C = 0.3 \ ; \qquad u_2 = G_2/C = 0.3$$

where G_1 and G_2 are the horizontal direction greens for junctions one and two. Using these controls, the second level controller computed the inflow trajectory $q_{12}(k)$ into junction 2 over cycle 2. This trajectory is shown in Fig. 9. On the first level, then, using this input and the given inflow trajectories q_{11} , q_{21} and q_{22} , the optimal controls for the second cycle were calculated to be

$$u_1 = u_2 = 0.3.$$

Fig. 9

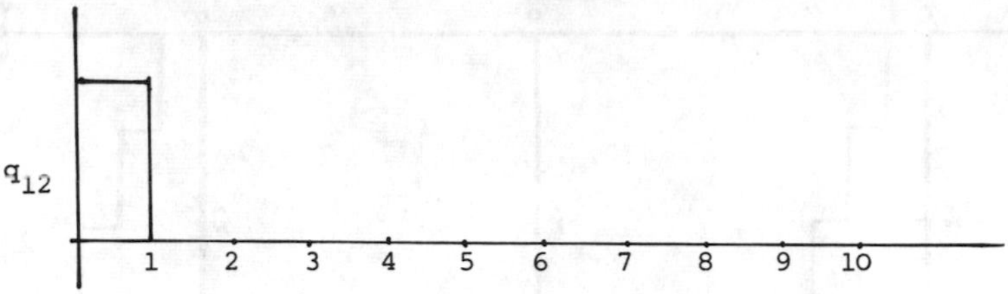

The optimal cost in the first cycle was 80, and each junction contributed a half of this. For the second cycle, the optimal cost was 101, of which the first junction contributed 40 and the second junction 61. Fig. 10 shows the optimal state trajectories over the 20 periods.

9. Conclusions

At the present time, on-line control of urban road traffic is virtually non-existent, so that any study of the problem, such as the one outlined in this paper, must be of a rather preliminary and tentative nature. However, such studies can be of practical use if only in clarifying the nature of the problem and highlighting those aspects which will need further detailed attention. Some of the points raised in the present paper are as follows:

a) The simple hierarchical approach to the optimisation of oversaturated systems has potential for application to on-line control due to its modest computational requirements. If parallel processing is used at the first level, the scheme could be used for the on-line control of the largest possible networks. It could also be used for offline optimisation in which case it could handle large networks without using parallel processors. More detailed practical studies of this problem are in progress and will be described elsewhere.

b) The hierarchical scheme for undersaturated systems has even smaller computational requirements since the period of optimisation is only one cycle and the second level does virtually no work.

Although, only simple one way networks have been considered in this paper, this has been primarily for ease of exposition and it is very easy to extend this work to cover more complex networks.

Acknowledgements

The author is indebted to his research supervisor Professor J.F. Coales for his encouragement and guidance, to Professor H. Tamura for many helpful discussions as well as for the use of his computer program, and to Dr. A.T. Fuller for reading the manuscript and his useful comments.

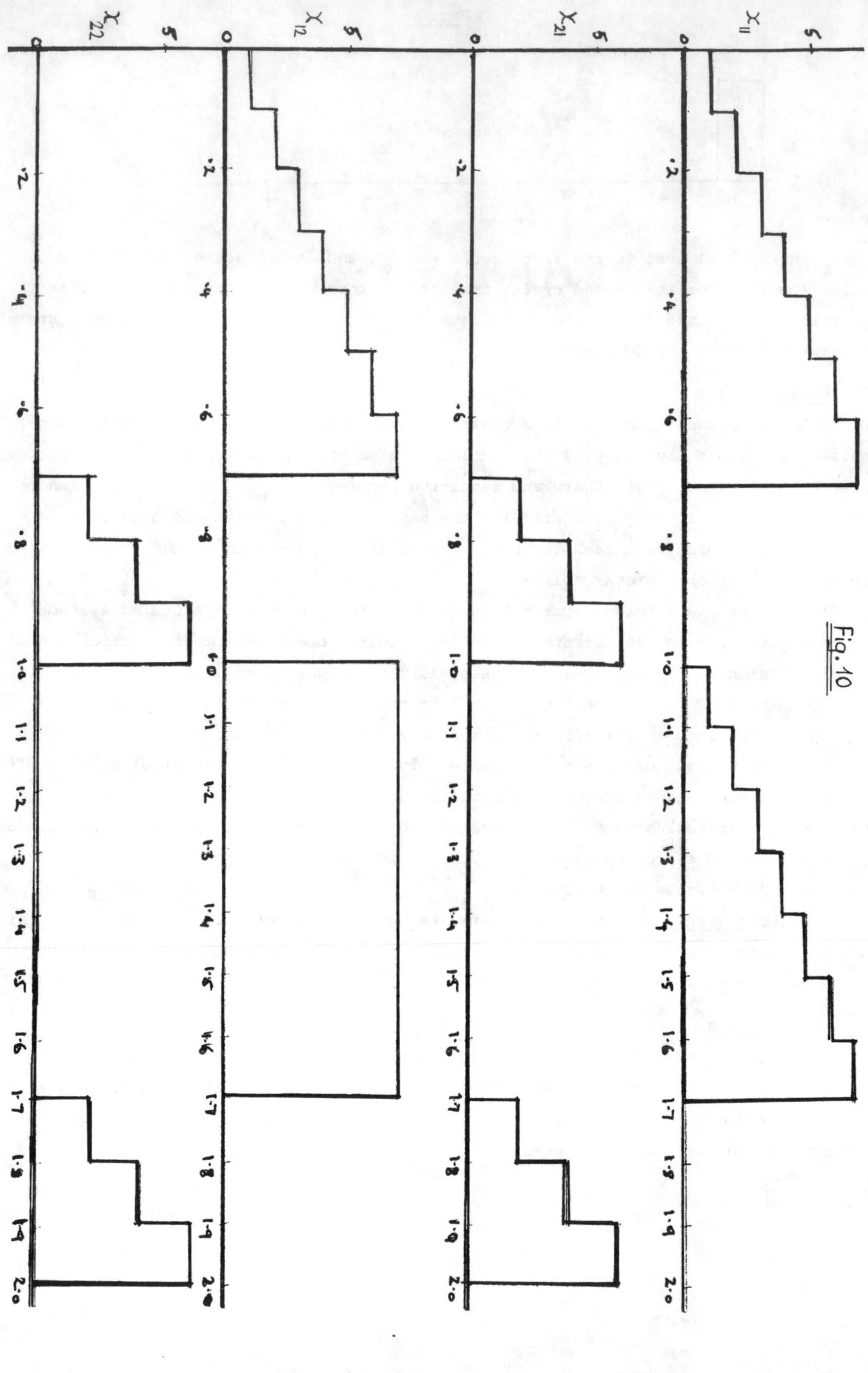

Fig. 10

References

1. Burhardt, K.K. and Kulikowski, R: "Optimum control of oversaturated traffic systems I: General Theory". Bull.Acad. Pol. Sci. Ser. Sci. Tech. (Poland), Vol. XVIII, No. 7, pp. 573-580, (1970).

2. Hillier, J.A.: "Glasgow's experiment in area traffic control", Part II, Traffic Engineering and Control, 7 (9), Jan. 1966, pp. 569-571.

3. Huddart, K.: of the Greater London Council. (Private discussion), 1972.

4. Singh, M.G., and Tamura, H.: "Hierarchical optimisation for oversaturated road traffic networks". Abstract submitted for the Second IFAC Symposium on "Traffic control and transportation systems", Cote d'Azur, 1974.

5. Tamura, H.: "Application of duality and decomposition in high order multistage decision processes", University of Cambridge Report, CUED/B-Control/TR49 (1973). Under consideration for publication.

6. Webster, F.V. and Cobb, B.M.: "Traffic signals", Road Research Technical Paper 56, HMSO, (1966).

APPLICATION OF OPTIMIZATION APPROACH TO THE PROBLEM OF LAND USE PLAN DESIGN

Kumares C. Sinha, Associate Professor of Civil Engineering

and

Alois J. Hartmann, Associate Professor of Civil Engineering

Marquette University
Milwaukee, Wisconsin, USA

ABSTRACT

This paper deals with an application of optimization technique to the problem of land use plan design: a spatial design of primary land uses in a given area and the associated linkage requirements for transportation and utilities. Using a random search procedure the discrete land use elements, called modules, are assigned to discrete land areas, known as cells. An optimal plan is obtained which satisfies the given design constraints as well as minimizes the private and public costs which include site development and linkage costs. This normative approach can be used to arrive at an ideal land use plan design which then can be considered as a standard to evaluate other alternatives. The validity of the proposed technique has been established through a small-scale controlled experiment. The large-scale application of the technique is illustrated by using the example of the Village of Germantown, Wisconsin. Apart from providing an optimal feasible plan the model also makes a sensitivity analysis for given design constraints. Furthermore, a plotting subroutine allows the model output to be given in the form of a map generated on-line. It is believed that such a model can be an extremely useful tool in the land use planning process.

INTRODUCTION

The concept of a land use plan design model is to prepare an ideal land use plan for an area at some target year -- an ideal plan that will minimize the total private and public costs as well as satisfy the community development objectives and design standards. Thus the land use plan design model follows a normative approach and provides an optimal target plan for an area. The usefulness of such a model is evident from the fact that it will enable practicing planners and engineers to arrive at a desired pattern of future land use through a more systematic and expeditious process than that offered by the conventional approach.

In using the land use plan design model approach the planning area under consideration is divided into a number of cells and the land use demand is expressed in terms of a series of discrete land use modules, such as residential neighborhoods, schools, commercial centers, parks, and so on. The basic operation of a land use design model consists of the placement of given modules in the specified cells of the planning area. The placement process is dependent upon the constraints that are associated with the land use plan of the area under consideration. These constraints form an essential part of the plan design process, for as a whole they control the feasibility of a plan. The total cost of a particular plan design is divided into two categories: site development cost, which includes the construction and maintenance costs of the module elements, and linkage cost, which consists of construction, maintenance and operation costs of facilities such as transportation routes, water and sewer lines, and connections for other public utilities between a pair of module units. The site costs are computed on the basis of the soil type and the type of module element, and they are expressed as dollars per acre of module size. The linkage costs are dependent on the types of module unit to be linked and on the comparative sizes of these units. The cost values represent present worth values of all cash flows for an interest rate of 6 percent considered during a period of 20 years.

OPTIMIZATION IN LAND USE PLAN DESIGN

The land use plan design model aims to provide an "optimal" land use plan, "optimal" meaning a plan with the lowest overall cost of development and operation that meets the specified design criteria. In this way, the problem can be considered as one of a class of "maximum-seeking" experiments to find the combination of factors which produce this "best" or lowest cost result. The factor combination producing the best result is termed the "optimal factor combination".

In the last few years several optimization procedures have been applied in land use plan design problem. In the initial effort a linear programming approach was utilized (1). However, land use plan design involves manipulation of discrete elements while the linear programming algorithm is generally capable of handling only continuous variable quantities. In addition, land use plan design also requires consideration of linkages, and in linear programming approach it is difficult to incorporate the linkage requirements. As an alternative approach in the land use plan design problem a procedure based on linear graph theory was then used (2,3).

The model algorithm prepared on the basis of linear graph theory consists essentially of a set decomposition technique. In the model operation, the planning area is successively divided into a series of subareas. Initially the algorithm provides for the placement of the modules into one of two halves of the planning area. The model then tests a series of successive adjacent subsets in an attempt to improve the initial allocation using a hill-climbing technique which searches for the best allocation. The best allocation is the one which produces the minimum combined site and linkage cost. Such an evaluation continuos until no improved partition can be obtained by shifting a unit element from one half of the partition to the other half. After a best partition of modules has been achieved, each module is located in one of the two halves of the planning area. The entire sequence of partitioning then continues within each of the halves of the proceding scanning process to generate another series of half areas when a new optimal partition is determined. This partitioning process continues until the area is subdivided to the degree of detail desired.

Although the model programs developed on the basis of set decomposition technique provided satisfactory application of the model, it was observed that this procedure fails to account for the possibility that a particular module element might have been better placed in a different topographic area after the initial partitioning had placed it earlier in a less desirable half-area. Moreover, in this procedure the model algorithm could consider only those linkage costs resulting from the latest division and not the cost of all the linkages required.

To eliminate the weaknesses associated with the use of set decomposition techniques, a new placement algorithm based on random search technique was then developed (4). In this procedure a set of experimental plans is developed through the combination of module-cell arrangement designed in a random fashion. The "best" plan is that experimental plan for which the random assignment of module-cell combinations produces the lowest total cost satisfying the design constraints.

RANDOM-SEARCH TECHNIQUE

In applying this approach, one can assume that there is an optimal zone of module-cell combinations, which contains a number of best alternative plans. This optimal zone can be defined a priori by establishing the level of plan accuracy that can be assumed as the proportion of optimal zone plans in the entire space of possible experimental plans. Another element that has to be decided a priori is the probability of "success", or the probability that at least one of the experimental plans is contained in the optimal zone. This probability can be expressed as

$$S = 1 - (1 - a)^n$$

where,

S = probability of success of the experiment,

a = level of plan accuracy, or the ratio of optimal zone to total number of possible experimental plans,

1 - a = probability that a trial plan will fail to be made inside the optimal zone, and

n = number of trials required to obtain the best plan with the plan accuracy a and the probability of success S.

By predetermining the values of S and a, the value of n can be obtained from the above equation. Thus, the number of experimental plans that must be prepared in order to have the desired probability of selection of a near-optimal design plan can be determined.

A detailed description of the random-search procedure including the results regarding the validity of the random placement algorithm has been discussed elsewhere (5). In the following paragraphs a description of the computer model as well as the results obtained from actual application of the model in a real-world planning problem are discussed.

OUTLINE OF THE ALGORITHM

Once the input data has been read in the operation of the algorithm is as follows: the program generates a distance vector which contains the shortest distance from any given cell to every other cell and this information is stored in order of increasing distance from cell in question. This information is needed to check generated plan design for possible distance constraint violations.

At this point all the temporary storage vectors in the program are zero and all the pre-existing module types and land usage representing the initial condition of the planning area are entered into the appropriate storage data. Two seed numbers are read in for use in random generation of cell numbers and module types.

A module type number is generated randomly, and this number is then checked to make sure there are more modules of this type available for placement. If so, a cell number is then generated using random number generator and this cell number has to be checked to make sure that the placement of this particular module type in this cell does not violate any module-cell constraint. If there is no module-cell constraint violation a check is then made to ensure that there are not more of this particular type of module placed in this cell than allowed.

If this cell is unacceptable a new cell number is generated and this process is continued until the module is placed. If there are no more modules of a given type available for placement, another module type number is randomly generated; and the process mentioned above is repeated, until all modules to be placed have been placed in some cell.

At this stage an experimental plan is prepared which may or may not be feasible in that it can violate distance constraints and/or compatibility constraints between modules. As the modules are placed, a running account of the site cost is kept which is later used to calculate the cost of the given plan. After placing all modules, the distance constraints are checked out. This is done by first checking what modules are in a given cell and making sure that the modules are compatible with one another. If this is so, a linkage vector is set up and each module is systematically connected with a module inside a given cell for which there is no linkage cost, or to the nearest module outside of the cell itself. As this linkage vector is filled, the distance between the module being checked and those it is to be connected to in other cells is checked against the given distance constraints, and if the constraint is violated, the plan is rejected as being infeasible. At this point, under operational mode II, the plan is checked to determine if some distance constraint makes this experimental plan infeasible. Should all distance constraints be satisfied, then the plan is assumed to be feasible and the linkage costs are summed up and this experimental plan is then stored for later consideration as to whether it is the lowest cost plan available. Once this is finished, another experimental plan is generated and the new plan is compared with the previous lowest cost plan to determine which has the lower cost; the plan with the lower cost is retained. This process is repeated until the required number of feasible plans have been generated. After the results have been printed out, the entire process is repeated as many times as specified in the input data. The steps involved in the operation of the computer algorithm are schematically presented in Figure I.

PROGRAM OPERATION

The purpose of the land use design model algorithm is to generate land use plans which have a minimal cost while satisfying given design constraints. The algorithm provides for two modes of operation; Type I is the land use design option, and Type II provides a sensitivity analysis. In the design option the prime end is the generation of an "optimum" design which satisfies present geographic, compatibility and distance constraints, while in the sensitivity analysis, the effects of a particular inter-module distance constraint on the development of feasible plans as well as on site and linkage costs are analyzed. Both Type I and Type II analysis modes allow for location of pre-existing modules and land use in the design area; although site costs for pre-existing modules are not included in the total plan cost, their linkage costs for connecting them with newly placed modules are included. Both modes of operation give the user the option to having the final design plan presented graphically.

Input Data

The input data can be divided into two basic classes; data defining the operation of the program and data to be manipulated. In order to prevent mixing up of the input data, each card carries an identification number, IDEN, which defines the character of the data being read in; if this number is not compatible with the read statement under which it is read, an error message is printed and the program operation is halted. The data is input as follows:

The first input card provides general information about the model operation and it defines the number of cells, NCELL; number of modules, NMOD; operational mode, ITYPE = 1, 2; number of cards containing descriptive information about the design, NTITLE; number of sets of random number seed values to be used, one set for each time a new sequence of plans is to be generated, NTIMES; the number of pre-existing module type - cell combinations to be read in, NINIT; the number of cells having pre-existing land use, NAREA; and need of a graphic plot of land use design, IWPLOT.

If NTITLE contains a value greater than zero, a series of cards containing decriptive material are read in and printed out on the problem output. If NTITLE equals zero, this read statement is ignored and the next set of data is read in. If IWPLOT is equal to 1, the program reads in parameters needed for graphical output of the design plan, otherwise this read statement is skipped.

At this point the program reads in the parameters needed for the land use design; the first set is the module data, XMOD. There is one card for each module type which contains the area requirements for this module type and the total number of modules (pre-existing plus new) of this type to be placed. Next comes the cell data, CELL, one for each cell, which contains the total area in cell and the coordinates of the cell. This is followed by a set of cards, one for each module-cell combination in which are listed the maximum number of a particular module that can be placed in the cell, MCC, and also the site cost of placing the module in this cell, COSTST. This is followed by module-to-module distance constraint data. The maximum distance (+) or minimum (-) between any two given module types are read in, CMMD in addition to the cost of linking these two modules, COSTLK. The constraints on the compatibility of two modules in the same cell are read in, IMMD, where a minus sign designates incompatibility.

Next comes data on current land use. The first set of information is concerned with which module types pre-exist in which cells, INIT, one card for each module type-cell combination. This is followed by information on area used in a cell by pre-existing structures or other land use, AREAUS; one card for each cell having preassigned land use. Finally a card defining the optimal region, a, and the probability of success, S, are read in.

Once the basic land use planning data has been stored, the nature of the remaining data to be input depends on the operational mode. In the case of mode I, land use plan design, one set of seed numbers is read in for each sequence of plans to be generated. In the case of mode II, sensitivity analysis, two cards are read in for each sequence of plans to be generated. The first card defines which module (MODF) to module (MODT) distance constraint is to be varied, the value of the constraint, CMMD, and the cost of this particular link, COSTLK; the second card contains a set of seed numbers.

Output Information

The form of the output depends on the form of the operational mode used. In mode I, design option program prints out number of feasible plans needed to satisfy the requirement on optimal region and probability of success. This is followed by the number of infeasible plans generated while satisfying the feasible plan requirement. The program then prints out in tabular form pre-existing module types in each cell of the land use plan area followed by printout of the optimal design. Finally, the total site cost, total linkage cost, and the total plan cost are printed out for the optimal plan. At this point the final plan will be plotted in bar graph form if the user desires to do so.

In mode II, a sensitivity analysis of the design constraints is carried out and the program prints out for each run the run number, the numbers of the modules for which the design distance constraint is being studied and the value of the distance constraint. In the process of generating the feasible plans required the program checks each infeasible plan design to check if failure was due to the distance constraint being studied. If the constraint is violated the cell numbers involved in the violation and the actual distance between them is printed out. Once the required number of feasible plans have been generated, this number and the total number of infeasible plans generated in the process are printed out. Finally, the total site, linkage, and total plan cost are printed out for the optimal plan; this plan can be plotted graphically if the users desire to do so.

EXAMPLE PROBLEM

Study Area

The Village of Germantown, selected as the study area for the application of the model, is located in southeastern Wisconsin. Its corporate limits nearly define a full U. S. Public Land Survey township of 36 square miles in area. The population of the area in 1970 was 7,000 and the projected population for the year 1990, derived from the regional forecast and used to provide model input data, is 27,400. At the present time only a small portion of this area is utilized and nearly all urban development is confined within the southeastern corner of the village. In the following paragraphs a brief description of the input data, output information, and sensitivity analysis conducted with the example problem are discussed. More detailed information about the model results are given in Reference 6.

Input Data

The available land area of the Village of Germantown was divided into 36 cells with each cell being one square mile, or 640 acres, in area. The land use requirements for the forecast year of 1990 were expressed in terms of ten module types. The module types used, area of each module, number of modules in the initial condition and the additional number to be placed are presented in Table 1. The site development costs for each module type as well as the intermodule linkage costs for each pair of modules were obtained from the Southeastern Wisconsin Regional Planning Commission. In addition, the input data included information about design constraints imposed on the model operation involving module-cell compatibility and inter-module distance constraints. The module-cell compatibility matrix combines two types of constraints: a site constraint, which excludes the placement of certain modules in certain cells, and a module limiting constraint, which specifies the maximum number of units of a certain module type that can be placed in a certain cell. More detailed information about the constraints can be obtained in Reference 2.

Output Information

A sample output from the model run with the Village of Germantown as the study area is presented graphically in Figures 2 and 3. In Figure 2 the initial land use development of the study area is depicted. The optimal design of the future land use pattern for the design year of 1990 is presented in Figure 3. The other necessary information such as total site and linkage costs associated with the optimal plan design is printed out. In addition, the output information about the module placement is obtained in tabular form which then can be transformed into conventional maps. However, the on-line graphical presentation of the model provides quick and ready output information about the resulting land use pattern which can be easily evaluated by the planner before proceeding to a new trial. For the plan presented in Figure 3, the total plan cost is $458,695,200; while the total site and linkage costs are $368,128,600 and $90,566,600, respectively.

The on-line maps of the model results indicate how many units of a particular module type are placed in any cell. It may be noted in the final plan that two units of module type 5, two units of module type 8, and one unit of module type 9 have been placed in cell 21. The maps also indicate how much of the total land area available in a cell is utilized by placing each type of module in it. This is done through the relative heights of the bars representing respective module types. If the height of a bar covers the entire vertical length of the cell in the plot, it is interpreted that the corresponding module type utilizes 100 percent of the available land area in this cell. For example, in the final plan the available land area in cell 6 is utilized entirely by the module type 1, which has two units located in this cell.

It may be noted that in some cells the total land area occupied by the modules is larger than the available land area in the cell. This situation resulted from the fact that the pre-existing land uses in each cell of the initial condition do not correspond to the specified discrete areal dimensions as employed in the land use plan design model. Consequently, some cells appear to have more modules than they can accommodate; whereas in reality, these cells have several pre-existing modules which are smaller than the standard module sizes used in the model.

Furthermore, as the model attempts to minimize the total cost, of which the site cost constitutes the largest portion, the modules are placed in those cells which would give lower site development costs. Consequently, most of the plans generated by the model do not show any strong clusterings around the existing development in the village center, but rather tend to follow a somewhat scattered pattern covering those cells which provided lower site cost.

Sensitivity Analysis

Using the Mode II of the model operation a sensitivity analysis can be conducted in order to observe the effects produced by changes in various design constraints. The intermodule distance constraints were varied, one module combination at a time, to find what effect an increase or decrease in the module-to-module distance constraint had on the feasibility of the plans generated as well as on the costs of the feasible plans. The role of cell-module compatibility constraint was also investigated by considering the effect of providing an environmental corridor in the plan. Some of the typical results are briefly discussed in the following paragraphs.

Effect of Intermodule Constraint -- The percentage of feasibility of a plan is directly related to the module-to-module distance constraints. The main module type which is involved in limiting distance constraints is type 1, the medium density residential land use. This module type must be within a specified distance of all of the other service modules in order to satisfy the predetermined design standards. Through the sensitivity analysis, it can be determined whether the design standards, which serve as the basis for the distance constraints, are too restrictive or if the number of units of certain module types should be increased. Experiments were conducted for all residential module to service module combinations. In each experiment, ten random best plans were obtained for each given set of distance constraints. For example, the intermodule distance constraint from module 1 to module 2 was varied from 2.0 to 4.0 miles and the constraint levels tested were at every half mile. The results of the trials conducted for this module to module linkage are displayed in Figure 4. It can be seen from the graph that at a level of 3.5 miles or more this particular distance constraint ceases to have a significant effect on the degree of infeasibility. A practical course of action in this case might be to set this distance constraint at 3.5 miles as a more stringent constraint would cause a high degree of infeasibility.

Effect of Preserving Land For Environmental Corridor -- A set of experiments was conducted to test the effect of restricting land development in certain parts of the area under consideration. The purpose of this set of experiments was to determine the change in the plan costs caused by the introduction of an environmental corridor through the plan area. The ten lowest cost plans were developed randomly, both with as well as without, the land preserved for the environmental corridor; the average and standard deviation of the total plan costs including the site and linkage costs were computed for both sets. A Student's "t" test was then conducted to check if the means of the sample plan costs were significantly different. It was observed, on the basis of the sample results, that the introduction of the environmental corridor does significantly affect the site costs; while the effect on the linkage costs were found not to be significant at the 95 percent level of confidence. Similar experiments may be conducted to test such policy decisions such as the preservation of prime agriculture land or prohibition of land development in a flood plain and so on.

CONCLUSIONS

An operational and flexible land use plan design model will aid in making decisions concerning the development of public and private policies regarding development and use of land in a systematic and efficient way. In metropolitan and regional planning process the model can be well utilized in establishing a "standard" or norm against which all proposed plans can be evaluated. Another important application of the model relates to the ready estimation of the cost of any suggested plan design constraints.

The land use plan design model in its present form and state of development has displayed only limited success in "real world" application. Although the model has been proven to be conceptually valid, there exist several deficiencies in the model which require further work. As further refinements are made in the model operation and its performance is improved, the land use plan design model can be an important tool in the operational planning process.

REFERENCES

1. Schlager, K. J., "A Land Use Plan Design Model", Journal of American Institute of Planners, May, 1965.

2. Southeastern Wisconsin Regional Planning Commission, "Land Use Plan Design Model-Model Development", Technical Report 8, Vol. 1, 1968.

3. Southeastern Wisconsin Regional Planning Commission, "Land Use Plan Design Model-Model Test", Technical Report 8, Vol. 2, 1969.

4. Sinha, K. C., Adamski, J. T. and Schlager, K. J., "Use of Random-Search Technique to Obtain Optimal Land Use Plan Design", Highway Research Record 422, Highway Research Board, 1973.

5. Ibid.

6. Sinha, K. C., "Development and Test of a Mathematical Model For Land Use Plan Design", Final Report, Department of Civil Engineering, Marquette University, 1973 (draft).

ACKNOWLEDGEMENT

The study was in part supported by the U. S. National Science Foundation Grant No. GK-17166, and this support is gratefully acknowledged. The necessary data was obtained from the Southeastern Wisconsin Regional Planning Commission, whose co-operation is highly appreciated.

TABLE I

MODULE DATA FOR LAND USE DESIGN MODEL EXAMPLE RUN: VILLAGE OF GERMANTOWN

Module No.	Module Description	Area (Acres)	Number in Initial Condition	Additional Number to Be Placed
1	Residential (Medium Density)	315.0	2	12
2	Local Commercial Center	6.4	3	5
3	Regional Commercial Center	90.0	0	1
4	Highway Commercial Center	13.8	1	1
5	Industry (Light)	315.0	1	2
6	Industry (Heavy)	315.0	0	2
7	Jr. High School	27.5	1	1
8	Environmental Corridor	150.0	34	0
9	Sewage Treatment Plant	50.0	1	0
10	Major Highway	5.0	16	0

Source: Southeastern Wisconsin Regional Planning Commission, Waukesha, Wisconsin

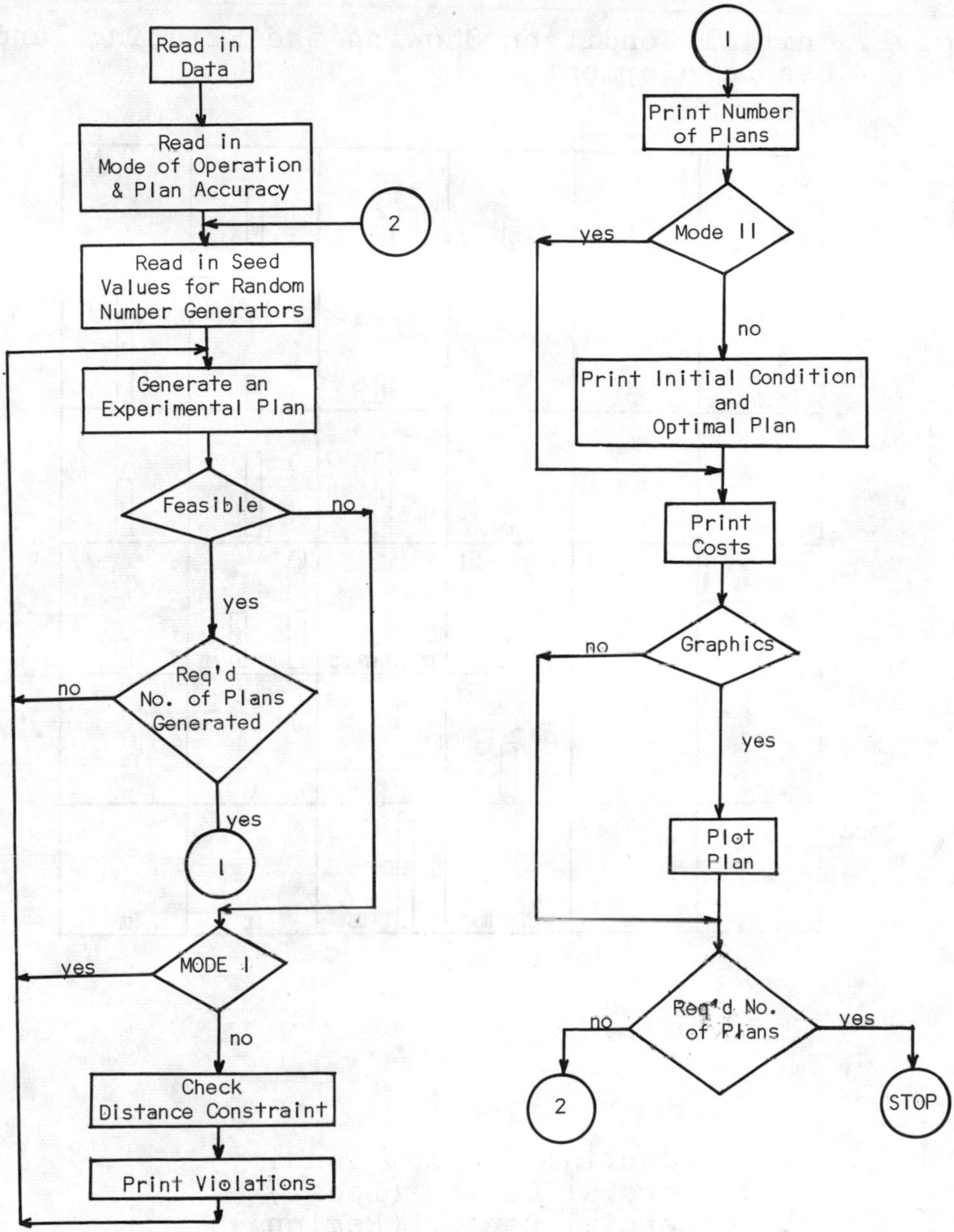

FIGURE I. COMPUTER FLOW CHART FOR MODEL OPERATION

Fig. 2. Initial Condition Showing the Existing Land Use Development

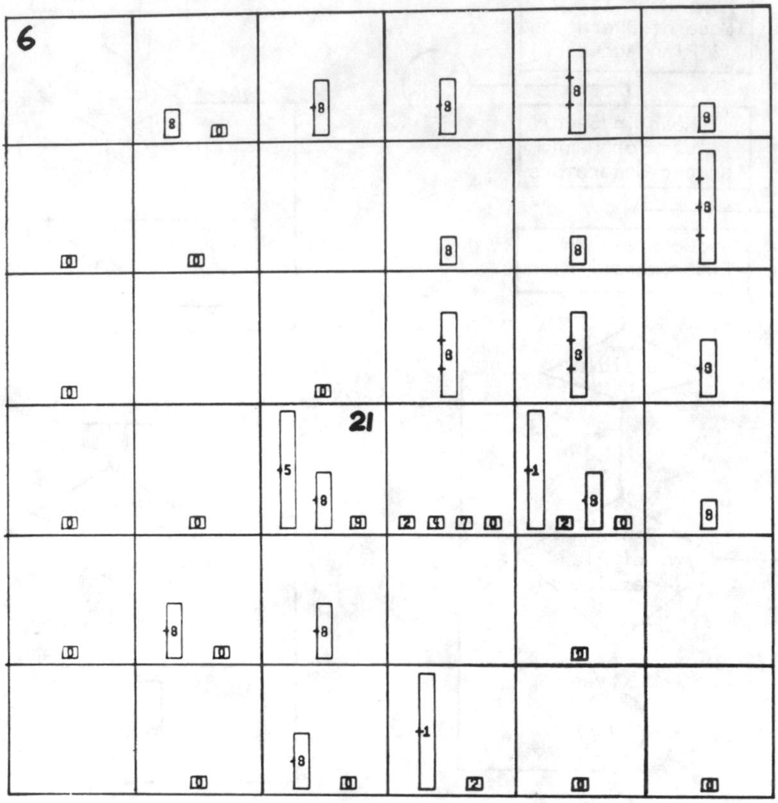

Notation

1. Residential (Medium Density)
2. Commercial Center (Local)
3. Commercial Center (Regional)
4. Commercial Center (Highway)
5. Industry (Light)
6. Industry (Heavy)
7. Secondary School (Jr. High)
8. Environmental Corridor
9. Sewage Treatment Plant
0. Major Highway

Initial Plan

Fig. 3. Final Land Use Plan Design for the Year 1990: Village of Germantown

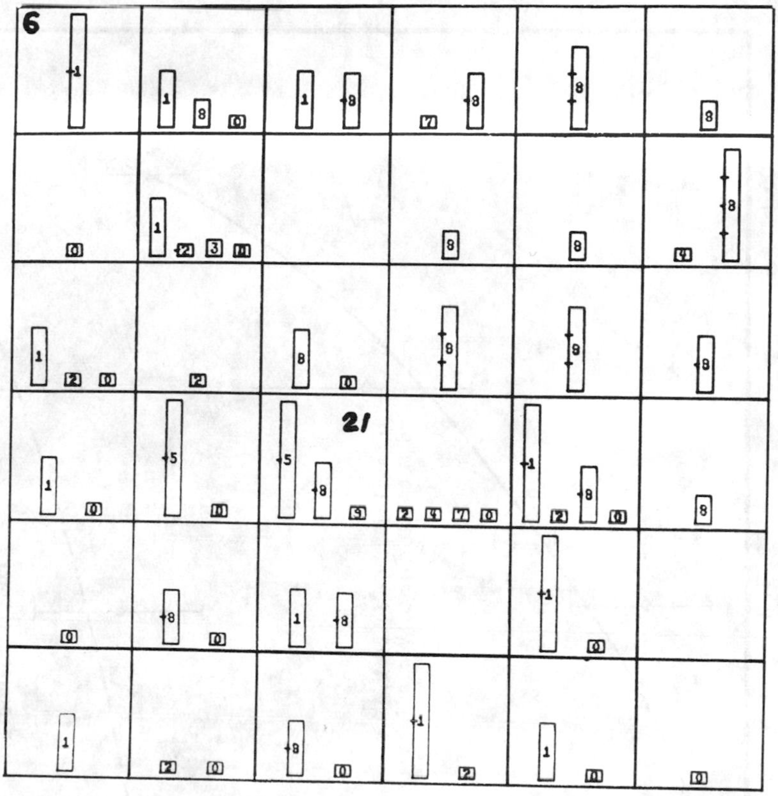

Notation

1. Residential (Medium Density)
2. Commercial Center (Local)
3. Commercial Center (Regional)
4. Commercial Center (Highway)
5. Industry (Light)
6. Industry (Heavy)
7. Secondary School (Jr. Hight)
8. Environmental Corridor
9. Sewage Treatment Plant
0. Major Highway

Final Plan

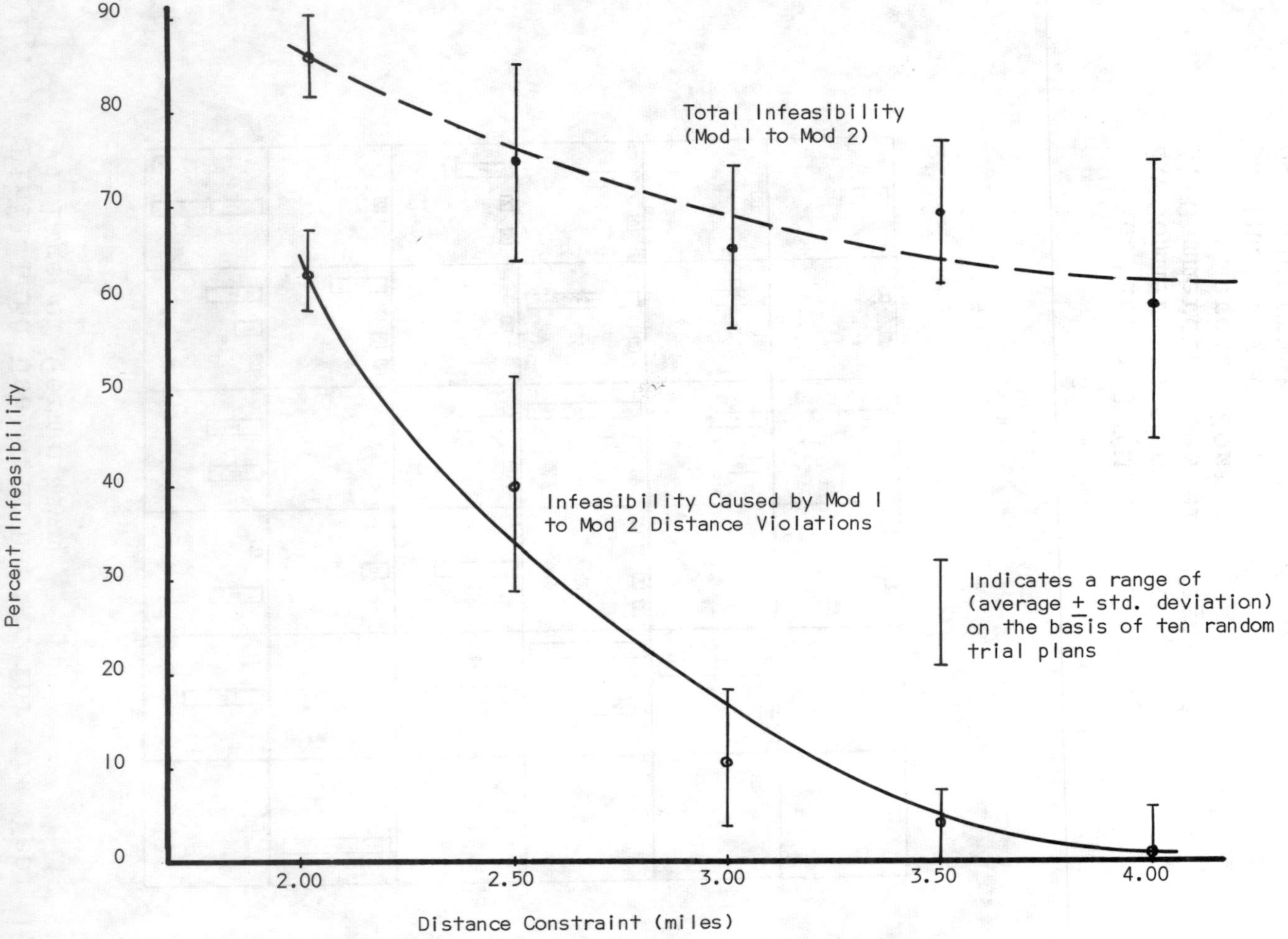

FIGURE 4. INFEASIBILITY CURVES FOR MODULE TYPE 1 TO MODULE TYPE 2

COMBINATORIAL OPTIMIZATION AND PREFERENCE PATTERN AGGREGATION

by

J. M. Blin* and Andrew B. Whinston**

(*) Northwestern University, Evanston, Illinois
(**) Purdue University, Lafayette, Indiana

I. INTRODUCTION

Studies of decentralized decision processes have mostly concerned themselves with markets for private commodities. Under some general convexity conditions on available technologies and consumer preferences, it has been shown that any competitive equilibrium is a Pareto-optimal state for an economy; furthermore, given a Pareto-optimal state, there exists a price system that will be consistent with a competitive equilibrium [1]. From the standpoint of decentralization theory, the significance of this result cannot be overemphasized. We can rest assured that a Pareto-optimal state is achievable via a decentralized competitive price mechanism. If we now consider the field of group decision making, and social choice theory more generally, it seems that this decentralization issue would be even more crucial. In fact, if we view the market place as an aggregation mechanism which combines various consumers' preferences and producers' technologies to yield a given allocation at a certain price, we should also consider another market place, viz that one which decides on public goods and collective issues in general.

Here again the aggregation issue arises. In fact, Arrow's fundamental work in this area [2] specifically addressed this question. If social decisions are to be made jointly and reflect the diversity of individual opinions with maximal accuracy, mechanisms must be devised to perform this formidable aggregation task. Some such mechanisms already exists and are commonly used. Voting mechanisms belong to this class of aggregation mappings and so do dictatorial decision rules -- although most people would view them as a degenerate case. But voting rules may fail at times -- just as markets do in the presence of indivisibilities and/or non-convexities. Such voting failures arise, for instance, in the form of intransitivities. To circumvent these problems, a study of alternate decentralized -- or centralized -- solutions becomes a pre-requisite for any further action. In this spirit a number of routes have been explored. Various centralized procedures have been proposed to mimic the traditional voting mechanism (See [4], [5], [6]). Alternatively a number of authors have recoursed to logrolling as a decentralized solution mechanism. Vote trading among legislators on various bills has been praised as an indirect but effective way of letting legislators' preference intensities count. From the early work of Arthur Bentley [3] to that of J. Buchanan and G. Tullock [8], James Coleman [9], Dennis Mueller [11], R. E. Park [12] and R. Wilson [14], the logrolling hypothesis has been proposed as a possible way of bypassing Arrow's General Impossibility Theorem. Arrow himself has remarked that this seemingly attractive solution breaks down if the bills before an assembly happen to be dependent. However there is no a priori compelling reason -- empirical or theoretical -- to assume such dependence. In the sequel, we shall propose a new conceptual framework for the consumers' choice space, which will then enable us to address the logrolling issue more effectively. A special case of this choice space will also be studied separately to show how majority voting can be represented by a centralized decision mechanism based on a discrete optimizing model.

II. A CONCEPTUAL FRAMEWORK OF ANALYSIS FOR THE CONSUMER'S CHOICE SPACE

We consider a finite set of social alternatives or, more briefly, a set of n bills

(1) $B = \{b_1, b_2, \ldots, b_i, \ldots, b_n\}$

Each individual (h) in the group S (society, assembly etc.) is assumed to have a certain hierarchical preference pattern over these bills. Specifically an individual may choose any of m_i courses of action (levels of preference) for each bill b_i. (Note that each m_i may differ from m_j for any $i \neq j$). Furthermore, each individual is assumed to have a most salient alternative say b_k, a second most salient alternative b_l etc... In other words each individual has a saliency (ordinal) scale for the various bills which represents the order of priorities that he assigns to the issues. This saliency scale is the basic preference structure for any individual. Once it is determined, the individual then decides on the outcome he prefers most for a given echelon on this scale, i.e. which course of action m_i^* he sees as best from his standpoint for the ith bill b_i. If there are three alternatives $(n = 3)$ and two courses of action on each bill (m_i = {Yes,No} or {Pass,Fail} for all $i = 1,2,3$) an individual hierarchical preference structure would then be written as

(2) $(\sim 3;2\ ;\ \sim 1)$ where $\sim \equiv$ not

This means that: (i) for this individual the third bill is most important, the second bill is next in importance and the first bill of least importance, and (ii) the preferences on each bill are of the (No;Yes;No) type. In regard to this saliency hypothesis, we might note that it is one possible way of introducing some further information somewhat like an intensity of preference. In effect we are making preference pattern into multidimensional entities. There are n dimensions (n bills) which are ordered differently by each individual, according to his own saliency scale. The preference patterns are simply represented by a subset of the lattice points in an n-dimensional space and these lattice points are linearly ordered by the saliency ordering of each individual.

The following example will serve as an illustration. Consider a set of three bills:

(3) $B = \{b_1, b_2, b_3\}$

Assume each of these three dimensions of individual preferences ia discrete scale viz:

(4) $(b_i;\sim b_i) \equiv (1;0)$ for $i = 1,2,3$

To generate the 3-dimensional preference patterns, we need to consider only 2^3 lattice points in $\mathbb{R}^3$. Moreover, the underlying saliency scale for an individual determines the ordering of these 3-dimensional patterns. For simplicity, suppose the saliency scale is simply the order $(\sim b_3 > b_2 > \sim b_1)$. Then the following tree enumerates all the preference patterns as the terminal nodes of the tree. For this individual the ordering of these patterns simply corresponds to the list of patterns read from top to bottom.

To characterize this saliency condition, we need only note that upon scanning the set of 2^n patterns vertically -- i.e. one dimension at a time -- the most salient issue for that individual corresponds to the first entry which remains invariant in the first 2^{n-1} patterns. Sequentially, the second most salient issue will correspond to the entry that remains invariant in the first 2^{n-2} patterns etc...

This characterization can be used to test for the existence of an underlying saliency scale in any set of individual preference patterns. From an experimental design standpoint, two approaches could be used

(i) On the one hand, we could directly ask the individual for his saliency scale on the issues then infer from it a theoretical ordering over his preference patterns, and finally compare it with his observed ordering. To simplify matters when a large number of issues are at stake, a good experimental procedure would consist in picking some pairs of patterns along the theoretical ordering and ask him to order them. To devise "saliency tests" would involve our allowing greater deviations from the theoretical ordering at the bottom of the scale than at the top. For instance, in the example below while we would insist on his ranking (010) > (110) (the top pair of patterns) we might want to dismiss answers such as (001) < (101) --

which is the reverse of the theoretical order between the bottom two patterns in our example. Reasons for allowing variable degrees of freedom along this ordering will be discussed separately.

(ii) On the other hand, we could ask the individual to rank order the set of 2^n n-dimensional patterns and look for evidence of a saliency scale by just applying the characterization of the saliency property which we have previously discussed. Here again, a paired comparison experiment over some (or all) the patterns could be used to derive a rank ordering (See [10], for instance).

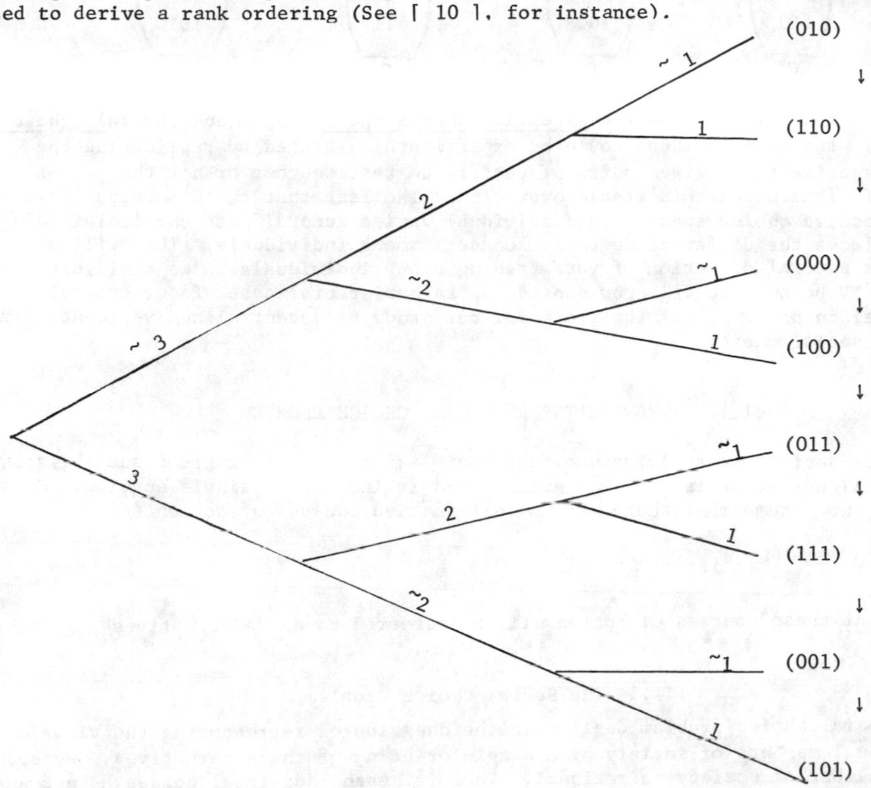

Turning back, now, to this notion of variable degrees of consistency which we may allow an individual as we move along a theoretical ordering of preference patterns, it should be noted that there exists a simple justification for this apparent inconsistency. Each individual may have a different degree of discrimination, a different threshold which each issue must pass in order to become effective in his preference structure. In other words there may exist two utility spaces that we must consider: the <u>theoretical utility space</u> U_n of full dimension (n) i.e. including each and every issue as a separate dimension; and the <u>perceived (or effective) utility space</u> U_k <u>for an individual</u> which is simply a subspace of the full space U_n. The dimension (n) of the full space is determined by considering each and every conceivable bill of interest to any member of the society. Clearly the logic of the saliency hypothesis requires that we also allow some individuals to be concerned with only a few issues, say k of them. As the concern threshold of an individual goes up k becomes smaller and vice versa. This also means that if we set the preference patterns P_t of an individual in an n-dimensional space, whereas his effective utility space U_k has dimension (k), we do not have a strict ordering over the set $\{P_t\}$ but rather a preordering with a set of equivalence classes $\{C_1, C_2, \ldots, C_\nu\}$. An equivalence class C_ν is the set of all patterns $\{P_{t_\nu}\}$ which are similar up to the first k^ν entries. A simple projection mapping Π from U_n to U_k will collapse each indifference class C_ν into a k-dimensional point, and the ordering will now be on these points. To illustrate these concepts, suppose that in our above example $k = 2$, i.e. only 3 and 2 are effective preference dimensions for a given individual whereas he is really indifferent as to the first issue. Then the set $\{C_\nu\}$ of equivalence classes reads

$$\underbrace{\begin{pmatrix} 010 \\ 110 \end{pmatrix}}_{C_1} > \underbrace{\begin{pmatrix} 000 \\ 100 \end{pmatrix}}_{C_2} > \underbrace{\begin{pmatrix} 011 \\ 111 \end{pmatrix}}_{C_3} > \underbrace{\begin{pmatrix} 001 \\ 101 \end{pmatrix}}_{C_4}$$

This distinction between the <u>effective choice space</u> (U_k) and the <u>full choice space</u> (U_n) for an individual could be experimentally tested by replicating the ordering experiment for given pairs of vectors to test whether or not the ordering of certain pairs is stable over time. The fact that the dimension (k) of the effective choice space of an individual varies according to the individuals simply reflects the different degrees of concern among individuals. This will also allow for a natural direction of vote trading among individuals as we shall discuss in section IV below. We will now consider a rather special case of our general choice model in order to set the stage for our study of decentralized vs. centralized methods of social choice.

III. A COMBINATORIAL SOCIAL CHOICE PROBLEM

In this section we shall assume that there exists only a single issue b so that the saliency scale is really degenerate as it includes a single echelon. On the other hand, we assume that there are m alternative courses of action for b:

(5) $A = \{a_1, a_2, \ldots, a_m\}$

In the sequel these courses of action will be referred to as "alternatives".

III.1 The Social Choice Problem

The social choice problem deals with the question of representing the views of the individual members of society by a single ordering of the alternatives. We assume that each member of society is rational. That is, each individual possesses a transitive preference ordering. The problem then amounts to combining the individual preference orderings to obtain a transitive social ordering. Two difficulties present themselves when we try to accomplish the above. The first difficulty arises in choosing a rule by which the aggregation should take place. The second problem is to guarantee that the resulting social ordering is transitive. The transitivity of a social ordering obtained from transitive individual orderings does not necessarily follow. This was demonstrated by Arrow [1] when the majority principle is used as the method of aggregation. In order to examine the problem of social choice we will find it convenient to place the problem within a specified frame of reference. In particular we shall use the method of paired comparisons in analyzing the problem of social choice. Briefly, the method of paired comparisons as applied to choice, individual or social, may be described as follows.

Each individual alternative is paired against every other alternative and the decision maker indicates his choice in each of the situations. If the decision maker is an individual consumer then the binary matrix T representing his preference ordering has an (i,j)<u>th</u> entry if and only if i is preferred to j. Clearly, each individual matrix T is such that

(6) $t_{ij} + t_{ji} = 1$

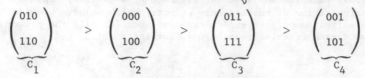

Now, a binary <u>voting process</u> can be simply characterized by some linear aggregation rule over the set of the individual preference matrices T. Simple majority voting, for instance, amounts to adopting as the (i j)<u>th</u> entry of the voting matrix A the arithmetic mean of the individual entries t_{ij} over all voters. Thus a <u>voting process</u> will associate to any set $\{T_h\}$ of individual preferences a (n x n) nonnegative matrix A with the following properties

(7) $a_{ij} \in [0,1] \ \forall \ i,j = 1,2,\ldots,n$

(8) $a_{ij} + a_{ji} = 1$

A <u>collective decision process</u> will associate to each such <u>voting matrix</u> A some collective preference matrix T with (0,1) entries and verifying equation (6) above. The problem is to choose a particular ordering of the alternatives that is closest or best fits the desires of the members of society. A more precise mathematical statement of this version of the aggregation problem will now be given.

First of all, it can easily be seen that the set $\mathcal{A}$ of all such matrices A forms a convex polyhedron in n^2 - space. It has also been demonstrated that the vertices of $\mathcal{A}$ correspond to the elementary preferences matrices T. Any social decision process can thus be represented as a vertex projection mapping d from an interior point $A \in \mathcal{A}$ obtained through majority voting, to a vertex T of $\mathcal{A}$ chosen on the basis of some optimality criterion. The well-known problem encountered with majority voting is that it does not ensure that the T matrix thus obtained corresponds to a transitive ordering.

III.2. An Associated Combinatorial Optimization Problem

Consider a given voting matrix A. The problem is then to rearrange the rows and columns of A in such a manner as to maximize a criterion function which measures the amount of agreement that exists in the society for a given ordering of the alternatives. Agreement shall be measured by the total number of votes that a particular ordering receives. The basic ordering principle is quite easy.

Place i ahead of $j \Leftrightarrow a_{ij} > a_{ji}$

This, of course, is simply the idea of majority rule.

Thus, the problem is to find that particular ordering of the alternatives such that for any candidate (i) the sum of all votes received by the candidates ranked ahead of (i) is maximal.

The mathematical nature of this problem is then to maximize a certain linear functional $\varphi: S \to R^1$ where S is the set of permutation operators ρ. Specifically the functional $\varphi(\rho)$ can be written:

(9) $\varphi(\rho) = \sum_{i \neq j} q_{ij} a_{\rho(i)\rho(j)}$

where

$$(10) \quad Q = \begin{bmatrix} 0 & 1 & \cdot & \cdot & \cdot & 1 \\ \cdot & & 1 & \cdot & \cdot & 1 \\ \cdot & & & & & \cdot \\ \cdot & & & & & \cdot \\ \cdot & & & & & \cdot \\ \cdot & & & & & 1 \\ 0 & \cdot & \cdot & \cdot & \cdot & 0 \end{bmatrix}$$

The optimization problem associated with a social choice problem can then be stated as

$$(11) \quad \max_{\rho \in S} \varphi(\rho)$$

This combinatorial optimization problem can be thought of as the centralized analogue of the voting mechanism in a decentralized society. In other words if a benevolent dictator or a planning agency wanted to "mimic" the democratic decision process while centralizing all the necessary information, its problem would be of the form (9) - (11). As noted previously, there exists a striking parallel between the decentralized versus the centralized approach to social choice on the one hand and the decentralized versus the centralized approach in a general equilibrium model of competitive markets for private goods on the other hand. To pursue this analogy, however, one should ask whether or not the equivalence between the solutions obtained under the two approaches always holds. More precisely do we have to impose certain conditions regarding the set of individual preference orderings to be aggregated in order to ensure this equivalence. As we know in the case of a system of competitive private goods markets, some convexity conditions are required. In the social choice case, however, is it true that some such conditions are needed, and, if so, what are they? This question will now be examined.

III.3 Centralized Versus Decentralized Choice

The basic result concerning the equivalence of solutions under the centralized and the decentralized methods of social choice can best be cast in the form of four theorems. Intuitively we suspect that the possible non-equivalence between the majority voting mechansim and the combinatorial optimization mechanism will occur when majority voting breaks down and leads to an intransitive social ordering. In cases where such an intransivity does not occur it seems reasonable to expect the two methods to be equivalent. This is, in fact, the case as shown below.

Theorem 1: If majority voting leads to a transitive collective ordering then this is an optimal solution to the associated optimization problem.

Proof: (i) We first note that a necessary and sufficient condition for any collective preference matrix to be transitive is that there exists some permutation matrix P such that

$$(12) \quad PTP^{tr} = \begin{bmatrix} 0 & 1 & \cdot & \cdot & \cdot & 1 \\ & & & & & \cdot \\ 0 & & & 1 & & \cdot \\ \cdot & & & & & \cdot \\ \cdot & & & & & 1 \\ 0 & \cdot & \cdot & \cdot & \cdot & 0 \end{bmatrix}$$

(ii) By assumption the A matrix is such that $a_{ij} > \frac{1}{2} \Leftrightarrow i < j \ \forall \ i,j=1,2,\ldots,n$ (where i and j have been relabelled according to the P permutation matrix).

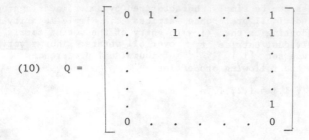

(iii) Clearly, then the identity permutation $p^* \in S$ maximizes the linear functional φ:

$$(13) \quad \varphi(\rho^*) = \sum_{i \neq j} q_{ij} a_{\rho^*(i)\rho^*(j)}$$

and

$$(14) \quad \varphi(\rho^*) > \varphi(\bar{\rho}) \ \forall \ \bar{\rho} \neq \rho^* \ ; \ \bar{\rho} \in S$$

This can be seen directly by taking ρ to be a simple transposition of k and ℓ:

$$(15) \quad \begin{aligned} \bar{\rho}(k) &= \ell \\ \bar{\rho}(\ell) &= k \end{aligned}$$

Then $\varphi(\rho) = \sum_{\substack{k \neq j \\ i,j \neq k,\ell}} q_{ij} a_{\rho^*(i)\rho^*(j)} + a_{\ell k}$ where $a_{\ell k} < a_{k\ell}$ by assumption.

From (i) and (ii) any other permutation different from P will also decrease φ. Q.E.D.

The converse of Theorem 1 is obviously not always true since there are cases where majority voting breaks down and leads to an intransitive ordering. A sufficient conditon which ensures that the converse proposition also holds will now be examined. This condition has been referred to as the "<u>single peakedness</u>" property. It amounts to assuming that the set of all individual orderings is restricted to a proper subset of itself consisting of all those meeting a certain regularity condition. If the alternatives are various levels of public spending, say, then we have a natural objective order to start with -- viz, that of the real field. An individual can choose any alternative as his preferred alternative but then he is to follow the objective order in a certain way in ranking the other alternatives.

Specifically once an individual chooses a preferred point from this set, then he is restricted in his other preferences: any other elements that lie on the same side of the preferred point in the objective order must also lie on the same side of the preferred point in his personal ordering. It has been shown by Black that if we restrict the domain of definition of the individual preference orderings to the single-peaked ones simple majority voting will always lead to the following result: there exists one alternative that obtains a simple majority over all others except the first one etc... We are now in a position to state

<u>Theorem 2:</u> Under the condition of single-peakedness, the optimal solution to the optimization problem can always be achieved by majority voting.

<u>Proof:</u> It follows directly from Black's result on single-peaked orders.

Let the optimal solution $\rho^* \in S$ that maximizes φ simply be the identity mapping.

Then we have: $a_{\rho^*(i)\rho^*(j)} > a_{\rho^*(j)\rho^*(i)} \ \forall \ i < j \ (i,j = 1,2,\ldots,n)$. But since $a_{\rho^*(i)\rho^*(j)} + a_{\rho^*(j)\rho^*(i)} = 1$ we must have $a_{\rho^*(i)\rho^*(j)} > \frac{1}{2} \ \forall \ i < j$. This is nothing else but the majority voting solution according to Black's result. Q.E.D.

As we know, majority voting over paired issues may lead to intransitive social orderings and thus leave the decision-maker with no simple rational way to single out a "best" course of action. The optimization model we have just developed will now be used to cast some light upon such phenomena. More specifically, it will show that the existence of a certain type of multiple optima for the optimization model is a sufficient condition for the existence of an intransitivity under majority voting. This class of multiple optima we are referring to is meant to eliminate the rather trivial case where we have in fact reached a tie on one or several pairs. Without loss of generality we shall assume that we have an odd number of individuals (or if it is even one individual is a tie-breaker, e.g. the chairman in some committees). The type of multiple optima we shall consider will be such that there exists at least two optimal orderings 0^* and 0^{**} and three alternatives (i,j,k) such that

$$0^* = (\ldots,i,\ldots,j,\ldots,k,\ldots)$$

(16) and

$$0^{**} = (\ldots,k,\ldots,i,\ldots,j,\ldots)$$

We can now state the following theorem.

<u>Theorem 3:</u> If the associated optimization problem (9) - (11) yields multiple optima of type (16) above, the majority voting procedure will lead to an intransitivity.

<u>Proof:</u> By assumption and given the properties of the $[a_{ij}]$ matrix we must have:

$$a_{ij} > \tfrac{1}{2}$$

$$a_{jk} > \tfrac{1}{2}$$

$$a_{ik} < \tfrac{1}{2} \Leftrightarrow a_{ki} > \tfrac{1}{2}$$

In other words (i) defeats (j) defeats (k) which in turn defeats (i). Q.E.D.

The fact that this condition is <u>not</u> necessary can be readily verified with a counterexample; for instance if there are 60 individuals with the following preference: abc(23),bca(17),bac(2),cba(10) the optimal solution to problem (5) - (7) is unique viz. (bca) but majority voting leads to (abca).

A necessary condition can be stated as follows:

<u>Theorem 4:</u> When the majority voting procedure leads to an intransitivity, the associated optimization problem (9) - (11) will yield multiple optima of type (16), only if there exists i,j,k such that

$$(17) \quad a_{ij} = a_{jk} = a_{ki}$$

Proof: We note first that from assumption (17) and property (6) for the class of A matrices we have

$$(18) \quad a_{ji} = a_{ik}$$

Now let $\rho^* \in S$ which maximizes φ be the identity mapping i.e.

$$(19) \quad \varphi(\rho^*) = \sum_{\substack{i'j' \\ i,j,k \neq i',j'}} q_{i'j'} a_{\rho^*(i')\rho^*(j')} + D$$

where $D = a_{ij} + a_{jk} + a_{ik}$.

Let $\bar{\rho}$ be such that

$$\bar{\rho}(i') = i' \qquad \text{while} \qquad \begin{cases} \bar{\rho}(i) = j \\ \bar{\rho}(j) = k \\ \bar{\rho}(k) = i \end{cases}$$

$$\bar{\rho}(j') = j'$$

Then (19) becomes

$$(20) \quad \varphi(\bar{\rho}) = \Sigma \; q_{i'j'} a_{\bar{\rho}(i')\bar{\rho}(j')} + D'$$

where $D' = a_{jk} + a_{ki} + a_{ji}$

and from (17) and (18) $D = D'$ so that $\varphi(\bar{\rho}) = \varphi(\rho^*)$ Q.E.D.

IV. HIERARCHICAL VOTE TRADING AND DECENTRALIZED GROUP DECISION-MAKING

In this section we shall now examine the logrolling hypothesis and its effect upon the group decision process, in the context of the individual choice model we have previously discussed. At the outset, we must clarify the vote trading rules that we will use. This is where the importance of the specific hierarchical choice model we have described, becomes apparent. In the absence of any hierarchy over the bills, the direction of vote trading between individuals remains ambiguous, since mere orderings do not carry any cardinal information that would represent consumer's preference intensities. Furthermore it is also unclear what actual pattern of vote trading would emerge -- if any -- since the incentive for vote trading exists only as long as a given majority has not yet been reached on a given issue -- assuming a majority voting decision rule.

In the present model, we shall first illustrate our discussion with the help of a simple example to clarify the trading rules that would prevail under the saliency hypothesis. Consider a society of three individuals: Mr. A., Mr. B. and Mr. C. The set of bills also consists of three issues (b_1, b_2, b_3) and each bill can be disposed of by adopting one of two alternative courses of actions $\{Yes, No\} \equiv \{1,0\}$. The complete preference structures of the three individuals are as follows:

$$(21) \quad \begin{aligned} &\text{Mr. A:} \quad (\sim 3 > 2 > \sim 1) = (0;1;0) \\ &\text{Mr. B:} \quad (\sim 2 > \sim 1 > 3) = (0;0;1) \\ &\text{Mr. C:} \quad (1 > 2 > 3) = (1;1;1) \end{aligned}$$

Given the saliency condition for the individual preference structures, the corresponding complete orderings of the preference patterns of these three individuals would read

(22)

Mr. A.	Mr. B.	Mr. C.
(0 1 0)	(0 0 1)	(1 1 1)
↓	↓	↓
(1 1 0)	(0 0 0)	(1 0 1)
↓	↓	↓
(0 0 0)	(1 0 1)	(1 1 0)
↓	↓	↓
(1 0 0)	(1 0 0)	(1 0 0)
↓	↓	↓
(0 1 1)	(0 1 1)	(0 1 1)
↓	↓	↓
(1 1 1)	(0 1 0)	(0 0 1)
↓	↓	↓
(0 0 1)	(1 1 1)	(0 1 0)
↓	↓	↓
(1 0 1)	(1 1 0)	(0 0 0)

(Note: For simplicity we have assumed that each individual effective choice space has full dimension i.e. $k = n = 3$).

Suppose the initial group decision rule is simple majority voting. The starting solution is then $(\sim 1;2;3) = (0\ 1\ 1)$.

Upon examining the three most preferred patterns as given in (21) above, it appears that some vote trades are feasible. Specifically the behavioral rules for a vote trade to be "feasible" are the following: the logic of the saliency condition implies that each individual tries just to get his way on his own most salient issue and in order to effect this goal he will trade hierarchically by giving away his vote (reversing the order of his preference if necessary i.e.selling a yes vote on this issue for which he favors a no vote) on the lowest-ranking bill on his saliency scale, then the next to last bill etc., as long as he can find a trading partner to effect the trade.

In this case the following trades would be carried out

$$(23)\quad \begin{array}{llll} \text{Mr. A.} \rightarrow \text{Mr. B. one} & \text{no vote on} & & 2 \\ \text{Mr. B.} \rightarrow \text{Mr. A. one} & '' \quad '' & '' & 3 \\ \text{Mr. A.} \rightarrow \text{Mr. C. one yes} & '' & '' & 1 \\ \text{Mr. C.} \rightarrow \text{Mr. A. one no} & '' & '' & 3 \\ \text{Mr. B.} \rightarrow \text{Mr. C. one yes} & '' & '' & 1 \\ \text{Mr. C.} \rightarrow \text{Mr. B. one no} & '' & '' & 2 \end{array}$$

The resulting group preference pattern then reads (1;~2;~3) or (1;0;0) which is different from the simple majority solution (~1;2;3) or (0;1;1). Now from the standpoint of a decentralized vs. a centralized group decision process, the problem could be formalized as follows.

The individual decentralized decision process consists of n maximization problems of the type described by equations (9) - (11) above i.e.

$$(24)\quad \begin{aligned} &\underset{\rho}{\text{Max}}\ \varphi^1(\rho) = \sum_{i \neq j} q_{ij} a^1_{\rho(i)\rho(j)} \\ &\underset{\rho}{\text{Max}}\ \varphi^2 \quad = \sum q_{ij} a^2_{\rho(i)\rho(j)} \\ &\quad\vdots \\ &\underset{\rho}{\text{Max}}\ \varphi^n(\rho) = \sum q_{ij} a^n_{\rho(i)\rho(j)} \end{aligned}$$

In the case of i, j = 1,2 i.e. only two courses of action are available as in our above example the [a] matrices for the subproblems would read

$$(25)\quad [a^1] = \begin{array}{c} \begin{array}{cc} 1 & \sim 1 \end{array} \\ \begin{bmatrix} 0 & 1 \\ 2 & 0 \end{bmatrix} \end{array} \begin{array}{c} 1 \\ \sim 1 \end{array} ; \quad [a^2] = \begin{array}{c} \begin{array}{cc} 2 & \sim 2 \end{array} \\ \begin{bmatrix} 0 & 2 \\ 1 & 0 \end{bmatrix} \end{array} \begin{array}{c} 2 \\ \sim 2 \end{array} ; \quad [a^3] = \begin{array}{c} \begin{array}{cc} 3 & \sim 3 \end{array} \\ \begin{bmatrix} 0 & 2 \\ 1 & 0 \end{bmatrix} \end{array} \begin{array}{c} 3 \\ \sim 3 \end{array}$$

and the solution would be the simple majority voting solution (0; 1; 1) as in section III.

Now from the standpoint of a centralized decision process to determine the same unique group pattern as obtained after vote trading, we could write a master problem of the same combinatorial optimization nature as the sub-problems. To do this we must first note that the dimension of the A matrix in the master problem would become equal to the number of points in the cartesian product of the row (and column) space of each individual subproblem.

For instance in our example let

$$(26)\quad \mathcal{P} = (0;\ 1) \times (0;\ 1) \times (0;\ 1)$$

Then $\mathcal{P}$ ~~has~~ 2^n points (patterns) each one of which corresponds to a row (and column) of A. More generally if $\{\mathcal{M}_i\}$ represents the set of courses of action available on the <u>ith</u> alternative, then

$$(27)\quad \mathcal{P} = \prod_{i=1}^{n} \{\mathcal{M}_i\}$$

The master problem which would lead to an equivalent (but <u>centralized</u>) solution as the vote trading mechansim would simply be

$$(28)\quad \text{Max}\ \Phi\ (\rho) = \sum_{i \neq j} q_{ij} A_{\rho(i)\rho(j)}$$

In our example for instance A would be:

$$
(29) \qquad A = \begin{bmatrix}
(100) & (111) & (110) & (101) & (011) & (010) & (001) & (000) \\
0 & 2 & 1 & 1 & 3 & 2 & 2 & 1 \\
1 & 0 & & & & & & \\
2 & & 0 & & & & & \\
2 & & & 0 & & & & \\
0 & & & & 0 & & & \\
1 & & & & & 0 & & \\
1 & & & & & & 0 & \\
2 & & & & & & & 0
\end{bmatrix}
$$

(The other entries of A can be readily computed from our example).

The optimal assignment for the first row is (100) which yields a score of 12 for the φ function. This solution is also the one obtained after vote trading. We conjecture that this similarity in the results obtained through a decentralized and then a centralized decision process is not a mere coincidence but is in fact true in general.

<u>Proposition 5:</u> Under the vote trading rules stated above the solution to the vote trading decision process is similar to the solution to problem (28) above.

To conclude we must emphasize the fact that this model provides a framework of analysis for hierarchical choice (which seems to be a common feature of actual citizens' preferences in actual group decision problems) and also integrates this hypothesis in the general context of a comparison between centralized vs. decentralized group decision processes.

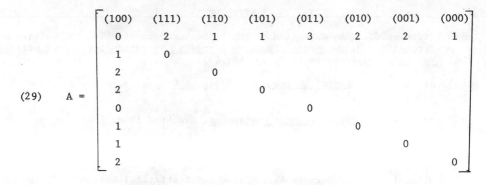

REFERENCES

[1] Arrow, K. J., "An Extension of the Basic Theorems of Classical Welfare Economics", Proceedings of the Second Berkeley Symposium on Mathematical Statistics and Probability, Berkeley 1950, pp. 481-492.

[2] Arrow, K. J., Social Choice and Individual Values, J. Wiley and Sons, N. Y., 1962, (2nd ed.).

[3] Bentley, A., The Process of Government, Evanston 1908, 1935.

[4] Blin, J. M., K. S. Fu, K. B. Moberg and A. B. Whinston, "Optimization Theory and Social Choice", Proceedings of the 6th Conference on Systems Science, 1973.

[5] Blin, J. M., "Preference Aggregation and Statistical Estimation", Theory and Decision (forthcoming).

[6] ________,Patterns and Configurations in Economic Science, D. Reidel Publishing Co., Dordrecht, Holland, Cambridge, Mass., 1973.

[7] Black, D., "The Decision of a Committee using a Special Majority", Econometrica, 16, 1948, pp. 245-261.

[8] Buchanan J., and G. Tullock, The Calculus of Consent, Ann Arbor, 1962.

[9] Coleman, J., "The Possibility of a Social Welfare Function", A.E.R., Dec. 1966, pp. 1311-17.

[10] David, H. A., The Method of Paired Comparisons, Ch. Griffin and Co., London, 1969.

[11] Mueller, D., Comment on [9], A.E.R., Dec. 1967, 57, pp. 1304-11.

[12] Park, R. E., Comment on [9], A.E.R., Dec.1967, 57, pp. 1300-1304.

[13] Saaty, T. L., Optimization in Integers and Related Extremal Problems, McGraw-Hill, 1970.

[14] Wilson R., "An Axiomatic Model of Logrolling", A.E.R., June 1969, Vol. 59, pp. 331-341.

A MICROSIMULATION MODEL OF THE HEALTH CARE SYSTEM IN THE UNITED STATES: THE ROLE OF THE PHYSICIAN SERVICES SECTOR*

Donald E. Yett
Leonard Drabek
Michael D. Intriligator
Larry J. Kimbell

Experience in the United States with Medicare, Medicaid, and other Federal programs clearly demonstrates the need to take account of factors affecting health services demand and supply in formulating and executing national health manpower policies. The research described in this paper is part of a large effort under way at the Human Resources Research Center (HRRC) to combine economic analysis, statistical estimation, and simulation techniques to develop a microsimulation model of the entire health care system. When it is completed, the HRRC model will facilitate improved health manpower planning by permitting forecasts of the complex interrelations between the demand and supply for health services and health manpower.

The paper is organized into two sections. The first presents an overview of the entire model. The second describes the physician services submodel in more detail, and plans for further research.

OVERVIEW OF THE MODEL

The HRRC Microsimulation Model consists of five major components or submodels. Each submodel is largely self-contained from a computer programming standpoint. That is, in developing the overall model, each submodel was coded and "debugged" in isolation from the rest of the model. Moreover, in some instances it is efficient to perform experiments by manipulating the relevant submodel before allowing the effects to be transmitted to the rest of the model. Figure 1 illustrates the overall model and the role played by each submodel.

The first submodel generates a population of consumers or individuals who demand medical services. The computer program generates annual estimates of the nation's population subdivided into cells according to the attributes age, sex,

*The research reported in this paper was supported, in part, by the Division of Manpower Intelligence, Bureau of Health Manpower Education under contract NIH 71-4065.

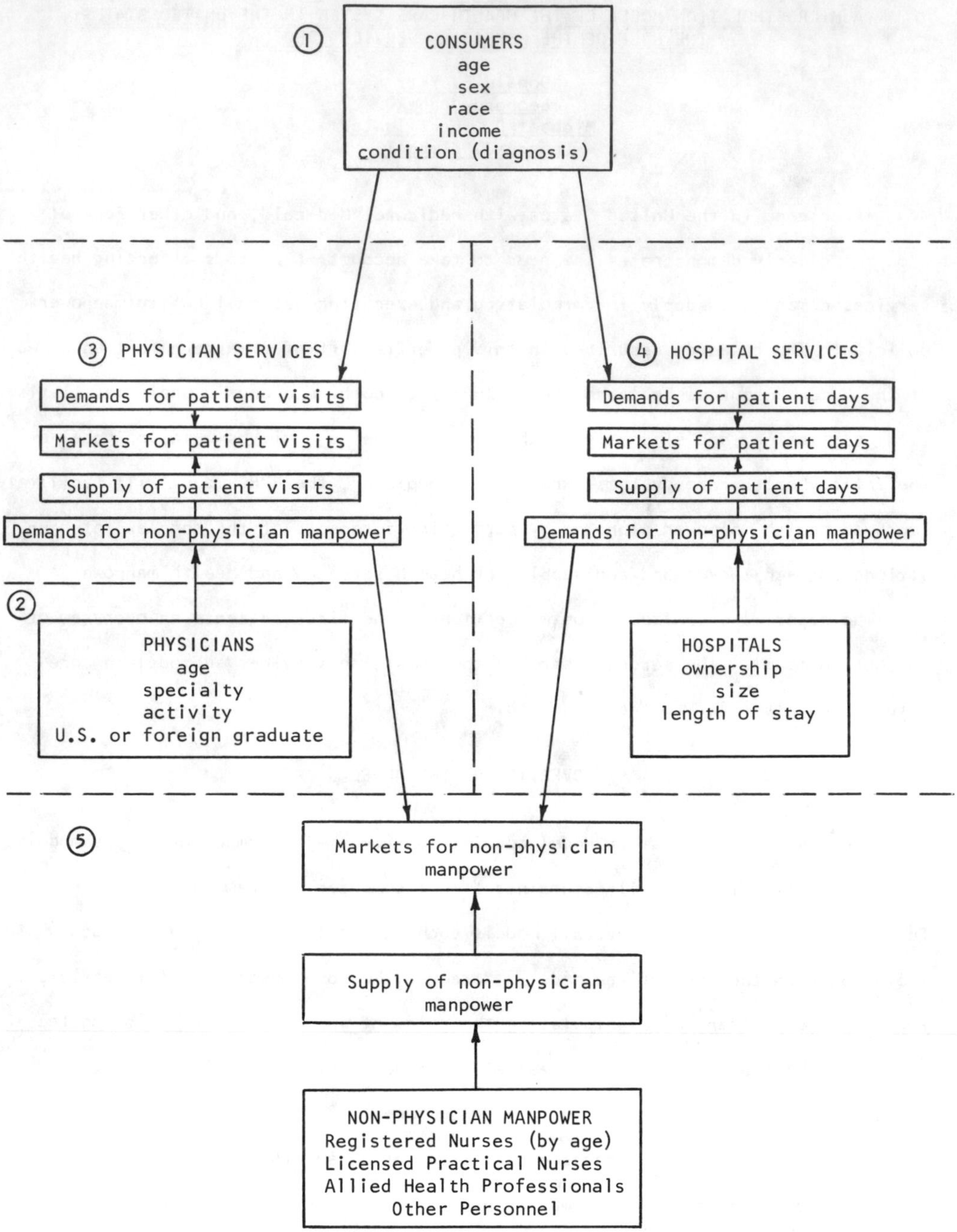

Figure 1 Block Diagram of the HRRC Microsimulation Model

race, income, and condition or diagnosis. The three major events which change the overall population are birth, death, and immigration.

The second submodel involves a similar computer program which generates a population of physicians, providing annual estimates of the stock of MDs subdivided into cells according to their age, specialty, and type of professional activity. Since U. S. and foreign medical school graduates differ dramatically in terms of their attribute distributions separate computer subroutines are used to generate the populations of the two types of physicians. The former are influenced by the volume of graduates from the nation's medical schools and physician deaths. The latter are influenced by net migration.

The two submodels which project the population of consumers and physicians over time are run first. Output from these submodels is then used as input to the other three submodels. This reflects our current specification that these populations affect the rest of the health care system but are not affected by the processes endogenous to the other three modules.

After the consumer and physician submodels generate their respective populations, the physician services submodel computes: (i) the demands by each consumer group for patient visits from physicians in private practice and hospital-based clinics; (ii) the supply of patient visits by MDs in office-based practice; and (iii) their demands for aides (i.e., non-physician manpower). The discrepancy between demand by consumers and supply by physicians leads to adjustments in the prices of patient visits, which, in turn, affect the solution of the solution of the submodel in the next period (year).

The fourth submodel computes the demands by each consumer group for patient days at short-term and long-term hospitals. More precisely, patient days is derived from separate equations for the admissions and average length of stay for each cell. Hospitals are characterized by ownership and bed size. Skilled nursing homes are included as a type of long-stay hospital. Hospital demands for non-physician manpower are based on patient days and outpatient visits to clinics and emergency rooms (determined by the previous submodel). The price of hospital care is primarily dependent upon labor and non-labor costs and the occupancy rate of hospitals. Over time the supply of hospital services, as measured by the number of hospital beds, adjusts to changes in the volume of patient demands.

The fifth submodel computes the supply of non-physician manpower. The routine for generating the stock of registered nurses is similar to the physician submodel, but the supply of RNs is influenced by the changes in their wage rates. Because of data limitations, the supply of licensed practical nurses (LPNs) and allied health professionals is presently treated as exogenous. The supply of other personnel is also exogenous to the model, but for a different reason. The supply of such manpower (e.g., clerks, secretaries, janitors, etc.) is largely exogenous to the health care system itself and, thus, is accurately portrayed as such in the model. The manpower demands by physicians and hospitals, together with the supply of manpower, are used to adjust wage rates for each occupation.

The solution of the third, fourth, and fifth submodels is repeated in the same sequence for each period (year) of the simulation.

THE PHYSICIAN SERVICES SUBMODEL

The physician services submodel treats the interactions between consumers and physicians and, accordingly, provides the linkage between the population of individuals and the population of physicians. It consists of three components treating: (1) the demand for outpatient physician services; (2) the supply of such services, and the derived demand for RNs, LPNs, technicians, and other office aides; and (3) the market interactions of supply and demand for services via fee adjustments.

Demand for Outpatient Physician Services

The demand for physician services depends on age, sex, race, family income (hereafter income), physician fees, and health insurance status. The population attribute subdivisions by which the demands vary are:

Age		Sex	Race	Family Income
0-5	6-16	Male	White	Under $5,000
17-24	25-34	Female	Other	$5,000-$10,000
35-44	45-54			Over $10,000
55-64	65-74			
75+				

The demands for specific sites are as follows:

Office Visits and Inhospital Visits by Physicians

General Practice (GP)
Internal Medicine (IM)
Pediatrics (PED)
Other Medical (O. MED)
General Surgery (GS)
Obstetrics/Gynecology (OBG)
Other Surgical (O. SUR)
Other Specialists (O. PHY)

Hospital Based and Other Sources

Hospital clinics (HOSCLI)
Hospital emergency rooms (HOSEMER)
Telephone, home, other (TELHOM)

Prices vary by sites and year. Health insurance is represented by coinsurance rates that vary by age and year.

The quantity of visits demanded (per person) in year t is:

$$(1) \qquad Q^d_{ijklmt} = R_{ijklm}\left[\left(P_{mt}\right)\left(C_{it}\right)\right]^{\beta_m}$$

where Q^d is the quantity demanded, P is price, C is the coinsurance rate (or the fraction of the price paid out-of-pocket), R represents the base rates of utilization, and β is the elasticity of demand. The subscripts are: age - k, sex - j, race - k, income - l, site - m, and year - t.

In order to implement each of the equations represented by (1), the base rates of utilization, R, were calculated from the Health Interview Survey (HIS) conducted by the U. S. National Center for Health Statistics.*

The estimates of the elasticity of demand with respect to price, β, were obtained from regression analysis using the following specification:

$$(2) \qquad Q^d_m = f(\text{Price, Income, Age, Sex, Conditions}),$$

where Q^d_m is the number of patient visits demanded per year from site m, Price is the cost of the most recent visit to the physician at site m, Income is the annual income of the person and his household, Age is the person's age in years, Sex is a

*These rates were computed by processing computer tapes of unpublished data. Summary tabulations of the survey data have been published. See: U. S. National Center for Health Statistics (1972).

binary which takes the value 1 if the person is female and 0 otherwise. Each of the four condition binary variables takes the value 1 if the person has at least one chronic condition, and has some activity limitation and is 0 otherwise. Multiple variables are included to measure the severity of the condition in terms of activity limitation. Equation (2) was estimated from HIS data using individuals as observations.

Table 1 presents the results for site 1 (i.e., visits to general practice physicians). The three sets of estimates it contains illustrate the importance of including condition in demand equations for medical services. Set C shows that the condition variables alone have a significant positive influence on demand. It is apparent from Set A that price has the expected negative effect on demand, while income has the expected positive effect. Set B shows that deleting the conditions variables results in a negative income elasticity. The results also show that while age is a proxy for conditions, it is much better to directly represent conditions.

Table 2 provides an example of output generated by the model. Since space limitations do not permit us to present the joint distribution of physician services demands, Table 2 gives a marginal distribution by age and site for 1970.

Supply of Physician Services

The aggregate supply of physician visits, by specialty, is the product of the number of such physicians in office-based practice, times the productivity of the average physician in each specialty. Productivity depends first on whether the physician is in group or non-group practice. (The non-group physician classification is dominated by solo physicians, and hereafter will be described as "solo" for brevity; logically the set contains two-man partnerships and "informal associations".) There are 14 specialties and the group-solo distinction, giving 28 types of physicians for which separate functions are maintained to treat the supply of services and the demand for aides.

Productivity for a given type of specialist depends on average annual hours of physician input and average number of nurses, technicians, and other aides employed (hereafter, "secretaries," who constitute the majority of other aides). The output of visits is related to the inputs of physician hours and the three types of aides

TABLE 1

DEMAND FOR PHYSICIAN OFFICE VISITS TO GENERAL PRACTICE PHYSICIANS

	SPECIFICATION							
	Set A			Set B			Set C	
	All Variables			No Conditions			Conditions Only	
	Coef.	T-Ratio	Elas.	Coef.	T-Ratio	Elas.	Coef.	T-Ratio
Age	.004	1.23	.021	.046	16.73	.246		
Sex	1.036	8.21		.851	6.58			
Price	-.023	-1.31	-.020	-.028	-1.52	-.024		
Income	.017	1.21	.018	-.018	-1.28	-.019		
Condition 2*	2.230	14.36					2.379	16.68
Condition 3	4.314	14.16					4.544	15.63
Condition 4	4.459	18.96					4.703	22.77
Condition 5	7.012	18.27					6.843	19.41
Constant		3.835			4.838			4.473
R-Squared		.1146			.0461			.1061
F Statistic		118.20730			88.24892			216.72010
Standard Error		5.23891			5.43649			5.26282
Number of Observations		7312.			7312.			7312.

*Condition 2 = 1 if 1+ chronic conditions, but no limitation of activity, 0 otherwise

Condition 3 = 1 if 1+ chronic conditions, with limitation but not in major activity, 0 otherwise

Condition 4 = 1 if 1+ chronic conditions, with limitation in amount or kind of major activity, 0 otherwise

Condition 5 = 1 if 1+ chronic conditions, and unable to carry on major activity, 0 otherwise.

TABLE 2

AGE X SITE YEAR = 1970

TOTAL VISITS

AGE	GP	IM	PED	O.MED	GS	OBG	O.SUR	O.PHY	HOSCLI	HOSEMER	TELHOM	TOTAL
0-5	37368.	2905.	55806.	2588.	3153.	1254.	7530.	8779.	20299.	11969.	44062.	195714.
6-16	47915.	10485.	24165.	10183.	5695.	690.	16013.	12816.	19258.	13440.	32747.	193406.
17-24	33547.	5527.	471.	9564.	4744.	19566.	8302.	11790.	12986.	6146.	16499.	129241.
25-34	38590.	9504.	1069.	10052.	8791.	25626.	13570.	23712.	15912.	7560.	19558.	173945.
35-44	37325.	13752.	351.	7560.	8944.	8400.	13739.	19685.	18392.	5042.	16821.	149911.
45-54	45613.	13872.	171.	7553.	11384.	5987.	15953.	29770.	16656.	3971.	16035.	166965.
55-64	46976.	15284.	0.	5501.	10839.	696.	16225.	27112.	10255.	3507.	14502.	150896.
65-74	36676.	13422.	0.	3210.	6686.	508.	11150.	16353.	7248.	2011.	13509.	110773.
75+	22607.	4723.	0.	4956.	2864.	155.	6830.	11404.	4920.	1258.	13187.	72902.
TOTAL	346616.	89574.	82032.	61167.	63000.	62882.	109311.	161420.	125926.	54904.	186920.	1343739.

ROW PERCENTAGE TABLE

AGE	GP	IM	PED	O.MED	GS	OBG	O.SUR	O.PHY	HOSCLI	HOSEMER	TELHOM	TOTAL
0-5	19.1	1.5	28.5	1.3	1.6	0.6	3.8	4.5	10.4	6.1	22.5	100.0
6-16	24.8	5.4	12.5	5.3	2.9	0.4	8.3	6.6	10.0	6.9	16.9	100.0
17-24	26.0	4.4	0.4	7.4	3.7	15.1	6.4	9.1	10.0	4.8	12.8	100.0
25-34	22.2	5.5	0.6	5.8	5.1	14.7	7.8	13.6	9.1	4.3	11.2	100.0
35-44	24.9	9.2	0.2	5.0	5.9	5.6	9.2	13.1	12.3	3.4	11.2	100.0
45-54	27.3	8.3	0.1	4.5	6.8	3.6	9.6	17.8	10.0	2.4	9.6	100.0
55-64	31.1	10.1	0.0	3.6	7.2	0.5	10.8	18.0	6.8	2.3	9.6	100.0
65-74	33.1	12.1	0.0	2.9	6.0	0.5	10.1	14.8	6.5	1.8	12.2	100.0
75+	31.0	6.5	0.0	6.8	3.9	0.2	9.4	15.6	6.7	1.7	18.1	100.0
TOTAL	25.8	6.7	6.1	4.6	4.7	4.7	8.1	12.0	9.4	4.1	13.9	100.0

COLUMN PERCENTAGE TABLE

AGE	GP	IM	PED	O.MED	GS	OBG	O.SUR	O.PHY	HOSCLI	HOSEMER	TELHOM	TOTAL
0-5	10.8	3.2	68.0	4.2	5.0	2.0	6.9	5.4	16.1	21.8	23.6	14.6
6-16	13.8	11.7	29.5	16.6	9.0	1.1	14.6	7.9	15.3	24.5	17.5	14.4
17-24	9.7	6.3	0.6	15.6	7.5	31.1	7.6	7.3	10.3	11.2	8.8	9.6
25-34	11.1	10.6	1.3	16.4	14.0	40.8	12.4	14.7	12.6	13.8	10.5	12.9
35-44	10.8	15.4	0.4	12.4	14.0	13.4	12.6	12.2	14.6	9.2	9.0	11.2
45-54	13.2	15.5	0.2	12.3	18.1	9.5	14.6	18.4	13.2	7.2	8.6	12.4
55-64	13.6	17.1	0.0	9.0	17.2	1.1	14.8	16.8	8.1	6.4	7.8	11.2
65-74	10.6	15.0	0.0	5.2	10.6	0.8	10.2	10.1	5.8	3.7	7.2	8.2
75+	6.5	5.3	0.0	8.1	4.5	0.2	6.2	7.1	3.9	2.3	7.1	5.4
TOTAL	100.0	100.0	100.0	100.0	100.0	100.0	100.0	100.0	100.0	100.0	100.0	100.0

by estimated production functions. Physician hours are exogenous. The number of each type of aide demanded depends on wages, prices, and visits. Since aides demanded and visits supplied are endogenous, this gives a system of four simultaneous structural equations for each specialty determining the quantities demanded for three types of aides and the quantity of visits supplied. The reduced form equations therefore yield nurses demanded, technicians demanded, secretaries demanded, and visits supplied, as linear functions of the wages of nurses relative to the output price (product wages); product wages of technicians, product wages of secretaries, and hours of physician input. Tables 3 and 4 give the predicted patient visits and employment patterns using these equation.

The development of the current version of the supply of physician services sector has drawn heavily upon pilot work on another study of major issues in private medical practice--including, inter alia, the comparison of group and solo form of medical practice.* The equations expressing the demand for aides are derived from estimates by Intriligator and Kehrer (1973). The production function estimates are adapted from estimates by Kimbell and Lorant (1972). These two sources of estimates have been used to synthesize our current version.

Fee Adjustment Procedures

The basic assumption underlying our fee adjustment procedures is that physician services markets do not equilibrate instantaneously and completely. Alternatively, we assume that they are typically in a state of disequilibrium, with gradual adjustments in fees in the direction of equilibrium. When there is growth in the quantities of visits demanded relative to quantities supplied, there will be an acceleration in the inflation of fees. When there is a slower growth in visits demanded than in visits supplied, there will be a retardation in the rate of fee inflation.

The basic form of the fee adjustment equations for the first seven type of specialists is:

---

*This project is being conducted jointly by the Human Resources Research Center of the University of Southern California and the Center for Health Services Research and Development of the American Medical Association, pursuant to Contract No. HSM 110-70-354 with the Health Services and Mental Health Administration, U. S. Department of Health, Education, and Welfare.

TABLE 3

QUANTITY OF VISITS SUPPLIED BY PHYSICIANS IN OFFICE BASED PRACTICE

TYPE	1960	1961	1962	1963	1964	1965	1966	1967	1968	1969	1970
GENERAL PRACTICE	513677	502285	491276	479649	468791	460029	450449	441269	433323	415226	408465
INTERNAL MEDICINE	101247	104102	107034	109916	114682	117490	120240	124162	128454	129064	135933
PEDIATRICS	60949	62578	64285	65968	68311	70098	71476	73069	75633	77618	79904
OTHER MEDICAL	55247	56725	58247	59778	61475	62734	64547	66238	68718	73915	77006
GENERAL SURGERY	90111	92116	94164	96252	99245	101335	102855	104158	105864	105676	109032
OBG	64293	66220	68173	70153	72922	74957	76668	77943	80391	82160	86517
OPTHAMOLOGY	30664	31551	32508	33415	34688	35574	36287	37287	38303	39453	40840
ORTHOPEDIC SURGERY	28157	29558	31028	32548	34604	36125	37655	39373	42928	45084	47361
OTHER SURGICAL	48570	49994	51445	52930	54869	56226	57608	58887	60460	62044	64561
ANESTHESIOLOGY	19830	20946	22158	23382	25266	28565	27835	29224	30239	31170	32341
PSYCHIATRY	15341	16170	17071	17937	18926	19942	21237	22497	23645	24832	26038
PATHOLOGY	11191	12301	13544	14837	17086	18427	20370	22663	24330	26069	27741
RADIOLOGY	45558	47916	60484	53030	57054	59392	62338	65670	68772	72344	75960
ALL OTHER PHYSICIANS	65613	67478	69509	70764	73110	74154	77095	78502	80473	91524	94473
TOTAL	1150448	1159950	1170926	1180559	1201019	1213048	1226660	1240942	1261533	1276179	1306172

TABLE 4

EMPLOYMENT OF RN'S,LPN'S,TECHNICIANS AND OTHER AIDES BY MD'S IN OFFICE

TYPE	1960	1961	1962	1963	1964	1965	1966	1967	1968	1969	1970
REGISTERED NURSES	57398	57691	58104	58474	59029	59346	59721	58597	57992	57310	57204
PRACTICAL NURSES	28270	28415	28618	28800	29074	29230	29414	28861	28563	28227	28175
TECHNICIANS	53630	54137	55274	55694	56487	56787	58572	59529	60090	61421	65185
OTHER PERSONNEL	214152	216484	218943	221271	225748	228719	231950	234786	239290	242718	249369
TOTAL	353450	356727	360939	364239	370338	374082	379657	381773	385935	389676	399933

$$Fee_{i,t+1} = \alpha_i \, Fee_{i,t} + \beta(V^D_{i,t} - V^S_{i,t})$$

where $Fee_{i,t}$ is the fee for specialty i in year t, $V^D_{i,t}$ is the aggregate quantity of visits demanded from specialty i in year t, $V^S_{i,t}$ is the aggregate quantity of visits supplied by specialty i in year t, $\alpha_i - 1$ is the fractional rate at which fees will grow when equilibrium obtains, and β is the adjustment factor which governs the speed of fee adjustment during periods of disequilibrium.

The α_i's were estimated from the mean rate of change observed in the physician fee component of the Consumer Price Index over the period 1960-1970. The speed adjustment coefficient, β, was specified at alternative values and sensitivity studies were performed to find the best calibration of this factor.

The Linkages Between the Physician Services Submodel and the Rest of the HRRC Model

The physician services submodel is linked to the rest of the HRRC microsimulation model by six channels: (1) the population of individuals generated by the consumer submodel drives the demographic variables in the demand for physician office visits; (2) the physician submodel provides the numbers of office-based physicians by specialty, which influences the supply of physician visits; (3) inhospital visits by physicians in office-based practice are taken from inpatient days generated in the hospital services submodel; (4) outpatient visits to hospital clinics and emergency rooms are generated in the physician services submodel for input to the hospital services submodel; (5) wages of RNs, LPNs, technicians, and secretaries enter the physician services submodel from the health manpower submodel; and (6) the physician office demands for these aides, generated in the physician services submodel, are inputs to the health manpower submodel.

Plans for Further Research

The present form of the complete HRRC model has been described in detail in a report to the U. S. Department of Health, Education, and Welfare.* This report also contains the results of a historical simulation run, suggestions for improving the model based upon these results, and a number of forecasting and policy simulation experiments which can be performed using the revised version of the model. These

*See: Yett, Drabek, Intriligator, and Kimbell (1973).

and, quite probably, additional experiments will be performed during the next phase of the project. Further revisions and expansions of the model will be made on the basis of the experience acquired. Gradually the current version of the model will evolve toward the prototype set forth in the conceptual design phase of the project.* It should be emphasized, however, that long before this is accomplished, the model will be capable of providing forecasts and policy simulations to assist U. S. health planners.

*See: Yett, Drabek, Intriligator, and Kimbell (1970) for a description of the prototype designated as "Mark I."

REFERENCES

Intriligator, M. D., and Kehrer, B. H. "Allied Health Personnel in Physicians' Offices: An Econometric Approach." Economics of Health and Medical Care. Edited by M. Perlman. International Economic Association, 1973. (Forthcoming).

Kimbell, L. J., and Lorant, J. H. "Production Functions for Physicians' Services." Paper presented at the Econometric Society Meetings, Toronto, Canada, December 29, 1972.

U. S. National Center for Health Statistics. "Physician Visits Volume and Interval Since Last Visit United States--1969," by C. S. Wilder. Vital and Health Statistics, Series 10, No. 75. DHEW Pubn. No. (HSM) 73-1062. Washington, D. C.: Government Printing Office, 1972.

Yett, D. E., Drabek, L., Intriligator, M. D., and Kimbell, L. J. "The Development of a Microsimulation Model of Health Manpower Demand and Supply." Proceedings and Report of Conference on a Health Manpower Simulation Model. Aug. 31 - Sept. 1, 1970, Vol. 1. Washington, D. C.: Government Printing Office, 1970.

Yett, D. E., Drabek, L., Intriligator, M. D., and Kimbell, L. J. The Preliminary Operational HRRC Microsimulation Model. Final Report on U.S. Public Health Service Contract No. NIH 71-4065. Los Angeles: Human Resources Research Center, University of Southern California, 1973.

A MODEL FOR FINITE STORAGE MESSAGE SWITCHING NETWORKS

by F. Borgonovo and L. Fratta
Istituto di Elettronica and
Centro Telecomunicazioni Spaziali - CNR
Politecnico di Milano, Italy

0. ABSTRACT

In this paper the analysis of particular message switching networks is performed taking into account the finite storage at nodes.

An upper and lower bound to the node blocking probability are computed when the exact values are not available because of the computational complexity.

1. INTRODUCTION

A communication network (CN) is a collection of nodes (communication centers) connected together by a set of links (communication channels). This paper consider a particular class of CN denoted as Message Switching Networks (MSN). Such networks accept message traffic from external sources and transmit this traffic over some route within the network to the destination nodes. This transmission takes place over one channel at a time, with possible storage of the message at each intermediate node. In fact if a message can not be transmitted out of a node because its output channel is not available it joins a queue and awaits its turn to be transmitted.

This kind of networks has been recently studied as a model for computer communication networks [Kleinrock, 1964, 1970] but all the analysis carried out consider that the nodes have infinite storage. This hypothesis simplifies the analysis but it does not take into account an important fact which happens in the real networks: the node "blocking". In fact, since the storage of each node is finite, it can become filled out and refuse the messages coming from external sources which then will get lost.

The goal of this work is to get some more insight into the analysis of MSN with finite storage nodes and to evaluate the blocking probability for some particular networks.

2. DEFINITION OF THE MODEL

The blocking probability of any node, i.e., the percentage of external traffic which is lost, may be exactly obtained from the knowledge of the statistical stationary behaviour of the network which depends upon the following three specifications:

a - topology

b - external traffic characteristics (message arrival process, message length distribution, message destination, etc.)

c - node and channel characteristics (node storage, channel capacity, queue discipline, etc.)

Unfortunately, a statistical analysis cannot be performed for general specifications a, b and c; for example, just the complexity introduced by the finite storage makes unfeasible the analysis even for networks which have been easily studied with the hypothesis of infinite storage. Such networks were supposed to have a high degree of connection and deal with poisson message arrival processes and exponential length distributions. With these hypotheses it was shown that the input messages at any node are still poisson distributed with independent exponential lengths and the analysis can be performed separately at each node [Kleinrock, 1964].

On the contrary when finite storages are considered the internal message arrival process is no more poisson distributed which implies a correlation among the nodes of the network and makes unfeasible their analysis separately. Furthermore when a message is forbidden to leave node i because of the saturation of the next node, j, on its path a feedback effect arises. Such an effect depends on the length distribution of the saturation times of node j which makes unfeasible the analysis. Moreover the whole network can become permanently blocked if a suitable procedure is not used at each node.

In the following we will then limit our analysis to a particular kind of network in which the feedback effect does not exist. Such networks are characterized by:

- topology: the N nodes are connected in a chain which may be closed (Fig. 1)
- external traffic characteristics: markovian message arrival process and constant message length of L [bits]
- node and channel characteristics: finite node storage S_i [messg] i=1, N channel capacity C_{ii+1} = C [bits/sec] for all channels and synchronous message transfer.

When synchronous behaviour is supposed, the time is devided into intervals of length $T=L/C$(°) in which messages are transmitted.

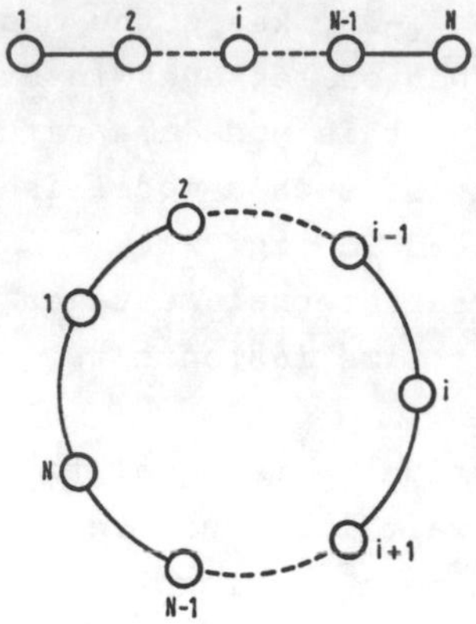

Fig. 1

This implies that no message entering node i from node i-1 will be ever rejected because even if node i was filled during (t_{K-1},t_K) when the message enters it at time t_K another message leaves the same node at the same time.

Then, the saturation of node i affects only the external traffic affered to itself.

Fig. 2 a) shows the model of the node i. The Input Register IR is utilized to receive from node i-1 at maximum one message at each interval T and at time t_K it routes the message, if any, either out of the node or into the storage.

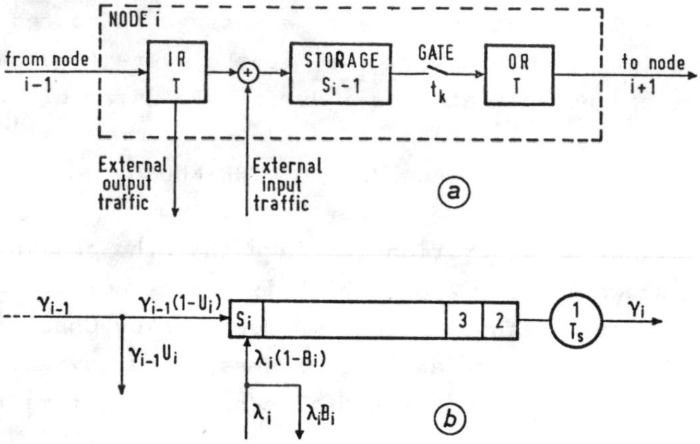

Fig. 2

(°) the k-th interval initiates at time t_K and terminates at time t_{K+1}.

The Output Register OR performs the transmission of one message during the interval T.

The gate allows the next message to enter the OR at time t_K and the storage of size S_i-1 , keeps the message in queue.

When the number of messages in Store i equals S_i-1 the external messages entering this node are refused.

The behaviour of such a model is equivalent to that of the queueing system represented in fig. 2b).

In this system internal messages are supposed to arrive and to leave exactly at time instants t_K, while the external messages arrive at random.

The service time T_S is equal to the transmission time T only for messages which wait in queue. In fact if an external message enter the empty node at time t^* during the interval $(t_{K-1}-t_K)$ it does not joins any queue but its service terminates at time t_{K+1}, that is its service time has to be augmented by the quantity t_K-t^*.

This is due to the presence of the gate in the model which also constrains to S_i-1 the maximum number for external messages that can enter the node during each interval T.

The following variables have to be defined:

γ_i = average number of messages leaving node i in t_K
λ_i = average number of external messages offered to node i in T
B_i = fraction of external traffic which is refused at node i
U_i = probability that an internal message leaves the network at node i.

As no internal message can be refused by node i, U_i represents also the probability that at any time t_K a message is routed out of the network and it follows that γ_{i-1} U_i is the average number of messages which leave the network at node i during the interval T.

3. ANALYSIS OF THE MODEL

From previous considerations we have that the analysis of the chain of queues, shown in Fig. 2 b), which is a model of our network can be carried out by the method of the imbedded Markov Chain [Saaty, 1961] when the external message arrival process is markovian.

In fact if we let the state of the network at time t_K to be represented by the set

$$\{n_1^{(K)}, n_2^{(K)}, \dots, n_N^{(K)}\}$$

where $n_i^{(K)}$ is the number of messages in the node i at the time

instant immediately following t_K (i.e., after the message transfer) the network state at time t_{K+1} is completely determined by the state at time t_K and the arrival process.

The behaviour of the network is known at time t_K once the set of $M = \prod_{i=1}^{N} (S_{i+1})$ joint probabilities

$$P[n_1^{(K)}, n_2^{(K)}, \ldots, n_N^{(K)}]$$

is computed.

By the hypothesis in the following set of M equations:

$$P\left[n_1^{(K+1)}=a_1,\ldots,n_N^{(K+1)}=a_N\right]=\sum_{b_1=0}^{S_1}\cdots\sum_{b_N=0}^{S_N} P\left[n_1^{(K+1)}=a_1,\ldots n_N^{(K+1)}=a_N / n_1^{(K)}=b_1\ldots n_N^{(K)}=b_N\right]\cdot P\left[n_1^{(K)}=b_1,\ldots, n_N^{(K)}=b_N\right] \quad 0\leq a_j \leq S_j \qquad (1)$$

for any $\{a_1,\ldots,a_N\}$ $\qquad j = 1,2,\ldots,N.$

The conditional probabilities $P\left[n_1^{(K+1)}=a_1,\ldots,n_N^{(K+1)}=a_N / n_1^{(K+1)}=b_1,\ldots,n_N^{(K+1)}=b_N\right]$ depend only upon both the arrival process and the probabilities U_i.

The stationary behaviour of the network is described by the probabilities

$$p\left[n_1,n_2,\ldots,n_N\right] = \lim_{K\to\infty} p\left[n_1^{(K)},n_2^{(K)},\ldots n_N^{(K)}\right] \qquad (2)$$

which are the solutions of the linear equation system (1') obtained for $K\to\infty$ from (1)

$$(1')\begin{cases} P\left[n_1=a_1,\ldots,n_N=a_N\right]=\sum_{b_1=0}^{S_1}\cdots\sum_{b_N=0}^{S_N} P\left[n_1=a_1,\ldots,n_N=a_N/n_1=b_1,\ldots n_N=b_N\right]\cdot \\ \quad\cdot P\left[n_1=b_1,\ldots,n_N=b_N\right] \qquad \text{for all sets } \{a_1,\ldots a_N\} \text{ except one} \\ \qquad\qquad 0\leq a_j\leq S_j \qquad j=1,2,\ldots,N \\ \sum_{a_1=0}^{S_1}\cdots\sum_{a_N=0}^{S_N} P\left[n_1=a_1,\ldots,n_N=a_N\right] = 1 \end{cases}$$

From (2) we may compute the probability, $P_i(0)$ that the service at node i is empty

$$P_i(0)=\sum_{a_1=0}^{S_1}\cdots\sum_{a_{i-1}=0}^{S_{i-1}}\sum_{a_{i+1}=0}^{S_{i+1}}\cdots\sum_{a_N=0}^{S_N} P\left[n_1=a_1,\ldots,n_i=0,\ldots,n_N=a_N\right]$$

from which we have

$$\gamma_i = [1-P_i(0)]$$

and for the flow conservation at node

$$\gamma_{i-1}(1-U_i)+\lambda_i(1-B_i) = \gamma_i \tag{3}$$

then

$$B_i = 1 - \frac{\gamma_i-\gamma_{i-1}\ (1-U_i)}{\lambda_i} \tag{4}$$

Note that in the general case we are not able to straitforwardly compute the probabilities U_i because the actual internal traffic γ_i is not known, and an iterative procedure has to be implemented starting from values of U_i corresponding to infinite storage.

Some particular cases exist where the computation of U_i can be easily performed. In the following we will consider one of these particular networks defined by a closed chain topology with N=3 (Fig. 3) and with the same traffic and storage capacity at all nodes. This implies that the behaviour of a node is the same for all nodes, then from now on we will delete index i from all quantities.

We will further assume that the traffic λ entering a node is random equally destinated to all the other nodes in the network. As the total traffic in each channel is

$$\gamma = \lambda(1-B)\ (1+ \frac{N-2}{N-1} + \frac{N-3}{N-1} \ldots + \frac{1}{N-1}) = \lambda(1-B)\ \frac{N}{2} \tag{5}$$

and from (3)

$$\gamma U = \lambda(1-B) \tag{6}$$

we have

$$U = \frac{2}{N}$$

The traffic intensity ρ, which measures the degree of utilization of the network, is defined as the total traffic offered at each node in T; that is, ρ, is equal to the total traffic that would be accepted by the node with infinite storage. Thus, as in this case B=0 from (5)

$$\rho = \lambda \cdot \frac{N}{2}$$

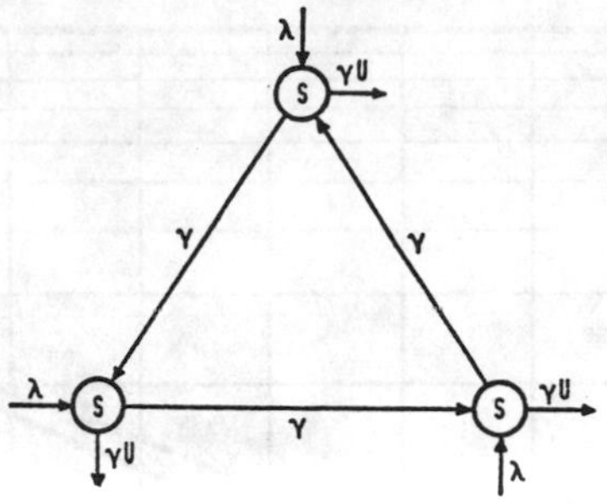

Fig. 3

Note that the channel utilization, i.e., the fraction of time that the channel is busy corresponds to the total traffic γ in the channel. Thus

$$\gamma = 1-p_o = \rho(1-B) \tag{7}$$

from which or equivalently from Eq.(4) we get

$$B = 1-\frac{\gamma}{\rho} \tag{8}$$

In Eq.(8) γ is unknown and we already pointed out that it can be computed by solving system (1').

Even if in this particular case the number of equations in system (1') can be reduced from $(S+1)^N$ to a fraction of it because of the symmetry of the problem it still remains too large for practical values of N and S and its utilization is limited to very small networks.

In Fig. 4 the values of the blocking probability B are reported as function of the traffic intensity ρ with storage capacity S=3,4,5 N=3 and poisson external arrival process with mean value $\lambda T = \rho \frac{2}{N} T \left[\frac{\text{messg}}{\text{sec}}\right]$.

4. APPROXIMATION METHODS

In the present paragraph we will propose two approximation methods which give a lower and an upper bound of the blocking probability for networks of the kind studied in the previous paragraph where the exact value can not be computed because of the large amount of computation required.

Both methods consist in studying a single node and they can not give an exact solution because the input traffic is unknown unless we

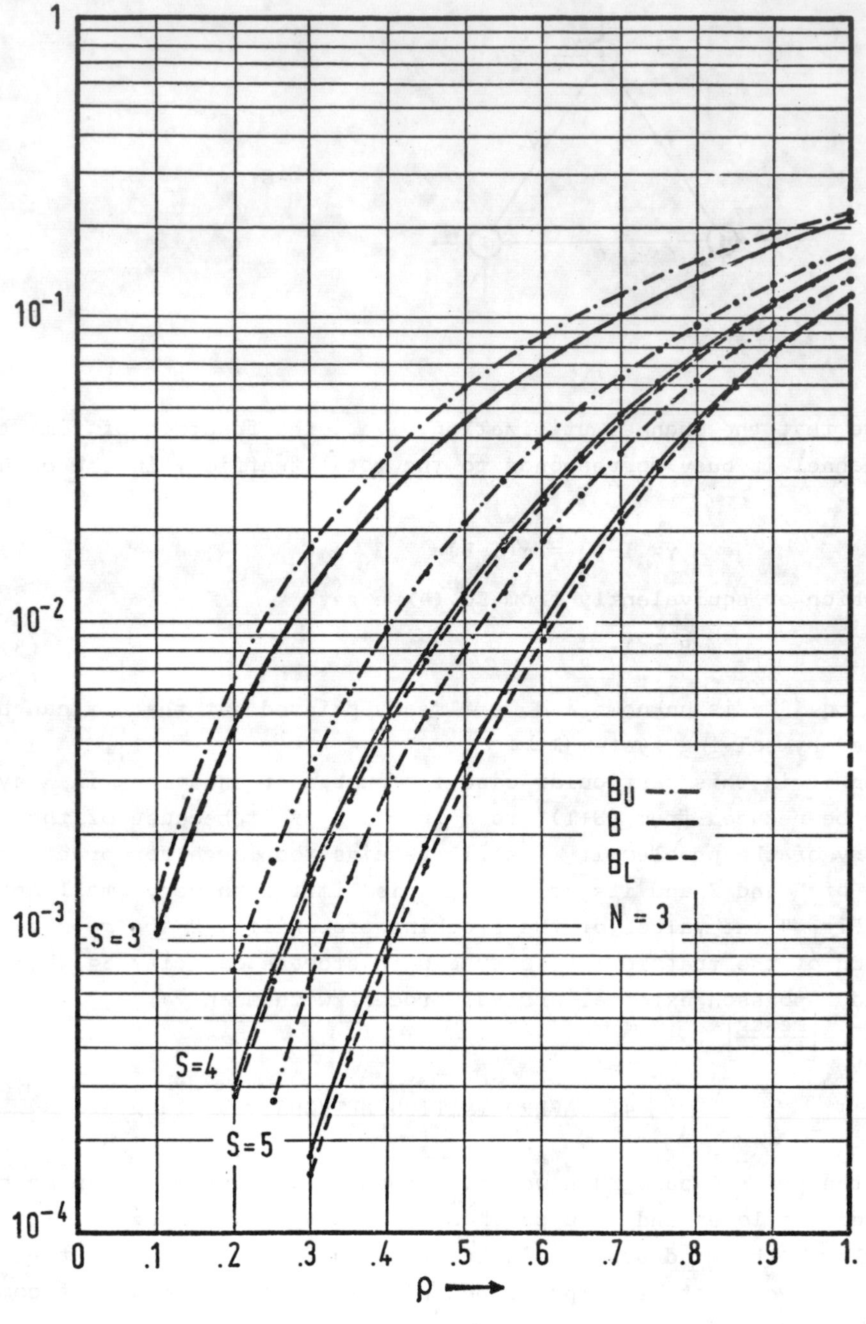

Fig. 4

consider the network in its entirety.

Lower Bound Method (LBM)

Such a method performs the analysis of node i by the imbedded Markov Chain assuming that the internal input traffic is merely random, that is at any interval time T a message arrives with probability $[1-P_{i-1}(0)]$ (1-U) and no correlation exists among different intervals.

With this assumption the node state probabilities $P_i(n)$ can be computed by solving the following system:

$$(9)\quad \begin{cases} P_i(n=a) = \sum_{b=0}^{S} P_i(n=a/n=b)\, P_i(n=b) & a=0,1,\dots,S-1 \\ \sum_{j=0}^{S} P_i(n=j) = 1 \end{cases}$$

The conditional probabilities P_i (n=a/n=b) depend on the arrival process, network variables S and N and on the probability $P_{i-1}(0)$. As the behaviour of all nodes is the same $P_{i-1}(0) = P_i(0)$, system (9) becomes non linear and may be numerically solved by an iterative procedure.

Once $P_i(0)$ has been computed, from Eq.(8) we get

$$B_L = 1 - \frac{1-P_i(0)}{\lambda}\,\frac{2}{N} \qquad (10)$$

which represents the lower bound of the blocking probability of the node.

A heuristic proof that $B_L \leq B$ is obtained recalling that the real internal arrival process at node i, say process A, has a correlation, introduced by the queue at node i-1, among the different intervals. Such a correlation, as it can be easily seen, yields that the ratio

$$\frac{P\left(\frac{x_{K+1}}{x_K}\right)}{P\left(\frac{x_{K+1}}{\bar{x}_K}\right)}$$

is greater for the process A than for the merely random process with the same average we imposed to compute B_L.

$P\left(\frac{x_{K+1}}{x_K}\right)$ represents the conditional probability that a message arrives (does not arrive) at time t_{K+1} if a message was (was not) arrived at time t_K;

$P\left(\frac{x_{K+1}}{\bar{x}_K}\right)$ represents the conditional probability that a message arrives (does not arrive) at time t_{K+1} if a message was not (was) arrived at time t_K.

This fact points out that in process A messages have the tendency to be grouped in packets and implies that if process A characterizes the arrival process of node i this node will assume both state S and state 0 with higher probability than in the case of merely random arrival process as it is shown in Table I.

P(n)	LBM	EXACT METHODS	UBM
P(0)	0.1421	0.1439	0.1587
P(1)	0.2328	0.2320	0.2105
P(2)	0.2267	0.2247	0.2054
P(3)	0.1934	0.1921	0.1910
P(4)	0.1614	0.1619	0.1770
P(5)	0.0436	0.0454	0.0574
$B=1-\frac{1-P(0)}{\rho}$	0.0969	0.0988	0.1145

TABLE I

Node state probability distribution for
$N=3 \quad S=5 \quad \rho=0.95$

Thus we expect to have in the real operation of the node a $P_i(0)$ greater than that we computed by LBM and consequently $B_L \leq B$.

The same kind of reasoning shows us the way how to find an upper bound to B.

Upper Bound Method (UBM)

This method as the previous one analyzes a single node where now the internal input traffic is the percentage (1-U) ofits internal output traffic as shown in Fig. 5.

Fig. 5

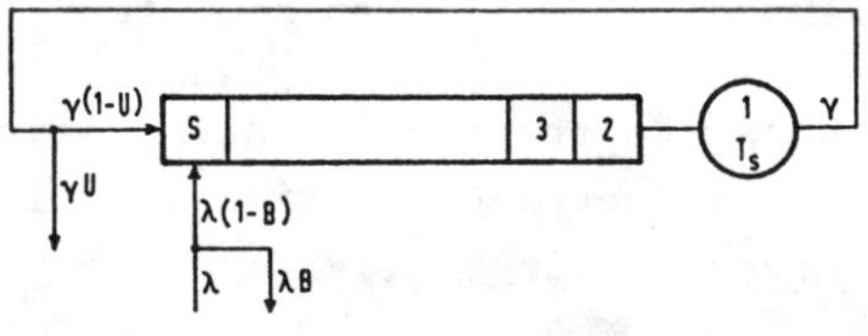

In such a scheme because of the traffic hypothesis the node will be furtherly constrained to be in both states S and 0 with higher probability than in the real case [TABLE 1]. In fact when the node is in state 0 no internal message will enter and the node will change the state only because of the external message traffic. Thus the probability to become busy is smaller than in the real operation of the node with a consequently increase of $P_i(0)$. The same holds for state S.

The state probabilities are computed by solving the equation system (9) which in this case becomes linear. Eq. 10 with the present value of $P_i(0)$ provides the upper bound B_U to the node blocking probabilities.

The complexity of the two approximation methods proposed is much smaller than that of the exact method, in fact they require the solution of a system of only S+1 equations independent from the number of nodes in the network.

Some computational results for poisson distributed external traffic are reported in Fig.4 & 6. Fig. 4 shows the very good performance of the lower bound.

When a direct comparison with the exact values is not possible (Fig. 6) we expect the lower bound is still very good for $\gamma \to 1$. This is supported by the fact that in this case both internal arrival processes (the real one and the merely random) trend to have the same behaviour.

5. CONCLUSIONS

In this paper we have proposed and analyzed a model for message switching networks with constant length messages. This is the case when messages are divided in "packets" as happens in some existing networks. With this condition the operation of chain topology network may be realized in such a way of never blocking the internal traffic and can be studied by an imbedded Markov chain method.

Two approximation techniques have been proposed to reduce the complexity of this exact analysis. The computational results have shown that the bounds are very good for high network utilization. This is particularly interesting because such a case is the worst for the analysis performed assuming infinite storage at nodes.

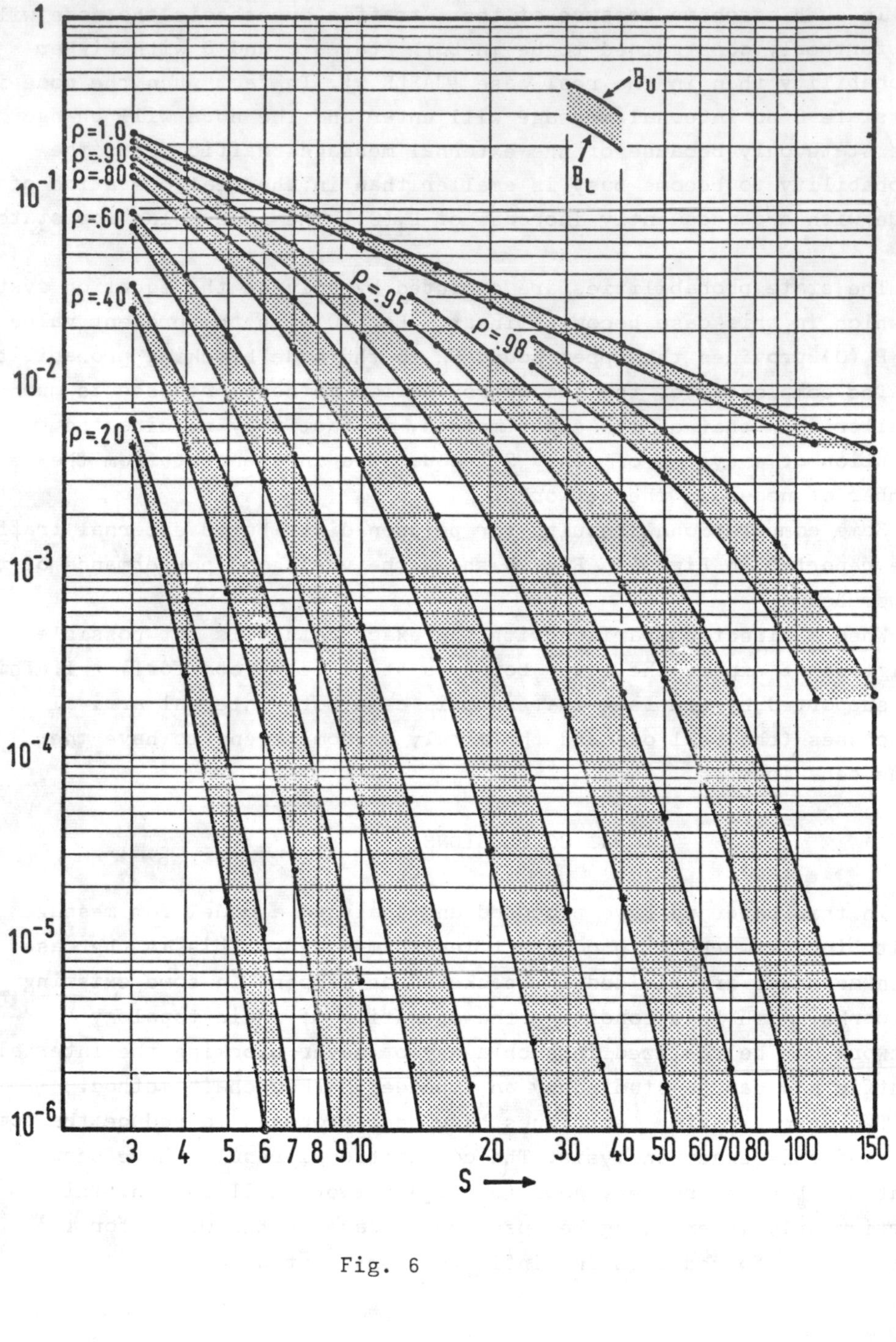
1
10^-1
10^-2
10^-3
10^-4
10^-5
10^-6
ρ=1.0
ρ=.90
ρ=.80
ρ=.60
ρ=.40
ρ=.20
ρ=.95
ρ=.98
B_U
B_L
3 4 5 6 7 8 9 10 20 30 40 50 60 70 80 100 150
S

Fig. 6

REFERENCES

Kleinrock, L.: "Communication Nets" McGraw-Hill, 1964

Kleinrock, L.: "Analytic and Simulation Methods in Computer Network Design" SJCC 1970

Saaty, T.L.: "Elements of Queueing Theory" McGraw-Hill, 1961.

ON CONSTRAINED DIAMETER AND MEDIUM OPTIMAL SPANNING TREES (°)

F. Maffioli

Istituto di Elettronica, Politecnico di Milano, Italy

0. ABSTRACT

When two weighting figures per arc must be considered in optimizing a network one has either to combine them in a single performance factor or to optimize the network with respect to one of them while respecting some constraint on the other. This work deals with trees, showing how the algorithm for finding the shortest spanning tree of a graph can be modified to handle some problems of this kind efficiently.

1. INTRODUCTION

Communication networks must often be studied considering more than one weighting figure for each element of the network. Let for instance $G = (N,A)$ be the graph representing the network, N being the set of nodes and A the set of arcs of G. Every arc (i,j) may be labeled by a length ℓ_{ij}, by a cost C_{ij}, by a probability of being operational p_{ij}, by a time delay τ_{ij}, etc. One of the problems most often encountered is that of finding a spanning tree of G (fig. 1) which has to be optimal in some sense.

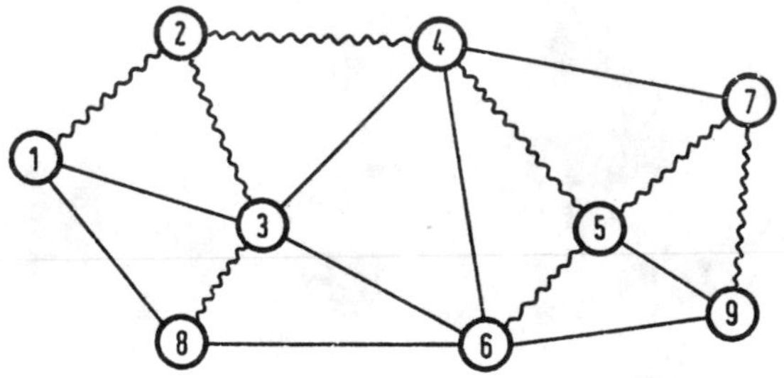

Fig. 1 - A graph and a spanning tree

(°) Partially supported by the Centro di Telecomunicazioni Spaziali del CNR, Milano, Italy.

If the tree has to be of minimum (or maximum) total weight with respect to only one weighting figure per arc the "greedy" algorithm may be used and is known to be quite efficient [Kerschenbaum - Van Slike (1972) Prim (1957)]. In this work the optimization of the tree will be carried out considering two weighting figures per arc, either combining them in a single performance factor or optimizing with respect to one of them while respecting some constraint on the other. The different methods of solution are nothing but suitable modifications of the "greedy" algorithm.

Most of this work could be considered as folklore, however it may help to point out how much has to be done even about the best known problems to handle practical situations of the kind for instance reported in Fratta et al. (1971).

2. MEDIUM OPTIMAL SPANNING TREE

The first problem considered here is the following.
Find the spanning tree T which minimizes $z = \sum_T c_{ij} / \sum_T p_{ij}$.

The suggested algorithm uses an idea already applied to the minimal cost to time ratio problem [Lawler (1972)]. Assume we have guessed the minimum value of z. Then

$$z = \min_k \{ \sum_{T_k} c_{ij} / \sum_{T_k} p_{ij} \}$$

where k indexes the trees of the given graph.
Define now

$$\bar{c}_{ij} = c_{ij} - z\, p_{ij}$$

and let us find

$$w = \min_k \{ \sum_{T_k} \bar{c}_{ij} \} = \sum_T \bar{c}_{ij} \,.$$

There exist three cases.

<u>w < 0</u>

$$\sum_T \bar{c}_{ij} = \sum_T (c_{ij} - z\, p_{ij}) < 0$$

then

$$\sum_T c_{ij} < z \sum_T p_{ij}$$

i.e.

$$z > \frac{\sum_T c_{ij}}{\sum_T p_{ij}}$$

and obviously the guessed value of z is too pessimistic since there exists a tree T for which a smaller value can be found.

<u>w > 0</u>

Similarly

$$z < \frac{\sum_T c_{ij}}{\sum_T p_{ij}}$$

and there is no tree meeting such a small value of z.

<u>w = 0</u>

The guessed value of z is optimal.

As it is suggested in Lawler (1972) a binary search can now be implemented to approach the correct z as close as one likes and hence to find the medium optimal spanning tree T.

3. EXISTENCE OF CONSTRAINED DIAMETER TREE

From now on we will consider only euclidean networks. However it should not be difficult to generalize meaningfully the problems to follow.

The first problem we shall consider is an existential one and shall be referred to as Problem 0. Note that in this paragraph we place no restrictions on using auxiliary nodes different from the given ones to build up shorter trees. The conditions we shall find apply therefore to general Steiner trees on the plane. Unfortunately this will be not the case in dealing with the algorithms to construct them. In fact it is well known that finding the Steiner tree even without constraints is an unsolved problem [Gilbert, Pollack (1968) and Shi-Kuo Chang (1972)].

PO) Assume n points and a tree connecting some of them are given on a plane. Does there exists a tree connecting them all and having diameter(1) not greater than that of the given subtree?

PO comes into play whenever, in considering the constraints of the problem in terms of one of the weighting figures labelling the arcs, this amounts to specify only part of the solution in the form of a subtree.

---

(1) The diameter of a graph is here by definition the (euclidean) length of the longest shortest path connecting any pair of nodes of the graph.

The simplest subtree is of course a simple arc and this is the case we shall consider first referring to the simpler corresponding problem as P'0. Let (p_1,p_2) of fig. 2 be such an arc, then the following lemmas are easily proved [Maffioli (1971)].

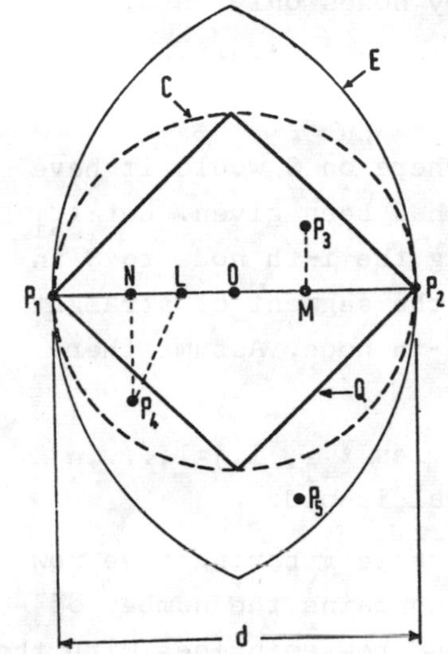

Fig. 2 - Solvability of P'0.

<u>Lemma 1</u> P'0 is solvable iff all points are inside C.

<u>Lemma 2</u> solving P'0 a point may be connected perpendicularly to the subtree iff it lies inside Q.

<u>Lemma 3</u> if (p_1,p_2) is the pair of most distant points, no other given point may be outside E.

Consider now P0 and refer to fig. 3, where the assigned subtree S is depicted with continuous lines.

On S, ABCDE is the longest path, that is the diameter of S. By successively rotating the branches of the longest path it is always possible, as in the example of fig. 3, to obtain a segment (A'E") whose length is equal to the diameter d of S and whose middle point, 0, lies on a branch of S. Let Γ be the circle having d as diameter and 0 as center.

<u>Theorem 1</u>. P0 is solvable iff all given points lie inside or on Γ. (For the proof see Maffioli (1971)).

P'0 is obviously a particular case of P0, so that theorem 1 implies lemma 1.

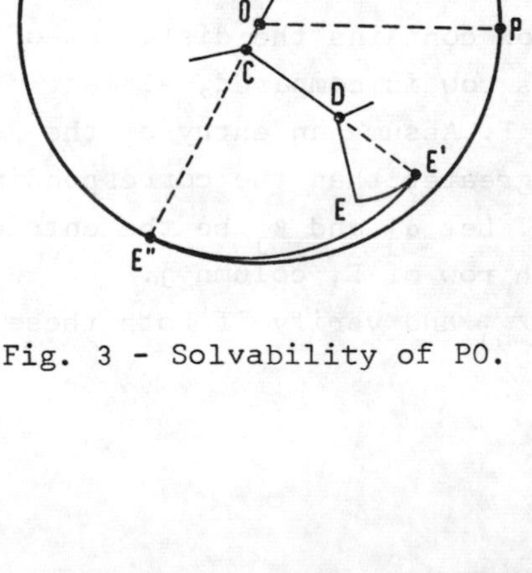

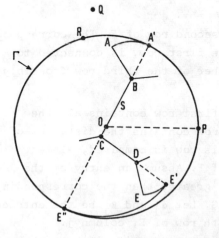

Fig. 3 - Solvability of P0.

4. AN ALGORITHM

To obtain efficient methods for finding diameter constrained shortest spanning trees we have now to restrict eventual auxiliary nodes to lie only on S (otherwise we would need to solve the Steiner problem efficiently). The following is a modification of the algorithm of Prim (1957), which solves P0 allowing auxiliary nodes only on S.

<u>Step 1</u>: check if theorem 1 is obeyed constructing Γ as indicated in fig. 3.

<u>Step 2</u>: for eachone of the given points find where on S would it have to be connected as if no other points had been given. Let ℓ_{Si} be the length of the segment connecting the i-th node to S in this way, and $\ell_{ij} = \ell_{ji}$ the length of the segment of straight line connecting the i-th node to the j-th node. Assume there are n given nodes outside S.

<u>Step 3</u>: form a matrix L, (n+1) x (n+1), of entries ℓ_{ij}, i,j=1,...,n and $\ell_{i,n+1}=\ell_{n+1,i}=\ell_{Si}$. The main diagonal is void.

<u>Step 4</u>: select the n+1-st row of L and form a table F having five rows and n columns. The first row from top contains the number of the nodes we are considering, the second row coincides with the n+1-st row of L, the third row contains all entries equal n+1, the integer assigned to identify S, the fourth row contains the distances of the node numbered as in the corresponding entry of the first row from one extreme of the longest path of S, the fifth row from the other extreme, assuming the node connected directly to S.

<u>Step 5</u>: select the smallest entry of the second row of F. The corresponding node, say j, as found from the first row, is connected to the node corresponding to the number of the third row (possibly S itself).

<u>Step 6</u>: another table, G, is formed, whose first row contains all the remaining nodes and whose second row contains the distances of the j-th node from the others. This row is compared, element by element, with the second row of F. Assume an entry of the second row of F is found which is greater than the corresponding element ℓ_{hj} of the second row of G. Let α_j and β_j be the entries respectively of the fourth and fifth row of F, column j. Adjoint to α_j and β_j the value of ℓ_{hj} and verify if both these sums are less than d.

If not examine the next entry of the second row of F; if yes put ℓ_{hj} in the second row of F in the h-th position and put j in the h-th position of the third row of F, then examine the next entry of the second row of G. The fourth and fifth row of F are also modified, since in the h-th position α_h and β_h must be substituted respectively with $\alpha_j+\ell_{hj}$ and $\beta_j+\ell_{hj}$. When all unconnected nodes have been analyzed the updating of F is terminated.

Step 7: cancel column j from F. If this was the only remaining column the algorithm terminates and all the nodes have been connected by the shortest possible tree T respecting the given constraint. If F is not empty, go to step 5.

Since the proofs of Prim (1957) and the fact that no auxiliary nodes are allowed outside S ensure that T will be the shortest tree, theorem 2 is sufficient to ensure that the above algorithm is optimal.

Theorem 2 [Maffioli (1971)] for any pair of given nodes the path connecting them on T respects the constraint of being shorter than or equal to d.

5. MINIMUM DIAMETER SHORTEST SPANNING TREE

Consider now the following problem.

P1) given a set of nodes on the plane find under which conditions a tree connecting these nodes using any number of auxiliary nodes will have minimum diameter.

A well known property of a set of points on the plane is the following: among n given nodes on the plane there exists always at least one triplet defining a circle which contains all the other nodes.

Usually there are many such triplets eachone defining a corresponding circles: let D be the smallest of these circles (fig. 4). The solution of PL is then obtained by the following theorem.

Theorem 3 for a tree connecting n given nodes on the plane to be of minimum diameter it is necessary (fig. 4)

a) to have $\overline{AB}$ as one of its branches, if all nodes are contained in C, the circle having the segment $\overline{AB}$ connecting the most distant pair of nodes as diameter;

b) if not all nodes are contained in C, to have the star ABPX as a subtree.

Proof: let A and B the pair of most distant given nodes. Let d be the length of the diameter of the tree we are searching. Then

$$d \geq \ell_{AB} \qquad (1)$$

If all given points lie inside or on C, by lemma 1 it is possible to construct a tree having AB as diameter. For such a tree inequality (1) will hold with an equal sign attaining the minimum possible diameter. If not all given points lie inside or on C, let A B and P be the triplet of points defining the circle D embedding all given nodes with minimum diameter. From theorem 1 it is possible to construct a tree connecting all given nodes to the star ABPX and diameter not greater than the diameter of the star. On the other side the star ABPX is the tree of minimum diameter connecting the three given nodes A, B and P. In this case therefore the diameter of the tree shall be equal to the diameter of D.

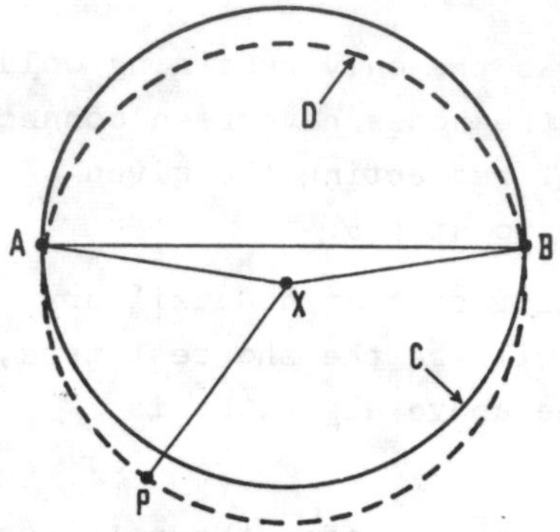

Fig. 4 - Shortest diameter subtrees.

Again in order to find an efficient algorithm let us now constrain the auxiliary nodes of our tree to be introduced only on AB or ABPX, and find the shortest among these trees obeying theorem 3. The algorithm follows.

<u>Step 1</u>: draw circle C. If all nodes are inside or on C go to step 3 considering $\overline{AB}$ as S. If not go to step 2.

<u>Step 2</u>: find D, the smallest circle defined by three of the given nodes and containing all the others. Let S be equal to the star ABPX.

<u>Step 3</u>: solve PO for S and the given nodes.

REFERENCES

-L. Fratta, F. Maffioli and G. Tartara, (1971) "Synthesis of communication networks with reliability of links exponentially decreasing with their length" Int. Conf. on Comm. 1971, Montreal (Canada).

-E.N. Gilbert and H.O. Pollack (1968) "Steiner minimal trees" SIAM J. Appl.Math. 16 No. 1 pp. 1-29.

-A. Kerschenbaum and R. Van Slike (1972)''Computing Minimum Spanning Trees Efficiently" Proc. of 1972 ACM Conference, Boston (Mass).

-E.L. Lawler (1972) "Optimal Cycles in Graphs and the Minimal Cost to Tima Ratio Problem", Memo No. ERL-M343, ERL, College of Engineering, UC Berkeley, (California).

-F. Maffioli (1971) "On diameter constrained shortest spanning tree", Int. Rep. 71-10, LCE-IEE, Politecnico di Milano (Italy).

-R.C. Prim (1957) "Shortest Connection Networks and some Generalizations" B.S.T.J. 26 pp. 1389-1401.

-Shi-Kuo Chang (1972) "The generation of minimal trees with a Steiner Topology" J. of ACM 19, No. 4, pp. 699-711.

SIMULATION TECHNIQUES FOR THE STUDY OF MODULATED COMMUNICATION CHANNELS

by

J.K. Skwirzynski
Research Laboratories
GEC-Marconi Co. Ltd.,
Great Baddow, Essex, UK

Description of some techniques for simulation of signal processing in communication channels by the use of a general programme for a digital computer. The method of simulation is explained and special techniques are developed for calculation of coherent and incoherent NPR in FDM/FM channels, for design of spectrum-shaping circuits which minimise the intersymbol interference in digital transmission and for calculation of transient responses of nonlinear phase-lock loops used for acquisition of carrier phase in PSK detection.

1. SIMULATION OF SIGNAL PROCESSING IN COMMUNICATION CHANNELS.

Direct simulation of signal processing by digital computer has become an indispensable tool in design of communication transponders, in the investigation of new or untried methods of modulation, in detailed study of noise and interference effects and in optimisation of signal or channel parameters.

In what follows, we shall describe some of our experience with a program for direct and complete simulation of communication channels. The simulation is direct in that we deal with typically generated sample signals and are able to observe these signals during their passage through stages of a transponder; starting with an input baseband signal, through a modulator and a transmission channel, to a demodulator and the output baseband signal. It is complete, in that the programme is suitable for study of any type of modulation (analogue, digital or hybrid) and that special algorithms are available for calculation of interference effects and of noise at each of the above stages.

An essential part of such simulation is the complex Fast Fourier Transform (FFT) algorithm [1] which serves for obtaining frequency spectra (i.e. discrete frequency samples of a spectrum) from time signals (i.e. discrete time samples of a signal) and vice versa.

The discrete nature of simulated signals and spectra points to basic limitations of this method. Thus, it deals only with repetitive waveforms, and therefore a sufficiently long period must be chosen in analysis so that any transients die away within it; otherwise the transient tails will distort the output waveform after the passage (for instance) through a filter. Let τ be the period of a waveform, f_o be a reference frequency and N the number of equally spaced samples in time or frequency. Then the spectrum generated by the FFT is also periodic and its frequency extent may be mapped on a circle to range from $f_o -(N-2)/2T$ to $f_o +N/2T$. If any component of the spectrum occurs outside this range then the aliasing effect takes place whereby the spectrum folds over itself on the circular domain and distorts its components in the range from $f_o -(N/2)/2T$ to $f_o +N/2T$ (for instance, the component at $f_o +3N/4T$ would coincide with the component at $f_o -N/4T$). To counter the aliasing, the number of samples of an input waveform has to be increased,while keeping the period τ fixed. It is therefore necessary to choose sufficient number of samples (N) and the period of the waveform (τ), so that a balance is maintained between the transient foldover in the time domain and the aliasing effect in the frequency domain.

The flexibility in choosing appropriate values of N has effectively increased with the development of an efficient FFT algorithm by Sigleton [2]. This no longer requires that N be a power of 2, but extends array sizes to such which have no prime factors greater than 23, and for which, if $N=2^rM$ (by prime decomposition), then $M < 210$.

We shall now indicate some of the facilities which are implemented in such a simulation programme *.

Signal Generation. Signals can be generated directly or by their spectra. Thus one would generate directly a digital sequence or a required waveform, such as the 'pulse-and-bar' signal used in television testing. A random number generator should have options for continuous and integer sequences, with uniform, normal and other distributions. In particular, uniformly distributed random integers would be used for setting up random bit sequences. Arithmetic subroutines are then needed for manipulating a generated random sequence, for instance for setting up several samples per bit, or for constructing strings of required pulses. Statistical analogue signals (e.g. when simulating a typical speech channel) are best generated by equivalent spectra in the baseband, typically by setting uniform amplitudes and random phases of particular tones (Rice model). Then the FFT will produce the corresponding random signal.

Modulation. Each type of modulation requires a particular treatment of a baseband signal, which can be arranged by using suitable programme routines. Here few examples should suffice.

Amplitude Modulation (AM) is linear and is carried out by forming the ratio-frequency (RF) spectrum by a special option of FFT, whereby the conjugate of the baseband spectrum is placed in the top half of the RF spectrum array. Then a

* A programme of this kind has been developed at the Research Laboratories of the GEC-Marconi Co. It is called MODSIM and uses Fortran IV language.

shift in frequency by N/2T gives the RF spectrum of lower sideband-carrier-upper sideband in the natural sequence.

Phase Modulation (PM) is simulated by mapping each baseband signal sample $V_r \rightarrow \exp(jV_r)$, where $j = \sqrt{-1}$, r = number of sample.

Frequency Modulation (FM) is carried out via PM, using spectrum integration and FFT. Thus instantaneous phase samples of the signal waveform are formed by integrating in the frequency domain (mapping $V_r \rightarrow V_r/j(r-1)$) * and then transforming to the time domain. When forming the RF spectrum of an FM signal it is essential to make provision for at least the Carson bandwidth.

Filters. Measured filter responses (i.e. amplitude in dB and phase) are entered as data and interpolation (cubic) is used to generate responses at required frequency samples of the filtered signal. Ideal spectrum truncation is considered as filtering.

Non-linear Devices. The action of nonlinear devices (such as a Travelling Wave Tube Amplifiers, TWTA's) must be simulated in the time domain, having specified in data the instantaneous transfer characteristic (complex). Again interpolation is used for mapping at required signal values. When transferring back to the RF spectrum it is essential to bear in mind the intermodulation products and resulting widening of this spectrum. Special routine is used for limiters; this is also uses for multi-level Phase Shift Keying (PSK) modulation maps.

Demodulation. Here again each type of modulation requires a particular treatment. One example is given; others are discussed in Section 2 below.

FM Detection. Here we require first the phase of the complex RF signal. Let $\phi(V_r)$ be this phase. Then we map:

$$V_r \rightarrow (-1)^r \left[\phi(V_r) - \frac{2\pi(r-1)NO}{N}\right] \qquad \text{(real sequence)}$$

where alternative sign reversals are incorporated to extract the N/2T frequency shift (see above) and NO = carrier phase shift over T (in cycles). The corresponding spectrum is then obtained and differentiated ($V_r \rightarrow j(r-1)V_r$). Special techniques are used to simulate performance of non-ideal detectors (particularly in the usual presence of AM at the output of the RF stage) and characteristics of noise threshold in FM discriminators. In the latter case noise (with a given statistics) is added at the input to the discriminator and it becomes necessary to establish the 'click rate' as a function of C/N (carrier to noise ratio). Now 'clicks' in FM are results of steps in phase, when the output signal vector sweeps an angle less than π, as the phase of the input signal sweeps a cycle. Such a step can be discovered by a suitably dense sampling of the RF

* Note that V_r may denote either a signal or a spectrum sample.

signal. We have established experimentally that to ensure a reliable count of 'clicks', for $C/N > 6dB$, it is necessary to sample at a frequency at least equal to 5 times the Carson band.

2. CALCULATION OF NOISE POWER RATIO (NPR) IN FDM/FM SYSTEMS.

An important question in the design Frequency Division Multiplex (FDM) systems, particularly when FM is used, is the estimation of NPR in individual channels. This quantity determines the necessary band spacing in a transponder and is influenced profoundly by both linear (e.g. IF filter responses) and non-linear (e.g. TWTA back-off) parameters of the transmission channel. The usual method of NPR measurement is based on the use of White Noise Test Meters. One could directly simulate such a process by introducing appropriate band-stop (BS) filters at the pre-modulation stage in a measured band. Yet this would have two draw-backs:

1) One generally requires an estimate of NPR over the whole base-band (and even beyond it for estimation of out-of-band NPR), thus necessitating repeated calculations for several positions of the BS filter. In addition, each value of estimated NPR would be affected by the assumed BS filter response.

2) This method does not directly distinguish between incoherent and coherent (intelligible) components of NPR, while the latter is an important paramotor used for optimisation of TWTA characteristics.

The proposed method is based upon repeating a simulation for several random input signals and averaging to give an NPR estimate. It is sufficient to consider the case of two FDM/FM bands modulated on separate and neighbouring carriers in the transponder (e.g. satellite) band. The simulation process is sketched in Fig.1. The carriers are labelled a and b. Let V_a and V_b be corresponding baseband tones of a Rice model. Then the resulting tone at the output (demodulated) in channel a can be written: $V_a' = C_a V_a + C_b V_b + V_n$ (1)

C_a and C_b are system gain functions and $(C_b V_b + V_n)$ is the total noise power on the V_a tone, with an intelligible component $C_b V_b$ and the incoherent component V_n.

In a well designed system $|C_a| \sim 1$. Our aim is to estimate the total NPR, i.e. $(C_b V_b + V_n)/V_a$ and the intelligible component C_b, as functions of the baseband frequency.

Two simulation runs are carried out. They are identical, except that the input tones in either of them are uncorrelated, being initialised by separate sequences from the random number generator. We label these runs by 1 and 2 respectively and can write for each tone in the baseband of channel a:

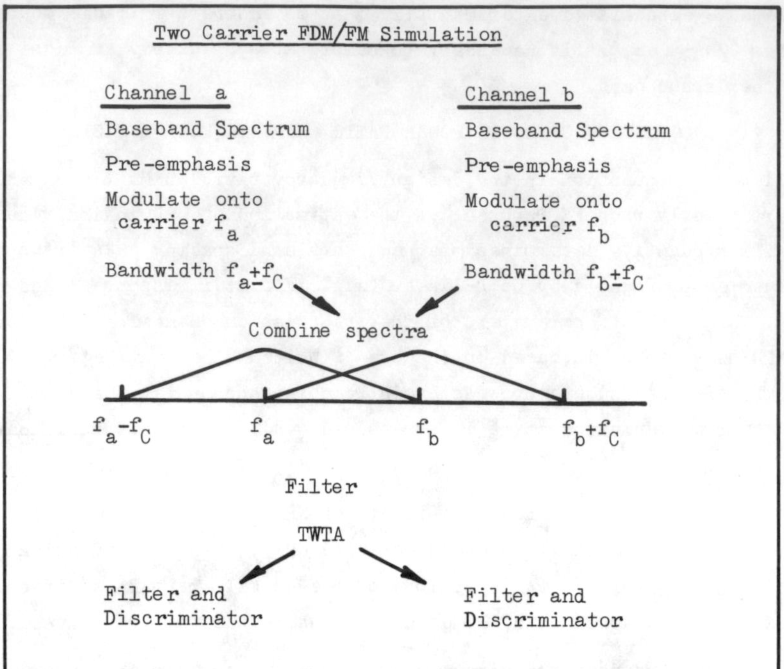

Fig. 1. Simulation Process for 2 Carrier FDM/FM.

$$V'_{a1} = C_a V_{a1} + C_b V_{b1} + V_{n1} \tag{2}$$

$$V'_{a2} = C_a V_{a2} + C_b V_{b2} + V_{n2}$$

Hence:

$$\frac{V'_{a1}}{V_{a1}} - \frac{V'_{a2}}{V_{a2}} = \frac{C_b V_{b1} + V_{n1}}{V_{a1}} - \frac{C_b V_{b2} + V_{n2}}{V_{a2}} \tag{3}$$

We write this more concisely: $Y_{12} = X_1 - X_2$ (3a)

The quantity Y_{12} is then calculated for several independently generated baseband signals (each of these is obtained from 4 random sequences) and averaged at each frequency of the baseband in channel a. Then:

$$E(|Y_{12}|^2) = E(|X_1|^2) + E(|X_2|^2) = 2\,E(|X|^2) \tag{4}$$

The quantity $E(|X|^2)$ is the required total NPR.

An extension of the above method enables us to identify the intelligible component. Again we consider two runs but this time with identical signals in the band a: $(V_{a1} = V_{a2})$ and with phase reversed signals in band b: $(V_{b2} = -V_{b1})$. Then:

$$V'_{a1} - V'_{a2} = 2\,C_b V_{b1} + (V_{n1} - V_{n2}) \tag{5}$$

Now when we average, the modulus of individual tones is the same for each run and therefore the incoherent noise component ($V_{n1} - V_{n2}$) will only arise from intermodulation products involving odd powers of V_b. Thus $(V'_{a1} - V'_{a2})/2V_{b1}$, when averaged over several runs, will give an estimate of the intelligible cross-talk C_b from carrier b to carrier a, for every tone V_a.

This method may be easily coded in a programme which has looping facilities for repeated runs.

3. SIGNAL SHAPING FOR MINIMUM INTERSYMBOL INTERFERENCE (ISI) IN DIGITAL CHANNELS.

In any type of digital transmission, whatever the type of modulation (e.g. PSK, FSK, PAPM, etc.), the main distortion effect, is the ISI and this can be minimised, for a given necessary bandwidth, by choosing suitable signal shapes. Let T be the symbol rate. Then a sequence of pulses can be written:

$$s(t) = \sum_i a_i h(t - iT) \tag{6}$$

where h(t) is the signal waveshape, that is the impulse response of the combined pre- and post-modulation filters, and a_i is the data sequence (thus for a 2-phase PSK, $a_i = 0, 1$). The requirement for the waveshape h(t) is that

$$h(rT) = 0 \quad , \quad r = 1, 2, \ldots \tag{7}$$

A well-known class of spectrum shapes, which lead to realisable filters, was introduced by Nyquist [3] for the minimum necessary bandwidth (1/T) of the ideal low-pass filter. A popular representative of this class is the 'raised-cosine' response given by:

$$\begin{aligned} H(f) &= 1 && , \quad 0 < f < (1-\alpha)/2T \\ &= \tfrac{1}{2}\,[1 + \cos \tfrac{\Pi}{\alpha}\,(Tf - \tfrac{1}{2}(1-\alpha))] && , \quad (1-\alpha)/2T < f < (1+\alpha)/2T \\ &= 0 && , \quad (1+\alpha)/2T < f \end{aligned} \tag{8}$$

where α is the 'roll-off' factor, a parameter defining the extension of the bandwidth beyond $f = 1/2T$. The ISI resulting from H(f) is zero, yet in practice this system assumes precise synchronisation of the timing at the receiver and zero time jitter. We measure the intersymbol ratio by

$$ISI = 10 \log \left[\sum_i |h(iT)|^2 / h(0) \right] \tag{9}$$

Let β be the timing error, measured as percentage of T. Then for $\alpha = 1/4$ in (8), ISI = 38 dB for $\beta = 1$ and ISI = 18 dB for $\beta = 10$. An ideal spectrum shape which minimises the effect of time jitter was proposed by Franks [4]. For this shape the ISI is better by 2.5 dB than for (8), when $\alpha = 1/4$.

Filter responses discussed above are assumed to have linear phase characteristics and such cannot be achieved in practice. That phase need not

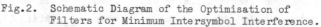

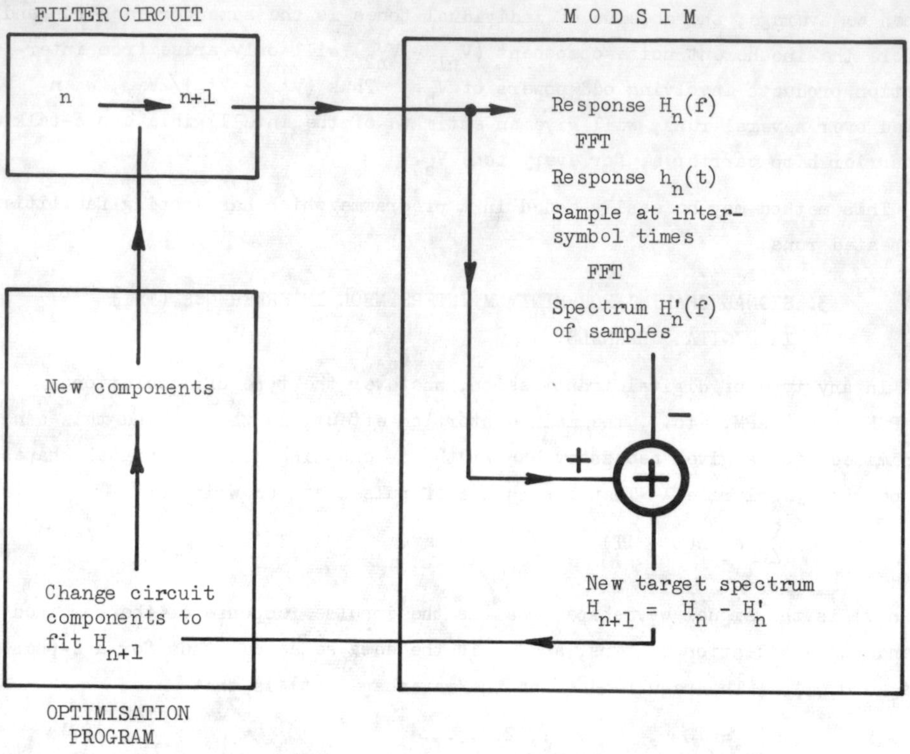

Fig.2. Schematic Diagram of the Optimisation of Filters for Minimum Intersymbol Interference.

be linear for zero ISI was already established by Nyquist [3], who showed that in the 'transition' band, $(1-\alpha)/2T < f < (1+\alpha)/2T$, the 'in-phase' response of an optimal spectrum shape needs to be 'skew-symmetric' with respect to the 6 dB value (at $f = 1/2T$), whereas the 'quadrature' response needs to be symmetric on both sides of this frequency. Such conclusion points to a possibility of deriving a practical filter response for minimal ISI.

Our method of design is based on iterative use of two coupled programmes;

1) The simulation programme (MODSIM) for obtaining impulse responses from actual filter characteristics;

2) The optimisation programme (OPTIMA) for adjusting values of circuit components to minimise a suitably defined cost function.

An optimisation cycle is sketched in Fig.2. We start with a filter circuit of suitable bandwidth, with an associated all-pass group delay equaliser. Let its response be $H_n(f)$ where n is the number of iteration. The impulse $h_n(t)$ is calculated and sampled at $t = rT$ (intersymbol times). The resultant sequence of samples produces a spectrum (complex) in the range $0 < f < 1/2T$. Let this spectrum be $H'_n(f)$. The new spectrum shape, $H_{n+1}(f) = H_n(f) - H'_n(f)$, is then used as a new target for the optimisation programme.

The optimisation method used is based on the so-called Levenberg algorithm [5 - 9]. A similar method of optimisation, but based on adjusting network response, rather than components, and on the Davidon-Fletcher-Powell algorithm, has recently been announced by Allemandou [10].

After four iterations, the ISI ratio (9) for a filter with $\alpha = 1/4$, with suitable optimal matching for a full duty cycle pulse, has given essentially the same ISI as the theoretical 'raised cosine' response for $\beta > 2$, whereas for $\beta = 1$, ISI = 33.5 dB. The asymptotic value of ISI, as $\beta \to 0$, is 38.5 dB.

4. DYNAMIC RESPONSES OF NONLINEAR PHASE-LOCK LOOPS USED FOR ACQUISITION OF SIGNAL PARAMETERS IN DIGITAL COMMUNICATION.

In digital communication optimal conditions for detection cannot be achieved without acquiring basic signal parameters which determine threshold boundaries in the complex carrier plane. These are: sampling times, carrier phase and carrier gain. The latter is required only for multi-level PSK modulation systems and for hybrid modulation methods (e.g. PAPM, see [11]).

Here we shall deal merely with an 'active' method of phase acquisition with the help of a phase-lock loop. This is not necessarily an optimal technique. When the carrier phase is shifted by a value near to the 'threshold angle' (i.e.near 45° for a 4-phase PSK modulation), a 'hang-over' effect may occur with a definite probability. To avoid such an effect, an 'active' method of phase acquisition is often associated with a 'passive' acquisition, based on frequency multiplication and subsequent averaging of the phase of a filtered signal [12].

In particular we propose to describe a new method for calculation of the dynamic (i.e. transient) response of a nonlinear phase-lock loop. The method will be applied to the 'Costas' loop [13], with a saw-tooth nonlinear characteristic. However, it is equally suitable for other loop circuits and in general, for solution of nonlinear integral equations of Fredholm type, with highly oscillatory driving terms.

The main feature of this algorithm is that the iterations (of Picard type) are conducted in the frequency domain, rather than directly in the time domain. Such a technique was proposed already by Neill [14], for analysis of intermodulation terms in a nonlinear transistor amplifier.

A schematic diagram of a Costas loop is shown in Fig. 3; it is used for phase acquisition of a 4-phase PSK signal.

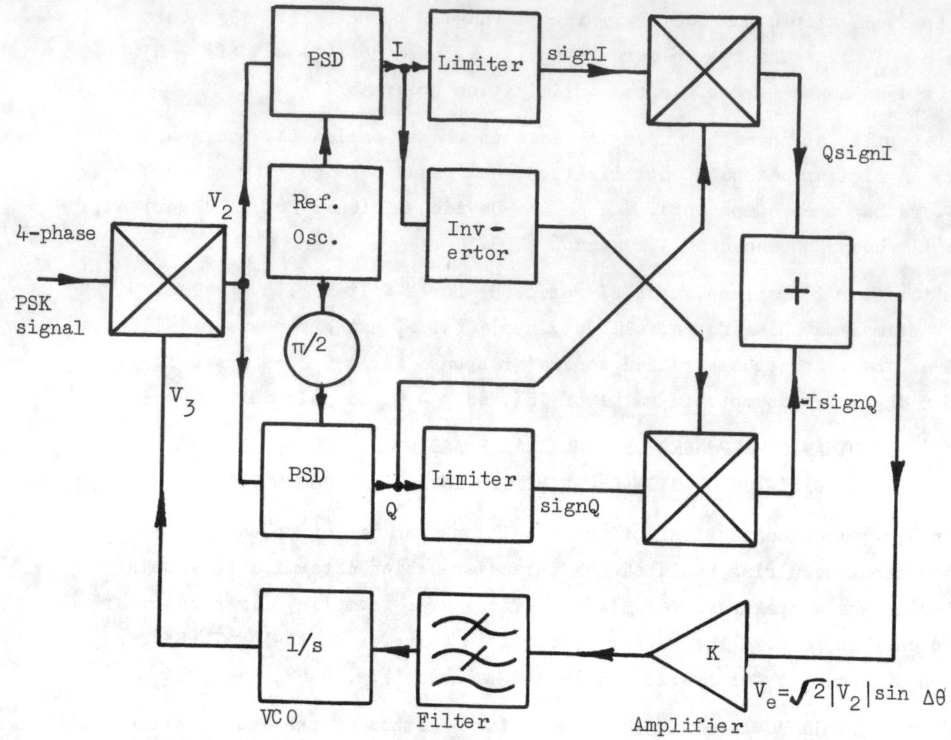

Fig. 3. Schematic Diagram of a Costas Phase-Lock Loop.

Since we are using the FFT algorithm for passing from the frequency to the time domain and vice versa on both sides of the nonlinear device, we initially generate a random PSK sequence, filter it, possibly degrade from minimum ISI conditions, possibly add a noise signal and then shift the phase of the first half of this sequence by $\frac{1}{2}\,\Delta\,\theta$ and the second half by $-\frac{1}{2}\,\Delta\,\theta$. Hence the signal is periodic with phase shift of $\Delta\theta$ at the centre of our observation 'window'. The two signals emerging from the Phase Sensitive Detectors (PSD) are in quadrature; we denote them by I and Q respectively. Then at the output from the adder we obtain: Q signI - I signQ = $2|V_2|\sin\,\Delta\theta$ (for hard limiting).

The simplified diagram of Coatas loop is shown in Fig. 4, with the nonlinearity represented by the box 'Costas'. Suppose we first open the loop as shown and apply the input signal θ_{os} to the nonlinearity, transform it to spectrum, modify by the open loop gain $G = KF(s)/s$ and thus obtain the spectrum B_o. If we identified directly $A_o = B_o$ we would close the loop and use essentially the direct

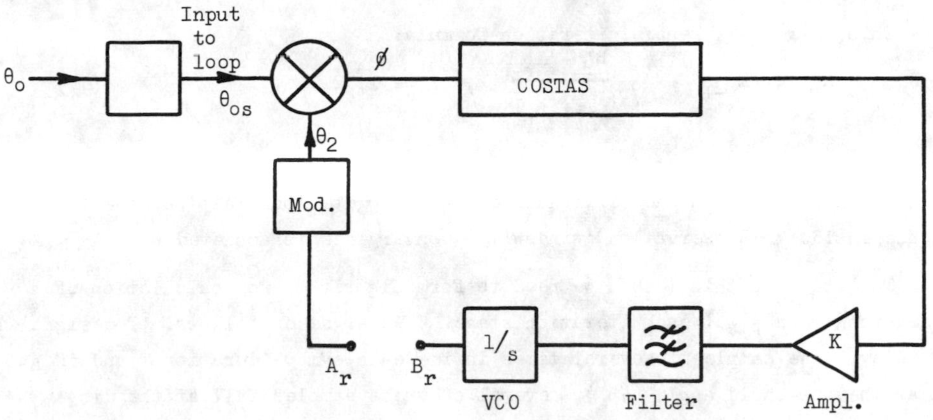

(a)

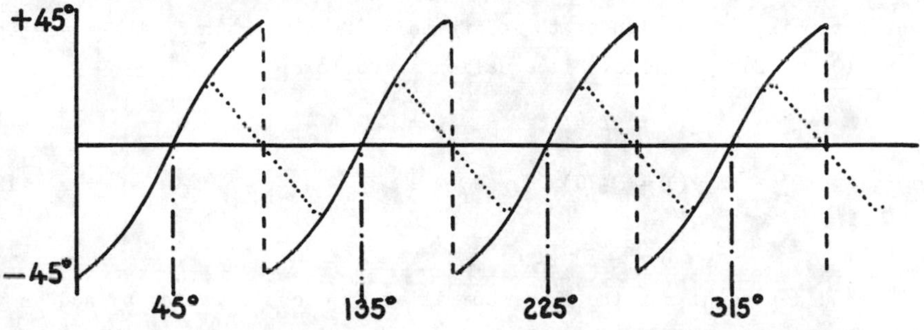

(b)

Fig. 4 Model for Solution of Costas loop.
(a) Block Diagram.
(b) Nonlinear characteristic 'COSTAS' (dotted curves give a 'soft limiter' characteristic).

contraction mapping algorithm. This cannot be done for we are not certain that the usual conditions for convergence hold [15]. However, $B_o \sim \theta_{os} G \sim \phi G$. Thus, to close the loop we propose $A_o = B_o/(1 + G)$. Then A_o is transformed to time, modulated and subtracted from θ_{os} to obtain new value for ϕ.

This suggests a general iteration formula:

$$A_r = A_{r-1} + \alpha_r \frac{B_r - A_{r-1}}{1 + \beta_r G}, \quad r = 1, 2, \ldots \tag{10}$$

We note that in this iteration A_r and B_r are spectra. Thus, on transforming to time we obtain at each stage (if required) the complete 'portrait' of the loop response inside the observation 'window'. Convergence is achieved when $A_r = A_{r-1}$.

To establish this method we need to form algorithms for calculation of parameters α_r and β_r at each iteration stage. We argue as follows: The signal is very noisy. We sample it several times inside each symbol duration T and if $\Delta\theta$ is near the threshold angle (45°), several of these samples fall at the discontinuities of the Costas characteristic, particularly if hard limiting is used. Thus the nonlinearity produces a random set of impulses, each of which modifies the effective loop response by a white noise spectrum. The parameter β_r is used to account for this, in particular to modify the effective gain in the iteration formula (10). Let us write this equation as follows:

$$A_r = A_{r-1} + \Delta A_r = A_{r-1} + \alpha_r \Delta A'_r \tag{11}$$

where $$\Delta A'_r = E_r/(1 + \beta_r G), \quad E_r = B_r - A_{r-1} \tag{12}$$

Let also $$\Delta B_r = B_r - B_{r-1} \tag{13}$$

Then β_r should account for the error committed at each iteration by suddenly arising impulses. Thus we assume: $\beta_r \sim \Delta B_r/G.\Delta A_{r-1}$ and hence the real parameter β_r is obtained by minimising

$$|| \beta_r G \Delta A_{r-1} - \Delta B_r ||^2 = \min \tag{14}$$

where $||\cdot||$ is the Euclidean norm in C^N.

Now since β_r is constant and does not vary over the spectrum we consider, equation (14) defines a 'noise' spectrum which we have generated by the above assumption. Let this noise be:

$$N_r = \beta_r G \Delta A_{r-1} - \Delta B_r \tag{15}$$

where β_r is obtained from (14). The corresponding noise gain is given by:

$$Z_r = || N_r ||^2 / || \Delta A_{r-1} ||^2 \tag{16}$$

Now α_r is found by combining two errors (the iteration error and the 'noise' error) and minimising:

$$P_r = Z_r \,||\, \Delta A'_r \,||\, \alpha_r^2 + (1 - \alpha_r)^2 \,||\, E_r ||^2 = \min \tag{17}$$

Fig. 5 shows the convergence of the error E_r for two values of $\Delta\theta$ (32° and 45°) and hard limiting. We see that in the latter case we need 12 iterations (10) before the error diminishes significantly. A typical portrait of

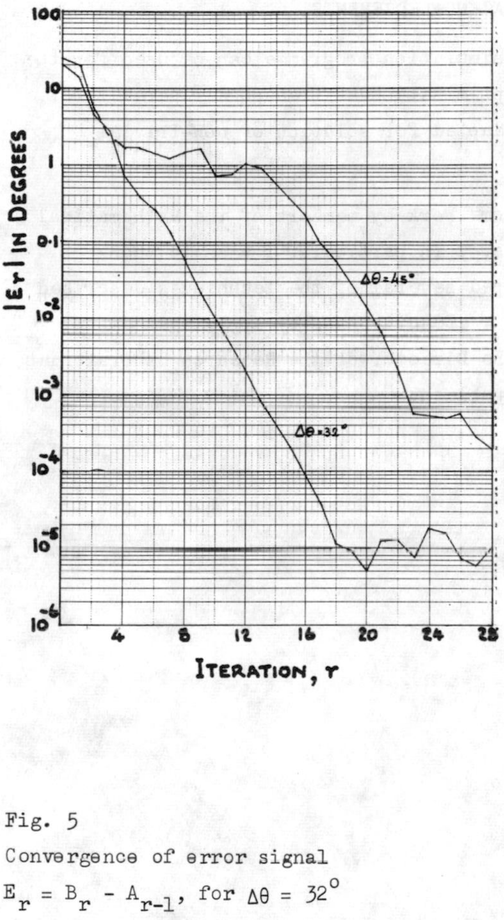

Fig. 5
Convergence of error signal
$E_r = B_r - A_{r-1}$, for $\Delta\theta = 32^\circ$
and 45°.

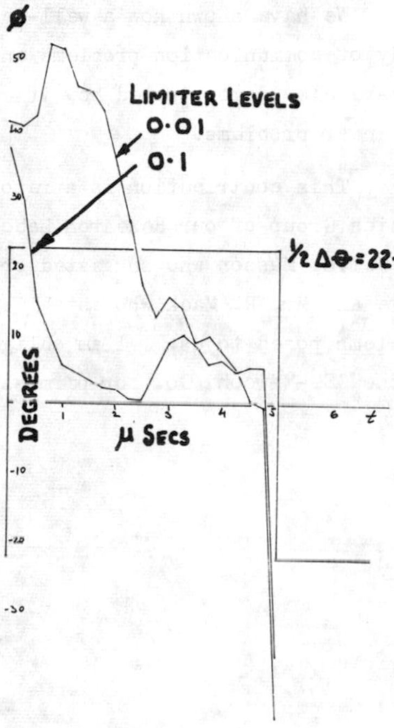

Fig. 6
Signal Ø (see Fig. 4a) for $\Delta\theta = 45^\circ$
and two values of limiter levels.
The phase shift $\pm \frac{1}{2}\, \Delta\theta = \pm\, 22.5^\circ$ is
also shown.

locking transient is shown in Fig. 6 for $\Delta\theta = 45^{\circ}$ and for two types of limiter in the Costas loop. Only the first half of the signal is shown, the other is very similar. The value of symbol duration was T = 0.25 μsec, sampled 5 times in each pulse. The case of hard limiter shows a tendency to 'hang-over'.

The algorithm described here is heuristic and for many cases we have tried it, it does break down occasionally, for a particular choice of the initial random sequence. We believe that these instances exhibit complete 'hang-overs' since for some of these it was possible to achieve convergence by increasing the number of pulses considered, yet maintaining initially the same sequence.

5. CONCLUSIONS AND ACKNOWLEDGEMENTS.

We have shown how a well-designed simulation programme can be used for the study of communication problems and for optimisation of transponder parameters. We have also demonstrated how it can be adapted for solution of non-trivial nonlinear problems.

This contribution is a report of some work by members of the Mathematical Physics Group of our Research Laboratories. In particular I wish to thank Mr. J.T.B. Musson who suggested or developed several of the techniques described above and Mr. R. Mack who established their effectiveness by applying them to problems posed to us. I am obliged to the Director of the Research Laboratories of the GEC-Marconi Co. for permission to publish.

REFERENCES

1. Special Issue on FFT. IEEE Trans. on Audio and Electroacoustics, Vol. AU-15(2), June 1967.

2. R.C. Singleton, An Algorithm for Computing the Mixed Radix Fast Fourier Transform, IEEE Trans. on Audio and Electroacoustics, Vol. AU-17(2),June 1969.

3. H. Nyquist, Certain Topics in Telegraph Transmission Theory, Trans. AIEE, Vol. 47, April 1928.

4. L.E. Franks, Signal Spaces: Applications to Signal Representation and Design Problems. In Network and Signal Theory, edited by J.K.Skwirzynski and J.O. Scanlan, Peter Peregrinus Ltd., London 1973.

5. K. Levenberg, A Method for the Solution of Certain Nonlinear Problems in Least Squares, Quart. Appl. Math., Vol. 2, 1944.

6. J.K. Skwirzynski, Optimisation of Electrical Network Responses. In Computing Methods in Optimisation Problems, Lecture Notes in Operations Research and Mathematical Economics, Springer Verlag, Berlin 1969.

7. J.K. Skwirzynski, Optimisation Techniques in Circuit Theory, In Progress in Radio Science 1966-1969, Vol. 3, Edited by W.V. Tilston and M. Souzade, International Union of Radio Science, Brussels 1971.

8. J. Kowalik and M.R. Osborne, Methods for Unconstrained Optimisation Problems, American Elsevier, New York 1968.

9. J.W. Daniels, The Approximate Minimisation of Functionals, Prentice-Hall, Inc., Englewood Cliffs 1971.

10. P. Allemandou, Quadripôles passe-bas de mise en forme d'impulsicns, Câbles et Transm., No. 4, Oct. 1972.

11. G.W. Welti, Pulse Amplitude-and-Phase Modulation. In Coll.Intern.sur les Télécommnications Numeriques par Satellite, Éditions Chiron, Paris 1972,

12. A. Ogawa and M. Ohkawa, A New Eight - Phase PSK Modom System for TDMA, ibid.

13. A.J. Viterbi, Principles of Coherent Communication, Mc Graw-Hill Book Co., New York 1966.

14. T.B.M. Neill, Spectral Analysis of Nonlinear Circuits, In Network and Signal Theory, see 4 above.

15. L.B. Rall, Computational Solutions of Nonlinear Operator Equations, John Wiley & Sons, Inc., New York 1969.

GESTION OPTIMALE D'UN ORDINATEUR MULTIPROGRAMME A MEMOIRE VIRTUELLE

E. GELENBE, D. POTIER, A. BRANDWAJN

(IRIA 78 - ROCQUENCOURT - FRANCE)

J. LENFANT

(Université de Rennes 35031 - RENNES-CEDEX)

I - INTRODUCTION

Les mesures réalisées sur le comportement des systèmes d'ordinateurs multiprogrammés à mémoire virtuelle paginée mettent en évidence la sensibilité des performances du système, comme le taux d'utilisation de la ressource unité centrale et le temps de réponse moyen de demandes de service, au nombre d'utilisateurs admis simultanément en mémoire centrale, ou degré de multiprogrammation. Citons par exemple les résultats des mesures de RODRIGUEZ-ROSELL obtenus sur le système cf. 67 de l'Université de Grenoble [7] : la figure 1 montre le pourcentage de temps réel passé par l'unité centrale en exécution des programmes utilisateurs en fonction du degré de multiprogrammation. Le taux d'utilisation de l'unité centrale détermine directement le débit du système, et donc le temps de séjour moyen des programmes dans le système.

La forme de la courbe représentée sur la figure 1 dépend étroitement des principes de fonctionnement des systèmes à mémoire virtuelle paginée et peut être expliquée simplement comme suit. Considérons le système idéalisé représenté sur la figure 2 et comprenant une unité centrale (U.C.) une unité de pagination et une unité d'entrée sortie, avec une file d'attente devant chacun de ces processeurs. On appelle "boucle mémoire" (B.M.) le système ainsi constitué.

Dans un système à mémoire virtuelle paginée, les programmes sont découpés en blocs de tailles égales, les pages sont contenues en totalité sur la mémoire secondaire (ici un disque de pagination). L'exécution d'un programme est initialisée en chargeant en mémoire centrale (M.C.) la page contenant la première instruction du programme et l'exécution se poursuit jusqu'à ce que l'U.C. fasse référence à une instruction en dehors de la page chargée en M.C. On a alors un défaut de page qui bloque l'exécution du programme et déclenche un mécanisme de recherche et de transfert de la page référencée de la M.S. vers la M.C. et le programme est de nouveau prêt à être exécuté sur l'O.C. Lorsque la M.C. est pleine une page est déchargée par l'algorithme de remplacement pour faire place à la nouvelle page. Le taux de défaut de page, qui détermine la fréquence des transferts M.S.-M.C., est fonction à la fois de l'algorithme de remplacement et de l'espace moyen M.C. alloué à un programme, et donc du degré de multiprogrammation. La figure 3 rapporte les résultats de [7] sur le taux de défaut de page.

Une seconde cause d'interruption de l'exécution d'un programme sur l'U.C., se

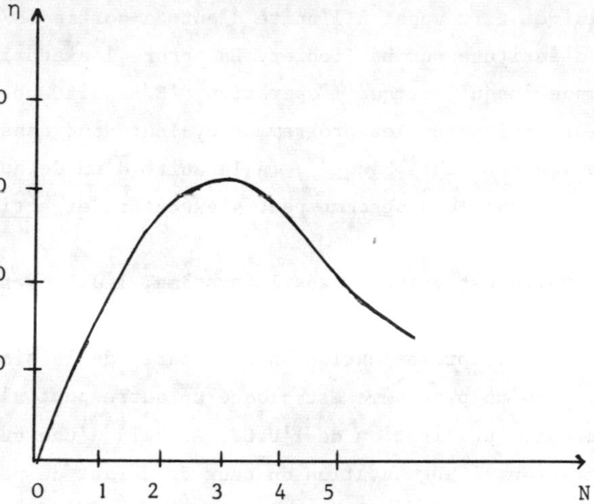

Figure 1
Taux d'utilisation U.C. en fonction du degré de multiprogrammation

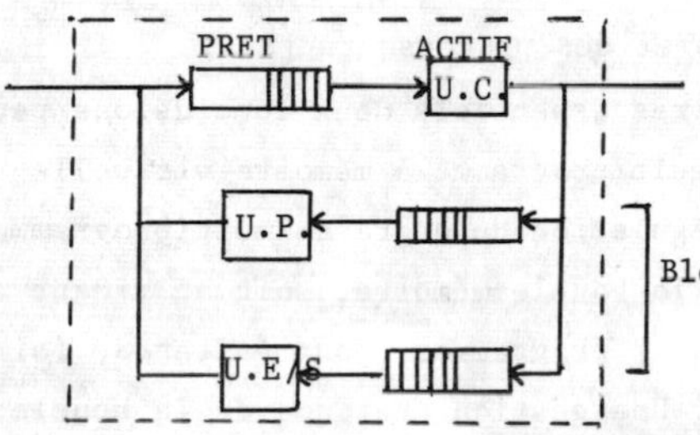

Figure 2
Boucle-mémoire idéalisée

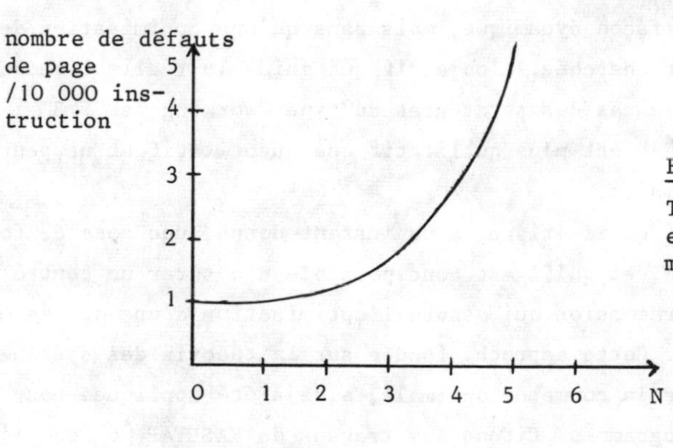

Figure 3
Taux de défauts de page en fonction du degré de multiprogrammation

produit lorsque celui-ci fait appel à l'unité d'entrée-sortie (E/S) pour une opération de lecture ou d'écriture sur un fichier. Là encore l'exécution du programme sur l'U.C. est interrompue jusqu'à ce que l'opération d'E/S soit achevée.

Au cours de leur traîtement les programmes cyclent donc dans la boucle mémoire en passant par la suite des états bloqué , (à la suite d'un défaut de page ou d'une demande d'E/S) prêt lorsque le programme peut s'exécuter, et actif lorsqu'il s'exécute sur l'U.C.

Si un seul programme est présent dans le système, l'U.C. sera inactive chaque fois que le programme sera bloqué par un défaut de page ou une opération d'E/S. En augmentant le degré de multiprogrammation on tire parti de la simultanéité des trois processeurs : pendant qu'un programme est bloqué un autre peut s'exécuter sur l'U.C. et donc accroître ainsi l'utilisation de l'U.C.. Au delà d'une certaine limite du degré de multiprogrammation, l'augmentation du taux de défaut de page et des opérations d'E/S provoque une saturation des unités de pagination et d'E/S si bien qu'avec une probabilité croissante l'ensemble des programmes va se trouver bloqué, tandis que l'U.C. reste inactive. C'est ce qui apparaît sur la figure 1 pour N supérieur à 4 où le système entre dans une zone de comportement catastrophique dite zone de "trashing", qui est caractérisée à la fois par une mauvaise utilisation de la ressource U.C. et par une augmentation considérable des temps de traîtement.

Ces résultats de mesures peuvent être retrouvés à partir de modèles analytiques de comportement des programmes et des processeurs [2] .

De ces rappels préliminaires essentiels deux conclusions peuvent être tirées pour la gestion d'un système multiprogrammé à mémoire virtuelle paginée. La première est qu'il faut opérer d'une régulation du degré de multiprogrammation, soit en limitant l'accès des programmes à la boucle-mémoire, soit en vidant de la B.M. avant la fin de leurs traîtements certains programmes, pour éviter de faire fonctionner la B.M. dans la zone de Trashing. Une gestion statique de la boucle-mémoire serait irréaliste parce que les caractéristiques des programmes évoluent au cours du temps, et provoquent des modifications dans le comportement de la B.M. qui doivent être prises en compte. Dans tous les systèmes existants la régulation du degré de multiprogrammation est réalisée de façon dynamique, mais sans qu'une optimisation des performances soit explicitement recherchée. L'objectif qui guide la réalisation du système de contrôle, comme dans le cas des stratégies du type "working-set" [3] , ou comme dans d'autres systèmes [1] est plus qualitatif que quantitatif et ne peut donc assurer des performances optimales.

La seconde conclusion est qu'il existe, à un instant donné, une zone de fonctionnement optimale du système, et qu'il est donc possible d'assurer un contrôle dynamique du degré de multiprogrammation qui assure l'optimisation d'une ou des mesures de performance du système. Cette approche fondée sur la théorie des systèmes et l'utilisation de la théorie de la commande optimale, a déjà été appliquée pour la régulation de systèmes monoprogrammés. Citons les travaux de KASHYAP [6] sur l'ordonnancement des programmes en U.C. sur le système CTSS, et de FIFE (4) sur le contrôle optimal des files d'attente. Nous le développons ici dans le cas d'un système multiprogrammé à mémoire virtuelle paginée en retenant pour critère de performance le temps de réponse moyen des demandes de service. Le système de contrôle proposé se compose d'un estimateur des paramètres d'un modèle de comportement de la B.M., et d'un algorithme de commande optimale qui, à partir des résultats de l'estimation, et au vu du nombre de programmes, en attente et du degré de multiprogrammation à un instant donné, décide si un nouvel usager doit être introduit.

Le calcul de la loi de commande est conduit à partir du modèle de comportement de la B.M. Ce modèle est volontairement choisi simple afin d'obtenir un système de contrôle qui puisse être facilement implanté sur un système réel ; il ne doit pas être considéré comme constituant un modèle du système, mais plutôt comme une base du calcul de la loi de commande optimale. On trouvera dans les références déjà citées (2) et (9) la description de modèles plus détaillés de B.M.

II - PROBLEME

2.1 - STRUCTURE DU SYSTEME IDEALISE.

Le système et son mécanisme de contrôle son représentés sur la figure 4. Les programmes émis par les utilisateurs à partir des consoles sont placés dans la file d'attente x dans l'ordre de leur arrivée, et leur accès au système de traîtement est commandé par l'interrupteur I. L'estimateur E observe le comportement de la boucle-mémoire B.M. et estime les paramètres du modèle du système utilisé pour le calcul des lois de commande optimale. Périodiquement le contrôleur C commande l'interrupteur suivant le nombre k de programmes en attente dans x et le degré de multiprogrammation i.

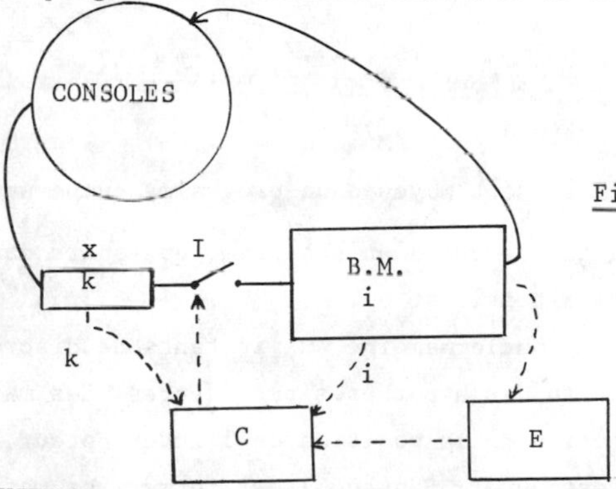

Figure 4

2.2 - MODELE DU SYSTEME.

On pose les hypothèses et les définitions suivantes :

H1 - l'interrupteur I est commandé périodiquement avec une période Δ . L'intervalle de temps Δ est pris pour unité de temps.

H2 - Lorsque I est fermé un programme au plus passe de la file x dans la B.M. pendant une période Δ .

H3 - Pendant une période Δ un programme au plus achève son traîtement et quitte la B.M.

Les hypothèses H1, H2, H3, définissent en partie la durée de la période Δ. Par H1 et H2 Δ doit être au moins égal à la durée de l'opération de commande de I et du passage du programme de la file x dans la B.M. L'hypothèse H3 n'est satisfaite que si la probabilité que plus d'un programme quitte la B.M. pendant une période Δ est négligeable, ce qui définit une borne supérieure de Δ.

H4 - A un instant donné l'état de la boucle-mémoire est entièrement caractérisé par son degré de multiprogrammation n, au plus égal à M.

L'hypothèse H4 réduit la description de la B.M. à un instant donné au nombre de programmes y circulant à cet instant sans préciser davantage l'état des différents processeurs et des files d'attente. Il est donc évident que la valeur de cette simplification ne peut être appréciée qu'en fonction des résultats obtenus. Dans le cas de résultats insuffisants l'état de la B.M. devrait être par exemple élargi en (n_1, n_2, n_3) où n_1 est le nombre de programmes en attente derrière l'U.C., n_2 le nombre derrière le disque de pagination, n_3 derrière l'unité d'E/S.

H5 - Le comportement de la boucle-mémoire est représenté par les quantités

$p_i = p_r$ (un programme donné quitte la boucle-mémoire pendant la période Δ quand $n = i$) $i=1,\dots,M$

$\pi_i = p_r$ (un programme quelconque quitte la boucle-mémoire pendant la période Δ quand $n=i$) $i=0,\dots,M$ (on a évidemment $\pi_o=0$)

π_i/Δ représente donc le débit de la boucle mémoire quand n=i. Les quantités p_i et π_i sont reliées par

(1) $$\pi_i = ip_i \qquad i=1,\dots,M$$

On pose

(2) $$a_i = \frac{\Delta}{p_i} \qquad i=1,\dots,M$$

et $$A = (o, a_1, \dots, a_M)$$

a_i représente le temps de traîtement moyen d'un programme quand n=i.

H6 - Le temps moyen de traîtement d'un programme qui entre dans la B.M. quand l'état de celle-ci est n=i est a_i.

La suite des états de la boucle-mémoire aux instants de discrétisation constitue une chaîne de Markov $\mathcal{C}$ en vertu des hypothèses précédentes. Les matrices de transition de la chaîne $\mathcal{C}$ sont fonction de la position de l'interrupteur I. Soit T la matrice de transition quand I est ouvert, S quand I est fermé. Les matrices T et S s'écrivent :

(3) $$T = \begin{pmatrix} 1 & \pi_1 & 0 & 0 & \vdots & 0 & 0 \\ 0 & 1-\pi_1 & \pi_2 & 0 & \vdots & 0 & 0 \\ \hline & & & & \vdots & & \\ 0 & 0 & 0 & 0 & \vdots & 1-\pi_{M-1} & \pi_M \\ 0 & 0 & 0 & 0 & \vdots & 0 & 1-\pi_M \end{pmatrix}$$

$$(4)\qquad S = \left(\begin{array}{cccc:cc} 0 & 0 & 0 & 0 & 0 & 0 \\ 1 & \pi_1 & 0 & 0 & 0 & 0 \\ 0 & 1-\pi_1 & \pi_2 & 0 & 0 & 0 \\ \hdashline 0 & 0 & 0 & 0 & \pi_{M-1} & 0 \\ 0 & 0 & 0 & 0 & 1-\pi_{M-1} & 1 \end{array}\right)$$

Appelons Q(t) un vecteur stochastique de dimension M+1 dont les composantes $q_i(t)$ sont définies par

$$q_i(t) = p_r\ (\text{état de la boucle-mémoire à l'instant } t \text{ soit } i)$$

Nous en déduisons :

(5) $Q(t+1) = TQ(t)$ si I est ouvert,

(6) $Q(t+1) = SQ(t)$ si I est fermé.

2.3 - ANALYSE DU SYSTEME SOUS COMMANDE ALEATOIRE DE L'INTERRUPTEUR.

2.3.1. - Analyse de la file d'attente

Considérons la file d'attente des usagers derrière I. Celui-ci étant commandé avec une probabilité de fermeture constante $1-\alpha$ à chaque période. Les programmes arrivent de l'extérieur suivant un processus de Poisson de paramètre λ , et sont placés dans la file dans l'ordre de leur arrivée. L'ensemble file d'attente-interrupteur continue un système M/G/1, qui a pour temps de service moyen le temps moyen entre deux fermetures consécutives de I soit $\Delta/(1-\alpha)$. On en déduit la probabilité p_o que cette file d'attente soit vide, et le nombre moyen $\bar{n}$ de programmes en attente aux instants de fermeture de I [8].

$$(7)\qquad p_o = 1 - \frac{\lambda\Delta}{1-\alpha}$$

$$(8)\qquad \bar{n} = \lambda\Delta\ \frac{1-\frac{\lambda\Delta}{2}}{1-\alpha-\lambda\Delta}$$

Le temps moyen d'attente w des programmes dans cette file en fonction de α s'écrit alors :

$$(9)\qquad w = \frac{\bar{n}}{\lambda} = \frac{1-\frac{\lambda\Delta}{2}}{1-\alpha-\lambda\Delta}$$

La condition d'existence d'une solution stationnaire est simplement $1-\alpha>\lambda\Delta$.

2.3.2 - Analyse de la B.M.

Définissons la fonction de commande de I par δ_i, i=0,...M, avec

$$\delta_i = p_r\{\text{ fermeture de I quand } n=i\ \}$$

Nous pouvons supposer que la file d'attente x n'est jamais vide en remarquant que si la file d'attente est vide quand n=i avec une probabilité θ_i, δ_i est remplacé par $\delta_i(1-\theta_i)$. A l'aide des coefficients δ_i les matrices τ et S se réduisent à une matrice unique R qui s'écrit

$$R=\left(\begin{array}{ccc:cc}
1-\delta_o & (1-\delta_1)\pi_1 & 0 & 0 & 0 \\
\delta_o & (1-\delta_1)(1-\pi_1)+\delta_1\pi_1 & (1-\delta_2)\pi_2 & 0 & 0 \\
0 & \delta_1(1-\pi_1) & (1-\delta_2)(1-\pi_2)+\delta_2\pi_2 & 0 & 0 \\
\hdashline
0 & 0 & 0 & 0 & 0 \\
0 & 0 & 0 & (1-\delta_{M-1})(1-\pi_{M-1})+\delta_{M-1}\pi_{M-1} & (1-\delta_M)\pi_M \\
0 & 0 & 0 & \delta_{M-1}(1-\pi_{M-1}) & (1-\delta_M)(1-\pi_M)+\delta_M
\end{array}\right)$$

Soit $U = {}^t(u_o,u_1,\ldots,u_M)$ la solution stationnaire de la chaîne $\mathcal{C}$. Une condition nécessaire d'existence de u est que $0< \delta_i <1$, $\forall i$, R étant singulière dans le cas contraire. La solution stationnaire u est définie par : $u_i=p_r\{n=i \text{ à l'équilibre}\}$ vérifiant

(10) $$U=RU$$

En développant (10) il vient

(11) $$u_o=u_o(1-\delta_o)+u_1(1-\delta_1)\pi_1$$

(12) $$u_j=(1-\pi_{j-1})\delta_{j-1}u_{j-1}+\{(1-\delta_j)(1-\pi_j)+\delta_j\pi_j\}\, u_j+\pi_{j+1}(1-\delta_{j+1})u_{j+1} \qquad j=1,\ldots M$$

soit

(13) $$0 = (1-\pi_{j-1})\delta_{j-1}u_{j-1}+(1-\delta_j\pi_j-\delta_j-\pi_j)u_j+\pi_{j+1}(1-\delta_{j+1})u_{j+1}$$

Définissons les fonctions génératrices U(x),D(x),P(x),Q(x)

$$U(x) = \sum_{i=0}^{\infty} u_i x^i$$

$$D(x) = \sum_{i=0}^{\infty} \delta_j u_i x^i$$

$$P(x) = \sum_{i=0}^{\infty} \pi_i \delta_i u_i x^i$$

$$Q(x) = \sum_{i=0}^{\infty} \pi_i u_i x^i$$

Nous en déduisons, à partir de (11) et (13), et après simplifications

(14) $$D(x)+P(x)\,\frac{1-x}{x} - \frac{Q(x)}{x} = 0$$

En identifiant les coefficients de x^i dans la relation (14) on obtient

(15) $$\delta_i u_i+\pi_{i+1}\delta_{i+1}\, u_{i+1} -\pi_i\delta_i u_i -\pi_{i+1}\, u_{i+1} = 0$$

(16) $$u_{i+1} = \frac{\delta_i(1-\pi_i)}{\pi_{i+1}(1-\delta_{i+1})}\; u_i$$

d'où

(17) $$u_i = \frac{\delta_{i-1}(1-\pi_{i-1})}{\pi_i(1-\delta_i)} \times\ldots\ldots\ldots\times \frac{\delta_o}{\pi_1(1-\delta_1)}\; u_o \qquad i=1,\ldots,M$$

u_o étant défini par

(18) $$\sum_{i=0}^{\infty} u_i=1$$

Les probabilités stationnaires u_i, $i=0,1,\dots,2,\dots$ peuvent être calculées numériquement en fonction des π_i et des fonctions de commande δ_i. Dans le cas d'une commande aléatoire constante $\delta_i = 1-\alpha$ les résultats s'écrivent :

(19) $$u_i = \left(\frac{1-\alpha}{\alpha}\right)^i \frac{(1-\pi_{i-1})\times\dots\dots(1-\pi_1)}{\pi_i\times\dots\dots\times\pi_1}\, u_o$$

2.3.3 - Analyse du système couplé : file d'attente boucle-mémoire.

Si on associe la file d'attente et la B.M. couplés par l'interrupteur I on réalise un modèle du système complet. Toutefois dans ce cas seule la longueur moyenne de la file d'attente va dépendre de la fréquence de fermeture de l'interrupteur suivant la formule (8). En effet à cause de l'hypothèse de fonctionnement stationnaire du système sous laquelle ont été conduites les analyses précédentes, et de la commande aléatoire indépendante de i, le flux de programmes sortant de l'interrupteur est égal au flux d'entrée. Plus précisément on a, pour la file d'attente

$$p_o = 1 - \frac{\lambda\Delta}{1-\alpha}$$

et donc, d'après la remarque faite au début de ce paragraphe, la probabilité réelle d'entrée d'un programme dans la B.M. est $(1-p_o)(1-\alpha)$ soit $\lambda\Delta$ indépendant de α.

L'analyse du système couplé peut être poursuivie en définissant les fonctions de commande $\delta_{k,i}$ de la façon suivante :

$$\delta_{k,i} = p_r\{\text{ fermeture de I quand n=i et k programmes sont en attente }\}$$

On décrit alors l'évolution du système couplé par une chaîne de Markov qui a pour état, à un instant donné, le couple (k,i) et on construit, en fonction de λ, des π_i, et des fonctions de commande $\delta_{k,i}$ la matrice de transition de la chaîne. Le calcul numérique de la solution stationnaire, dans le cas où elle existe, donne le nombre moyen de programmes dans le système à l'équilibre, et donc le temps de réponse moyen accordé à ces programmes. Ce modèle analytique est utilisé pour comparer différentes politiques de commande $\delta_{k,i}$ et vérifier les résultats obtenus au terme de la procédure d'optimisation décrite dans le paragraphe qui suit (de même que précédemment la solution stationnaire de la chaîne n'existe que si $0 < \delta_{k,i} < 1$).

III - CALCUL DE LA LOI DE COMMANDE OPTIMALE ET DE L'ESTIMATEUR

3.1 - CALCUL DES TEMPS DE REPONSE MOYENS.

Considérons d'abord le cas d'un programme dans la file x immédiatement derrière I, et appelons δ ($\delta \geq 1$) le nombre de périodes écoulées avant la fermeture de I. Pendant $\delta - 1$ périodes le programme attend derrière I et entre dans la B.M. pendant la δ-ième période. Soit Q(t) le vecteur d'état de la B.M. quand le programme arrive derrière I, le vecteur d'état de la B.M. en $t+\delta$, quand le programme y a pénétré, s'écrit $Q(t+\delta)$ avec

(20) $$Q(t+\delta) = ST^{\delta-1}Q(t).$$

On peut alors obtenir l'espérance mathématique du temps de traitement T_t du programme considéré à l'aide du vecteur A, en vertu de l'hypothèse H6

(21) $$T_t = AQ(t+\ ) = AST^{\delta-1}Q(T).$$

Finalement, le temps de réponse T_r total obtenu par le programme est

(22) $$T_r = \delta + T_t = \delta + AST^{\delta-1}Q(T)$$

Dans le cas d'un programme la politique de commande optimale est obtenue en résolvant

$$\underset{\delta \geqslant 1}{\text{Minimiser}} \ \{ \delta + AST^{\delta-1}Q(t)\}$$

pour tous les états possibles de la B.M., c'est-à-dire pour tous les vecteurs Q(t) de la forme

$$Q(t) = \begin{bmatrix} 0 \\ 1 \\ \vdots \\ 0 \end{bmatrix} i \qquad i=0,\dots M$$

On obtient ainsi la loi de commande $\delta(i)$, $i=0,\dots M$. Il est cependant évident que cette loi de commande, valable dans le cas où un seul programme est considéré, n'est plus optimale si on cherche à minimiser le temps de réponse moyen de plusieurs programmes dans la file. Soit K le nombre de programmes dans la file à l'instant t, les programmes étant numérotés de 1 à K comme sur la figure 5. Evaluons les temps de réponses T_k de chacun de ces programmes en fonction des quantités δ_k, nombre de périodes séparant

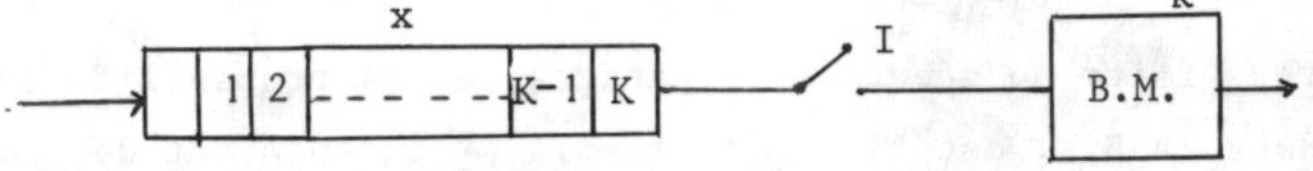

Figure 5

la fermeture de I pour laisser entrer le programme k+1, de celle pour laisser entrer le programme k. En raisonnant comme précédemment on peut écrire :

$$T_K = \delta_K + AST^{\delta_K-1} Q(t)$$

$$T_{K-1} = \delta_{K-1} + \delta_K + AST^{\delta_{K-1}-1} Q(t+\delta_K)$$

(23) $$T_k = \delta_k + \delta_{k+1} + \dots + \delta_K + AST^{\delta_k-1} Q(t+\delta_k+\dots+\delta_K)$$

$$T_1 = \delta_1 + \delta_2 + \dots + \delta_K + AST^{\delta_1-1} Q(t+\delta_1+\delta_2+\dots+\delta_K)$$

Posons $\qquad Q(t+\delta_k+\dots+\delta_K) = Q_k$, il vient

$$T_K = \delta_K + AST^{\delta_K-1} Q_K$$

(24) $$T_k = \delta_k + \delta_{k+1} + \dots + \delta_K + AST^{\delta_k-1} Q_k$$

$$T_1 = \delta_1 + \delta_2 + \dots + \delta_K + AST^{\delta_1-1} Q_1$$

avec

(25) $$Q_{k-1} = ST^{\delta_k-1} Q_k$$

Le problème s'écrit alors

$$\mathcal{P}(k) \left| \begin{array}{l} \underset{\substack{\delta_k \geqslant 1 \\ k=1,\dots K}}{\text{Minimiser}} \ \sum_{k=1}^{K} T_k = \sum_{k=1}^{K} (k\delta_k + AQ_{k-1}) \\ Q_{k-1} = ST^{\delta_k-1} Q_k \end{array} \right.$$

3.2 - CALCUL DE LA LOI DE COMMANDE OPTIMALE.

Le problème S(k) peut être résolu indirectement par programmation dynamique en introduisant les fonctions $f_k(Q)$ définies comme suit :

$$f_k(Q) = \underset{\delta_k \geq 1}{\text{minimum}} \{k\delta_k + AQ' + \tau f_{k-1}(Q')\}$$

$$f_o(Q) = 0 \quad ; \quad \tau < 1$$

$$Q' = ST^{\delta_k - 1} Q$$

On calcule ainsi récursivement la suite des fonctions δ(k,.). On appelle politique optimale liée au problème S(k) l'ensemble S(k) des fonctions δ(k,.), k=1,...K.

$$S(k) = \{\delta(k,.),\ k=1,\dots K\}$$

Remarquons qu'en raison de la technique itérative du calcul on a

$$S(k) \supset S(k-1) \supset \dots \supset S(1)$$

qui exprime qu'en calculant la politique optimale liée à S(k), on calcule également les politiques optimales des problèmes S(k), k=1,...k-1. Pratiquement S(k) fournit la loi de commande optimale de I lorsque la longueur de la file est inférieure ou égale à k. En effet δ(k,Q) représente le nombre de périodes à attendre avant de fermer I lorsque k programmes sont en attente et que le vecteur d'état de la B.M. est Q. Plus précisément, l'état de la B.M. étant totalement observable on ne considère que les politiques δ(k,i) où i est le degré de multiprogrammation.

La loi de commande optimale est donc condensée dans la table δ(k,i) qui peut être utilisée de deux façons. Soit en utilisant toute l'information contenue dans la table, c'est-à-dire, après avoir observé k et i, en laissant ouvert I pendant δ(k,i)-1 périodes, pour de nouveau observer les nouvelles valeurs de k et i et répéter le processus. Soit en utilisant simplement l'information δ(k,1)=1 ou δ(k,i) > 1, et en répétant l'observation de k et i, ainsi que la commande de I à chaque période. La seconde méthode qui permet de mieux suivre l'évolution et le contrôle est celle que nous retiendrons, et nous pouvons en conséquence condenser la table δ(k,i) en une table de 1 et de 0, 1 indiquant que I doit être fermé, 0 que I doit rester ouvert, comme représenté sur la figure 6

k \ i	0	1	2	3	4	5
1	1	1	1	1	0	0
2	1	1	1	0	0	0
3	1	1	0	0	0	0
4	1	1	0	0	0	0
5	1	1	0	0	0	0

Figure 6

Example de table de commande de I

On peut démontrer à l'aide d'un théorème de point fixe que quand k tend vers ∞, les fonctions δ(k,i) convergent vers des fonctions δ(∞,i), et que cette convergence est obtenue en un nombre fini d'itérations de l'algorithme de calcul. Ce résultat exprime simplement que dans le cas d'une file d'attente infinie minimiser le temps de réponse moyen des programmes est équivalent à maximiser le débit de la B.M. et donc à opérer une régulation en fonction uniquement du degré de multiprogrammation.

3.2 - CALCUL DE L'ESTIMATEUR.

Le calcul de la loi de commande optimale nécessite la connaissance des coefficients de débit π_i, i=1,...M, à partir desquels les autres paramètres p_i et a_i peuvent être obtenus par les équations (1) et (2). La fonction de l'estimateur est donc de fournir au contrôleur une estimation des paramètres π_i à partir de i observation des fréquences de sorties de la B.M., et de l'état de la B.M. au moment des sorties.

Il s'agit de calculer la valeur $\tilde{\pi}_i$ la plus probable du paramètre π_i, i=1,...M, qui est appelée estimateur du maximum de vraisemblance [5]. Le calcul de l'estimateur est effectué en supposant connue la fonction de densité des observations, excepté pour certains de ses paramètres. La fonction de vraissemblance est obtenue en substituant les valeurs des observations dans la fonction de densité. Il reste alors à calculer le maximum de la fonction de vraisemblance par rapport aux paramètres inconnus. Dans l'exemple qui nous occupe le paramètre π_i est le paramètre d'une loi binominale puisque, en raison des hypothèses faites, la probabilité de R sorties sur Q observations est

$$\pi_i^R (1-\pi_i)^{Q-R} .$$

La fonction de vraissemblance dans le cas de q observations et de r sorties s'écrit donc :

(26) $$L(\pi_i) = \pi_i^r (1-\pi_i)^{q-r}$$

En différentiant $L(\pi_i)$ le maximum de $L(\pi_i)$ est obtenu pour

(27) $$\tilde{\pi}_i = \frac{r}{q}$$

qui est l'estimateur du maximum de vraisemblance du paramètre π_i, $i=1,\ldots M$. Cet estimateur a été utilisé avec succès au cours des simulations.

IV - RESULTATS NUMERIQUES ET SIMULATIONS.

Le calcul de la table de commande optimale de l'interrupteur I a été réalisé dans le cas d'un système de degré de multiprogrammation maximum M=6, avec les coefficients π_i donnés dans le tableau ci-dessous (Fig. 7).

i	0	1	2	3	4	5	6
π_i	0	.042	.067	.075	.067	.043	.003

figure 7

pour une valeur de la période Δ = 30ms. Les résultats sont rassemblés dans le tableau suivant (Figure 8)

k \ i	0	1	2	3	4	5	6
1	1	1	1	0	0	0	0
2	1	1	0	0	0	0	0
3	1	1	0	0	0	0	0
	1	1	0	0	0	0	0

Figure 8

La table de commande optimale de I a été utilisée dans une série de simulations du système en faisant varier le débit d'entrée λ et en comparant les temps de réponse moyens obtenus dans le cas de l'application de la loi de commande optimale $\delta(k,i)$et de la loi limite $\delta(\infty,i)$ (fig. 9). Le temps de réponse moyen ainsi mesurés sont rapportés sur la figure 10. Pour les politiques $\delta(k,i)$ et $\delta(\infty,i)$ les résultats obtenus sont identiques pour les valeurs extrêmes du débit d'arrivée, correspondant d'une part au cas où le degré de multiprogrammation est très rarement supérieur à 1, c'est à dire au cas d'une sous-utilisation du système, d'autre part au cas de la saturation où le débit d'aarivée approche le débit maximum de la B.M. Dans la zone intermédiaire d'utilisation normale du système l'application de la loi de commande optimale δ (k,i) conduit à une amélioration sensible de l'ordre de 10% des performances du système.

V - CONCLUSION

Les résultats cités plus haut montrent l'intérêt de mettre en oeuvre un mécanisme de contrôle dynamique qui règle le degré de multiprogrammation en fonction de la charge du système et du comportement de la B.M. L'étape suivante du développement de cet outil doit permettre de préciser les conditions de son implantation à l'aide des simulations détaillées du système complet qui décrivent précisément les mécanismes de pagination et d'E/S afin d'évaluer les valeurs de Δ , période de commande de l'interrupteur,et de la période d'estimation qui réalisent le compromis optimal entre le coût des opérations de commande et d'estimation, et l'amélioration des performances qui en découlent.

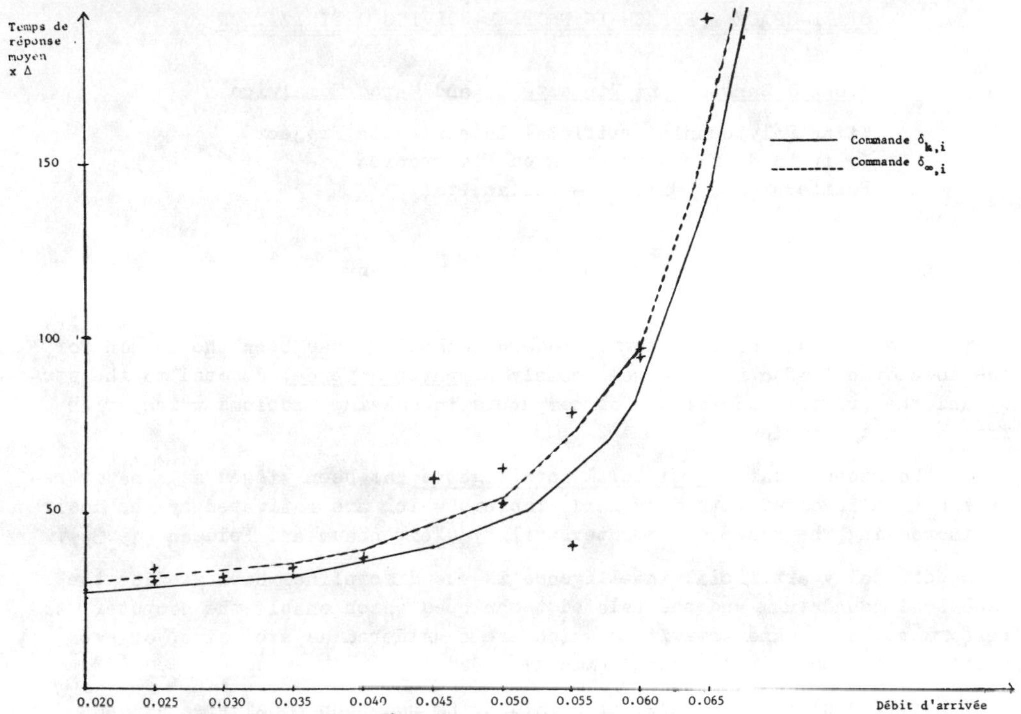

Figure 10

REFERENCES

(1) Betourne C., e.a. "Process management and ressource sharing in the multiaccess system ESOPE". Comm. ACM, 13, (1970.

(2) Brandwajn A., Gelenbe E., Potier D., LENFANT J., "Optimal degree of multiprogramming in a virtual memory system" Technical report LABORIA-IRIA, 1973.

(3) Denning P.J. "The working set model for program behavior", Comm. ACM. 11 (1968).

(4) Fife D.W., "An optimisation model for time-sharing", AFIPS, SJCC (1966).

(5) Jenkins G.M., and Watts D.G., "Spectral Analysis and its applications", Holden day (1969).

(6) Kashyap R.L., "Optimisation of stochastic finite state systems" IEEE Trans. AC 11, (1966).

(7) Rodriguez-Rosell J., and Dupuy J.P. "The evaluation of a time-sharing page demand system" AFIPS, SJCC, 759-765 (1972).

(8) Saaty T.L., "Elements of queuing theory" McGraw-Hill, New York (1961).

(9) Sekino A., "Performance evaluation of multiprogrammed time-shared computer systems". Ph. D. Thesis, MIT project MAC report MAC-TR-103, (1972).

STATE-SPACE APPROACH IN PROBLEM-SOLVING OPTIMIZATION

Alberto Sangiovanni Vincentelli and Marco Somalvico

Milan Polytechnic Artificial Intelligence Project
Istituto di Elettrotecnica ed Elettronica
Politecnico di Milano - Milan, Italy

I. INTRODUCTION

The great impact of computers in modern technology has been the reason for the development of a new science, namely computer science, devoted to the study and the progress of the use of computers in solving problems which arise from human exigencies.

In the recent years, artificial intelligence has been viewed as a major research area involved with many hard problems which are motivated by the desire of increasing the range of computer ability (Feigenbaum and Feldman (1963)).

Specifically artificial intelligence is the discipline which studies the technical foundations and the related techniques which enable the computer to perform mechanisms and activities which are considered as exclusive, or even not available to human intelligence (Minsky (1968)).

In artificial intelligence, problem-solving has been considered as one of the most important research subjects, worth of extensive efforts and carrier of significative results (Nilsson (1971)).

The goal of problem-solving is the availability of efficient methods which provide the machine of the capability for obtaining, within a mechanical process, the solution of a problem which has been proposed to the computer in an appropriate way (Slagle (1971)).

The fundamental issue in problem-solving is the necessity of providing the computer with a problem representation, i.e., a precise framework which contains all the informations which are required for a mechanical construction of the problem solution (Amarel (1968)).

Also the efficiency of a search procedure is heavily influenced by the way of carrying on the representation task.

In particular the heuristically guided search strategies are based on the possibility of coupling the information contained inside the representation with a new information, the heuristic information, which has the power of greatly increasing the efficiency of the search process (Hart, Nilsson, and Raphael (1968)).

The process of representing a problem involves necessarily a passage between

---

This work has been partially supported by the Special Program for Computer Science of the National Research Council.

two different worlds, namely the world of the intuitive notion that the man has of the problem, and the world of the formal and precise description of the problem, that will be given to the computer.

This passage is achieved by means of an appropriate selection of only one part of the information which is pertinent to the intuitive knowledge of the problem; the selected information is then arranged in a structured form, thus yielding the representation of the problem.

However, even the most structured and rich representation, which might be obtained within this passage, will always present a difference from the unformal knowledge about the problem.

This distance, which reminds the similar gap between physical phenomenon and physical law, can only be narrowed, but can never be wholly deleted.

The previous considerations have been the basic motivations for the development of a "theory of problem-solving" denoted to the understanding of the different aspects of the problem-solving process, and centered around the focal point of problem representation.

This theory of problem-solving, which is being developed at the Milan Polytechnic Artificial Intelligence Project (MP-AI Project), is intended to achieve the following goals (Mandrioli, Sangiovanni Vincentelli, and Somalvico (1973)):

- formalization of problem representation methods;
- formalization of solution search techniques;
- formalization and selection of "good" problem representations;
- automatic evaluation and use of heuristic information;
- generalization and operation of learning processes;
- structured organization of a problem in subproblems as a basis of automatic programming.

The previous research work that has been done in this direction (Sangiovanni Vincentelli and Somalvico (1972), (1973 (a))), has been centered on the syntactic description of state-space approach to problem-solving (SSPS), i.e., the classical problem representation made up of states and operators.

Application of SSPS to computer assisted medical diagnosis has been investigated as well (Sangiovanni Vincentelli, and Somalvico (1973 (b))).

The purpose of this paper is to propose a formal problem representation, in SSPS, called "semantic description", which constitutes a framework useful to structure a rich content of informations about a given problem.

The semantic description is shown to be equivalent to the syntactic description.

Furthermore a method is here presented, which is based on the semantic description, and which makes possible to extract in an automatic way, i.e., by computation, the heuristic information useful to guide the search in the state space.

More precisely, the method consists in a procedure which associates to a main problem, an auxiliary problem whose solution, easy to be found, i.e., computed, yields an estimate for the main problem.

The estimate is essentially the formal quantification of heuristic information which, according to the well known Hart-Nilsson-Raphael algorithm, allows one to perform an efficient heuristically guided search.

In Section II, the syntactic description of SSPS, according to the previous work of the authors, is briefly reviewed.

In Section III, the semantic description of SSPS is presented and its equivalence with the syntactic description is illustrated.

In Section IV, the implication of the semantic description on the notion of auxiliary problem is exposed, the outline of the method for computing an estimate is proposed as well.

In Section V, some conclusive remarks, and the research problems, which are open for further investigation, are presented as well.

II. SYNTACTIC DESCRIPTION OF STATE-SPACE APPROACH TO PROBLEM-SOLVING

Computer science is basically involved in the human activity of understanding and solving a problem with the help of computers (Feigenbaum, and Feldman (1963)).

In artificial intelligence research, problem-solving is devoted to the goal of a complete automatization of this human activity (Minsky (1968)).

Therefore the computer needs a description of the problem through a formalization, of a certain type, of the problem domain, i.e., of the knowledge about the problem.

In this formalization process, man performs a selection of the information, on the problem domain, which he estimates as sufficient to the computer to automatically solve the problem which is considered.

Thus in problem-solving we are faced with two different aspects (Nilsson (1971)):

- representation, i.e., the precise organization of the information which we select from the problem domain, and which we provide to the computer as the description of the problem;
- search, i.e., the technique which operates on the representation, and realizes the process of investigation whose goal is the construction of the solution of the problem.

The state-space approach to problem-solving (SSPS) constitutes, together with the problem-reduction approach, one method of defining the representation and the search aspects of problem-solving.

The SSPS is important and widely adopted since it provides some very intuitive and simple notions, together with efficient techniques, which assist the man during the task of constructing the representation of the problem, and of performing the search of the solution.

When we define the representation of a problem in SSPS, we have to provide the notion of state space, composed by a set of states and a set of operators, and the notion of solution of the problem, constituted by a sequence of operators.

We shall now briefly recall the basic notions of a description of SSPS, called syntactic description, which has been the result of previous research work of the authors (Sangiovanni Vincentelli, and Somalvico (1972),(1973(a)),(1973(b))).

Definition 2.1. A problem schema M is a couple $M = (S,\Gamma)$, where S is a non empty (possibly infinite) set of states, and Γ is a set of functions, called operators, s.t. :

$$\Gamma = \left\{\gamma_i \,\middle|\, \gamma_i : A_i \longrightarrow S,\ A_i \subset S\right\} \tag{2.1}$$

□

Theorem 2.1. The set Γ of operators, yields a function Γ_1, s.t. :

$$\Gamma_1 : S \longrightarrow P(S) \tag{2.2}$$

$$\Gamma_1 : s_j \longmapsto A_j \ , \quad A_j \subset S \tag{3.3}$$

Proof. For each $s_j \in S$, it is possible to determine a set A_j, s.t.:

$$A_j = \left\{ s \,\middle|\, (\forall s)(\exists \gamma_i)\ ((\gamma_i \in \Gamma) \wedge (\gamma_i : A_i \longrightarrow S) \wedge (s_j \in A_i) \wedge (s = \gamma_i(s_j)))\right\} \tag{2.4}$$

Moreover, by Definition 2.1, the set A_j is unique, since each γ_i is a function; and by (2.1) and (2.4), $s \in S$), and therefore, $A_j \subset S$.

Thus :

$$\Gamma_1 : s_j \longmapsto A_j, \quad A_j \subset S \tag{2.5}$$

□

Theorem 2.2. The function Γ_1 yields a function Γ_2, s.t.:

$$\Gamma_2 : P(S) \longrightarrow P(S) \tag{2.6}$$

$$\Gamma_2 : A_i \longmapsto A_j, \quad A_i, A_j \subset S \tag{2.7}$$

Proof. For each $A_i \subset S$, we shall determine an unique $A_j \subset S$.

Infact, in correspondence of each $s_k \in A_i$, by Theorem 2.1, it exists an unique set $\Gamma_1(s_k) \subset S$.

Therefore, we obtain an A_j, s.t. :

$$A_j = \bigcup_{(s_k \in A_i)} \Gamma_1(s_k) \tag{2.8}$$

Thus:

$$\Gamma_2 : A_i \longmapsto A_j, \quad A_i, A_j \subset S \tag{2.9}$$

□

Definition 2.2. We define the n-step global operator Γ^n, s.t.:

$$\Gamma^n : S \longrightarrow P(S) \tag{2.10}$$

in the following way :

$$\Gamma^1 = \Gamma_1 \tag{2.11}$$

$$\Gamma^2 = \Gamma_1 \circ \Gamma_2 \tag{2.12}$$

i.e., Γ^2 is the concatenation of Γ_1 with Γ_2, and, in general :

$$\Gamma^n = \Gamma_1 \circ \Gamma_2 \circ \Gamma_2 \circ \ldots \circ \Gamma_2 \tag{2.13}$$

i.e., Γ^n is the concatenation of Γ_1 with Γ_2 taken n-1 times.

□

The problem schema M represents the "skeleton" of a problem in the SSPS.

We can now introduce the concepts of initial and final states, to obtain the complete notion of problem.

Definition 2.3 . A problem P is a triple P=(M,i,f) where M is a problem schema, i is an element of S called initial (or source) state, and f is an element of S called final (or goal) state.

We can extended the notion of problem as a triple P =(M,i,K), where we consider, in place of f, a set $K \subset S$, called set of final states.

□

An other notion, which is intermediate between problem schema and problem, is presented in the following definition.

Definition 2.4 . A goal problem F is a couple F = (M,f), where M is a problem schema, and f is an element of S called final (or goal) state.

We can extend the notion of goal problem as a couple F = (M,K), where we consider, in place of f, a set $K \subset S$, called set of final states.

□

The notion of goal problem is interesting , because when we consider some problems with the same final state f (or with the same set of final states K), but with different initial states i, we can deduce that all these problems share among themselves the same goal problem.

We can therefore conclude that a problem has been formalised within the SSPS, whenever the previously defined elements have been specified.

More precisely, in order to obtain a state-space formulation of a problem, we have to introduce :

1) the states and their correspondence with "patterns" or "situations" which exist in the problem;
2) the set of operators and their effects on state descriptions;
3) the initial state;
4) the final state (or the set of final states), or the properties which characterize a state as a final one.

We want to label this way of formalizing the SSPS, as syntactic, since the informations about the real problem, which have been strictly confined into this mathematical framework , are mainly centered on the existence of states, and on the transformations between two states, while the informations about the nature of a state are not included.

Namely the knowledge, from the real problem domain, about the "structure" of a state is ignored; the notion of state is itself of atomic nature, since, it is represented by the mathematical concept of the element of a set.

Thus both the meaning of states and operators cannot anymore be related to the real description of the problem, but can only be considered on their algebraic nature.

More specifically, in the description that we have presented, we have not taken in account the particular "meaning" (i.e., structure) of the sets S and Γ.

Once we have setted up the problem in this formulation, we must find the solution; let us recall an other definition.

Definition 2.5. An n-step solution of a problem P, is a sequence of functions $\mathcal{G}^n$, s.t.:

$$\mathcal{G}^n = \langle \alpha_1, \alpha_2, \ldots, \alpha_j, \ldots, \alpha_n \rangle \tag{2.14}$$

where each function α_j, satisfies the following two conditions :

(i) $$(\forall\, j \in \bar{n})(\exists \gamma_i)((\gamma_i \in \Gamma) \wedge (\alpha_j = \gamma_i)) \tag{2.15}$$

where :

$$\bar{n} = \{1, 2, \ldots, j, \ldots, n\} \tag{2.16}$$

(ii) $$i \xmapsto{\alpha_1} s_1 \xmapsto{\alpha_2} s_2 \;\cdots\; s_{j-1} \xmapsto{\alpha_j} s_j \;\cdots\; s_{n-1} \xmapsto{\alpha_n} f \qquad (f\ K) \tag{2.17}$$

□

Theorem 2.3. An n-step solution $\mathcal{G}^n$ yields one and only one sequence of states, $S[\mathcal{G}^n]$, s.t.:

$$S[\mathcal{G}^n] = \{s_1, s_2, \ldots, s_{n-1}\} \tag{2.18}$$

$S[\mathcal{G}^n]$ is called n-step solution sequence, and it is composed of n-1 states called intermediate solution states.

Proof. If $\mathcal{G}^n$ is an n-step solution, because of Definition 2.5 and relations (2.14) and (2.17), it individuates a sequence of n-1 states; the sequence is unique since in (2.17) each α_j is, because of (2.15), an operator γ_i, and, therefore, because of Definition 2.1, each α_j is a function.

□

We can use the mathematical notion of directed graph in order to handle the SSPS, both in its aspects of representation and search.

We associate a vertex to each state and an arc to each operator.

When we apply an operator γ_i to a state $\bar{s}$, we obtain a new state $s^* = \gamma_i(\bar{s})$.

In the graph model we shall have an arc (corresponding to the operator γ_i) which joins $\bar{s}$ and s^*.

If we define some "costs" associated with the operators, we could be interested in obtaining an "optimum solution", i.e., a solution $\bar{\mathcal{G}}^n$ whose total cost, i.e., the sum of the costs of the $\alpha_j \in \bar{\mathcal{G}}^n$, is minimal.

More precisely, we may introduce the following definitions.

Definition 2.6. An operator $\gamma_i \in \Gamma$ is associated with a cost C_i, which evaluates the application of γ_i between a state s and a state t, i.e.:

$$s \xmapsto{\gamma_i} t \tag{2.19}$$

□

Definition 2.7. The cost of a solution $\mathcal{G}^n$.

$$\mathcal{G}^n = \langle \alpha_1, \alpha_2, \ldots, \alpha_n \rangle \tag{2.20}$$

is defined as :

$$C(\mathcal{G}^n) = \sum_{1}^{n}{}_j\, C_{\alpha_j} \tag{2.21}$$

□

Definition 2.8. Given a problem P and all its solutions $\mathcal{G}^n$, we call optimum cost $\bar{C}(\mathcal{G}^{\bar{n}})$ of the solution to P, the minimum of the costs $C(\mathcal{G}^n)$, and we call optimum solution $\bar{\mathcal{G}}^n$ the related $\bar{n}$-step solution to P. □

In the graph model the optimum solution is represented by a minimum cost path from i to f.

The search aspects of the automatic problem-solving consist mainly in obtainig a solution, possibly an optimum solution.

The construction of the solution (or of the optimum solution) involves the use of some search strategies.

Since the dimension of the state space, for real problems, is usually of big magnitude, the problem arises of limiting the occupation of memory and the computation time.

Thus, normally, the graph associated to the state space, is not stored in a way that is called explicit description, i.e., with the complete list of all vertices and of all arcs.

In effects, the graph is stored in a way, which is denoted implicit description, and which consists of the initial state i, of the set of operators Γ, and of the final state f (or the set of final states K).

Therefore the search strategy always consists of two processes which are developed in parallel :

1) incremental construction of the path (i.e., of the solution);
2) incremental explicitation of the graph (i.e., by the application of the n-step global operator Γ^n).

Because of this incremental procedure, these search strategies are called expansion tehniques.

The most conceptually simple expansion search strategy is the breadth first method, which consists in computing step by step Γ^n and testing if some of the new states which are obtained is f (or is an alement of K).

This strategy is very costly because of the memory and of the computation time which are required.

New techniques of expansion are based on the application of Γ^n at each step (i.e., of Γ_2 for $n > 1$) not to all the new states obtained at the previous step, but only to one new state, selected with some criterion (therefore Γ_2 is practically equivalent to Γ_1).

Please note that in this way the expansion process implies an ordering of the states which are expanded.

We say that a state s is expanded, when we apply Γ_2 to s (or Γ_1 to s).

Depth first method is one expansion search strategy which guides (i.e. which, orders) the expansion process in this way: it is expanded first, the first new state which was obtained in the expansion of the previous step.

This method does not assure that the solution which is obtained is an optimum solution.

An other expansion technique is the Dijkstra-Dantzig or uniform cost algorithm.

In this method a state s is associated with an evaluation function f(s).

This function is computed for each state on the basis of the costs of the operators α_i which constitute a solution for a problem which has s as final state (and i as initial state).

More precisely, $f(s)$ is the cost of the optimum solution for such problem.

The value of $f(s)$ for each state s guides the choice of which new state has to be expanded.

It is expanded the state s for which is minimum the value $f(s)$.

Although this method is more efficient than the previous ones, it is not yet satisfactory for the automatic solution of large scale problems.

All these methods use "blind search strategies" since they are based only on the information which is completely contained inside the representation of the problem.

The problem arises of utilizing some additional information, called the "heuristic information", for guiding the search.

This information lies outside the given representation and consists of knowledge which is extracted from the "semantic domain" of the problem (Nilsson (1971)).

A classical way of introducing heuristic information, proposed by Hart, Nilsson, and Raphael, consists in assigning to each state s a new evaluation function $\hat{g}(s)$.

This function represents an "estimate" of the cost $g(s)$ of a solution to the problem, with f as final state, and with s as an intermediate solution state.

More precisely the new evaluation function is the following one:

$$\hat{g}(s) = f(s) + \hat{h}(s) \qquad (2.22)$$

where $f(s)$ is the evaluation function of the Dijkstra-Dantzig algorithm and $\hat{h}(s)$ is a lower bound of the cost $h(s)$ of the optimum solution for a problem which has s as initial state (and f as final state).

Therefore we have :

$$\hat{h}(s) \leq h(s) \qquad (2.23)$$

$$\hat{g}(s) \leq g(s) \qquad (2.24)$$

This method, under assumption (2.23), is admissible, i.e., it finds, whenever it exists, a solution which is an optimum one.

An important question is, in this case, how to provide a technique for obtaining an evaluation function $\hat{g}(s)$ (actually, $\hat{h}(s)$).

Usually this estimate requires human ingenuity based on an appropriate inspection and processing of the semantic domain of the problem.

Therefore the goal of a complete automatization of the problem solving procedure is incompatible with the use of an heuristic information, which is not contained inside the computer , and which thus implies invention from the man.

For the above considerations, it seems important to develop new techniques which enable in some way the computer to obtain, within an automatic procedure, the heuristic information itself.

This task requires the availability to the computer of new information, outside the SSPS representation, from which, with an extraction process, the evaluation function can be computed.

A new way of conceiving the SSPS is therefore required, which we shall call semantic description, in order to distinguish it from the previously exposed syntactic description.

The richer information, which is proper of the semantic description, is embedded on the SSPS representation, by exploding the notion of state, and by associating to each state a structure in which the new information is inserted.

We shall describible how to utilize the semantic description in order to obtain the computation of the evaluation function.

This technique, which shall be exposed in the following Sections, constitutes a new progress in the direction of automatic problem solving.

III. SEMANTIC DESCRIPTION OF STATE-SPACE APPROACH TO PROBLEM-SOLVING

In this Section we will expose a new formal framework in which it is possible to arrange the new information necessary for the computation of the evaluation function.

Since this new formalisation is always related to the SSPS, we will illustrate it with two goals :

1. it is sufficient for describing a problem in SSPS ;
2. it is equivalent to the syntactic description of SSPS.

The main idea on which the semantic description is based, is to associate to a state a structure which contains the new information.

Therefore we will expose the formal schema which is apted to describe the structure of states in SSPS.

This formal schema is designed with the purpose of providing these properties:

1. it is sufficiently powerful in order to contain a great amount of information from the real problem domain, i.e., the semantics of the problem;
2. it presents great flexibility in order to be utilized for the description of a wide class of problems, arising from very different semantic domains;
3. it is a fruitful basis on which efficient procedures can operate in order to reach the goal of computing the evaluation function, and, in general, the fundamental goals of automatic problem solving.

The schema that is proposed here for a formal structure of a state, is a sequence of attribute and value comples.

For this reason, we shall call the problems, whose structure can be setted up in this way, attribute-value problems (AV problems).

More precisely we may now introduce the following definitions and properties.

Definition 3.1. In a problem, we identify a set A of elements called attributes:

$$A = \left\{A_1, A_2, \ldots, A_i, \ldots, A_n\right\} \qquad (3.1)$$

□

Definition 3.2. Each attribute A_i is associated with a value set V_i, i.e. :

$$V_i = \left\{v_1^i, v_2^i, \ldots, v_j^i, \ldots, v_{n_i}^i\right\} \qquad (3.2)$$

where each v_j^i is called value for the attribute A_i. □

Definition 3.3. An attribute - value couple (AVC) for a parameter A_i is a couple $c_i = (A_i, v_j^i)$ where A_i is an attribute, ans v_j^i is a value for the attribute A_i. □

Theorem 3.1. An attribute A_i individuates a set C_i of AVC's, called AVC set for A_i.

Proof. By Definitions 3.1, 3.2 and 3.3, we construct C_i in the following way :

$$C_i = \left\{ c_i \;\middle|\; (c_i = (A_i, v_j^i)) \wedge (v_j^i \in V_i) \right\} \quad (3.3)$$

□

We arrive now to the main definition in which the notion of structure of a state is introduced.

Definition 3.4. A structured state (SS) $\bar{s}$ (i.e., a state with structure), is an n-tuple of AVC's, s.t.:

$$\bar{s} = \langle c_1, c_2, \dots, c_i, \dots, c_n \rangle \quad (3.4)$$

where each c_i is an AVC for each parameter A_i of A, i.e. :

$$(\forall A_i)\,((A_i \in A) \wedge (c_i = (A_i, v_j^i))) \quad (3.5)$$

□

We may now introduce the notion of state space (structured).

Theorem 3.2. The SS set (i.e., the set of structured states) $\bar{S}$ is the cartesian product of the AVC sets for all the attributes A_i of A, i.e. :

$$\bar{S} = C_1 \times C_2 \times \dots \times C_i \times \dots \times C_n \quad (3.6)$$

Proof. From the definition of cartesian product, an element $\bar{s}$ of $\bar{S}$ is, because of (3.6), an n-tuple, made up with elements of each AVC set C_i.

From (3.3) we obtain therefore the SS $\bar{s}$, s.t. the (3.4) holds. □

Please note that, within this description, we can clearly understand what does it means that two states s' and s" are different: it means that they have some AVC's (e.g., c_i' and c_i'') which are different, i.e., some attribute A_i takes two differente values (e.g. , v_j^i and $v_j^{i}{}''$).

Now, in order to set up the notion of problem, we have to introduce somehow the notion of transformations from an SS $\bar{s}$ to an other SS s* (which in the syntac tic description were formalised with the operators of Γ).

If any transformation between any two states would be possible, we would be faced by a very trivial situation in which any problem (i.e., any choice of initial and final states) would be solved with just an one-step solution.

Problems of this nature, called universal problems, would be represented by a complete graph, because all the transformations are possible, i.e., any two vertices are connected by an arc.

In this case, the set of the operators, in the syntactic description, would be represented by the set of all the functions on S, i.e.:

$$\Gamma_u = \left\{ f \;\middle|\; f : S \rightarrow S \right\} \quad (3.7)$$

where Γ_u is the universal operator set.

In a real problem, on the other hand, we are faced with some restrictions with respect to the possible transformations; therefore the graph associated with the problem is an incomplete graph, and the set of operators Γ would be less powerful than Γ_u.

The idea which is embedded in the semantic description is practically this one: to describe in the formal way, the constraints which are imposed on the possible transformations between two SS's.

Moore precisely, it will be exposed how to describe, with expressions which deal with attributes and values, i.e.,with the structure of a state, the limitations which make up a real problem (i.e., an incomplete graph) starting from an universal problem (i.e., a complete graph).

It is interesting to observe that these expressions will present two different mathematical aspects, namely, the algebraic one, and the logical one.

More precisely we may now introduce the following definitions and properties.

Definition 3.5. A legal condition (LC) L_i is a binary relation on S, i.e. :

$$L_i \subset S \times S \tag{3.8}$$

or

$$L_i = \{(s', s'') \mid P_i(s', s'')\} \tag{3.9}$$

where P_i is a predicate, i.e. :

$$P_i(s', s'') : S \times S \longrightarrow \{T, F\} \tag{3.10}$$

(T and F stand for true and false). □

Please note that in Definition 3.5, while (3.8) implies the algebraic nature of a legal condition, (3.9) and (3.10) illustrate its logical aspect: in particular the predicate P_i can be expressed within some logical calculus (e.g., the first-order predicate calculus) operating on the attributes and the values as variables and constants.

A legal condition is therefore a mathematical expression, drawn from the intuitive notion of a problem, which enables one to describe some of the limitations existing on the transformations between SS's.

Therefore a problem can be made up with a certain number of these expressions.

Definition 3.6. An LC set L is the set :

$$L = \{L_o, L_1, L_2, \ldots, L_h, \ldots, L_r\} \tag{3.11}$$

where :

$$L_o = S \times S \tag{3.12}$$

and L_h, $1 \leq h \leq r$ are all the LC's.

When there are not LC's, we absume the existence of the special LC L_o which yields the universal problem.

With the next definition we introduce an important notion which takes, in the semantic description, a place equivalent to that of Γ in the syntatic description. □

Definition 3.7. The constraint C of a problem is a bynary relation on S (i.e., $C \subset S \times S$), s.t. :

$$C = \bigcap_{h=0}^{r} L_i \tag{3.13}$$

□

The constraint C is therefore the unification of all the formalised information which makes up a real problem from an universal problem.

This concept provides a new way of defining the notion of problem which is equivalent to that one given in the syntactic description.

Definition 3.8. An AV problem schema $\bar{M}$ is a couple $\bar{M} = (\bar{S}, C)$ where $\bar{S}$ is an SS set, and C is a constraint. □

Definition 3.9. An AV problem $\bar{P}$ is a triple $\bar{P} = (\bar{M}, \bar{i}, \bar{f})$(or $\bar{P} = (\bar{M}, \bar{i}, \bar{K})$) where $\bar{M}$ is an AV problem schema, $\bar{i}$ is an initial SS and $\bar{f}$ is a final SS (or $\bar{K}$ is a set of final SS's). □

We can now illustrate the main result, which shows, essentially, the equivalence between the semantic description and the syntactic description.

Theorem 3.3. The set of operators Γ and the constraint C are equivalent.

Proof. First of all we recall from algebra that, given a function f on a set Q, i.e. :

$$f : Q \longrightarrow Q \tag{3.14}$$

$$f : q' \longmapsto q'' \tag{3.15}$$

we can associate f, with a binary relation G(f) on Q, called the graph of f, s.t.:

$$G(f) = \left\{(q', q'') \,\middle|\, (q', q'' \in Q) \wedge (q'' = f(q'))\right\} \tag{3.16}$$

It exists clearly an equivalence between S and $\bar{S}$, i.e., between s and $\bar{s}$, because each SS $\bar{s}$ (i.e., each n-tuple of AVC's) constitutes the formal structure of each state s.

Now, if we consider the set of operators Γ, because of Definition 2.1., and (2.1) we may obtain the constraint C, in this way :

$$C = \bigcup_{\gamma_i \in \Gamma} G(\gamma_i) \tag{3.17}$$

Please note that the C which is obtained from Γ is unique.

On the other hand, if we consider the constraint C, which is a binary relation, we may set up a set of operators Γ in this way :

$$\Gamma = \left\{\gamma_i \,\middle|\, (\gamma_i : A_i \longrightarrow S,\ A_i \subset S) \wedge \left(\bigcup_{\gamma_i \in \Gamma} G(\gamma_i) = C\right)\right\} \tag{3.18}$$

In this case the choice of the operators γ_i of Γ consists in the covering of a binary relation with functions (i.e., the operators γ_i), which may be done in many different ways, all equivalent to eachother. □

We may therefore state these two final results.

Theorem 3.4. An AV problem schema $\bar{M}$ is equivalent with a problem schema M.

Proof. From Theorem 3.3 we obtain directly that $\bar{S}$ is equivalent to S, and C is equivalent to Γ.

Therefore, from Definitions 2.1 and 3.8, we obtain that $\bar{M}$ is equivalent to M. □

Theorem 3.5. An AV problem $\bar{P}$ is equivalent with a problem P.

Proof. From Theorem 3.4 we derive that $\bar{M}$ is equivalent to M. Moreover, from Theorem 3.3., since $\bar{S}$ is equivalent to S, we have also that $\bar{i}$ is equivalent to i, and $\bar{f}$ is equivalent to f (or $\bar{K}$ is equivalent to K).

Therefore, from Definitions 2.3 and 3.9., we conclude that $\bar{P}$ is equivalent to P. □

IV. THE COMPUTATION OF THE EVALUATION FUNCTION

We will briefly outline a method, based on the semantic description of SSPS, which provides the computation of the evaluation function $\hat{g}(\bar{s})$.

This technique is based on the idea of computing the estimate $\hat{h}(\bar{s})$, by solving an <u>auxiliary problem</u> in which $\bar{s}$ is the initial state, and the solution is easy to be found, and provides a lower bound of the cost $h(\bar{s})$ of the optimum solution for the main problem.

Since all the auxiliary problems share the same goal problem, we shall introduce the notion of auxiliary goal problem.

<u>Definition 4.1</u>. Let $\bar{F} = (\bar{S}, C, \bar{f})$ (or $\bar{F} = (\bar{S},C,\bar{K})$) be an AV goal problem, obtained from an AV problem $\bar{P}'$, $\bar{F}'=(\bar{S}', C', \bar{f}')$ (or $\bar{F}'=(\bar{S}', C', \bar{K}')$) is called an <u>auxiliary AV goal problem</u> for $\bar{F}$ iff:

(i) $\bar{S} = \bar{S}'$ (4.1)

(ii) $\bar{f} = \bar{f}'$ (or $\bar{K} = \bar{K}'$) (4.2)

(iii) $L' \subseteq L$ (4.3)

where L' and L are the LC sets from which C' and C are obtained.

We shall indicate :

$$\bar{F}' \sqsubseteq \bar{F} \tag{4.4}$$

□

In other words, we say that $\bar{F}' \sqsubseteq \bar{F}$, if they have the same SS set, the same initial and final SS's, and if the LC set L' is a subset of the LC set L.

<u>Theorem 4.1</u>. If $\bar{F}' \sqsubseteq \bar{F}$ then :

$$C' \supseteq C \tag{4.5}$$

<u>Proof</u>. The proof is obtained directly from Definition 3.7 and from (4.3) because of the property of set intersection. □

Now we have the main result which provides the computation of the estimate.

<u>Theorem 4.2</u>. If $\bar{F}' \sqsubseteq \bar{F}$ then for all $\bar{s} \in \bar{S}$, the optimum cost of the solution to the AV problem $\bar{P}'=(\bar{F}',\bar{s})=(\bar{M}',C',\bar{s},\bar{f}')$ (or $\bar{P}'=(\bar{M}',C',\bar{s},\bar{K}')$), is a lower bound $\hat{h}(\bar{s})$ with respect to the AV problem $\bar{P}=(\bar{M},\bar{s},\bar{f})$ (or $\bar{P}=(\bar{M},\bar{s},\bar{K})$).

<u>Proof</u>. Because of Theorem 4.1, $C' \supseteq C$; therefore the graph $G=(\bar{S},A)$ and the graph $G'=(\bar{S}',A')$ respectively associated to the AV problem schemata $\bar{M}$ and $\bar{M}'$ are such that, the set of arcs A is a subset of the set of arcs A' (see Definition 3.7 and Theorem 3.3), i.e. :

$$A' \supseteq A \tag{4.6}$$

Then, if $\bar{\mathcal{G}}^n$ is an optimum solution for the problem $\bar{P}$, (i.e., a minimum path in G from $\bar{s}$ to $\bar{f}$ (or $\bar{f} \in \bar{K}$)), $\bar{\mathcal{G}}^n$ is also a solution $\mathcal{G}'^n$ for the problem $\bar{P}'$ (i.e., also a path in G', but not necessary a minimum one).

Since $h(\bar{s})$ is, by definition, the optimum cost of a solution for $\bar{P}$, then the optimum cost of a solution for $\bar{P}'$, which is not necessarily a solution for $\bar{P}$, is a lower bound for $h(\bar{s})$, i.e., it is the estimate $\hat{h}(\bar{s})$.

□

Theorem 4.2 constitutes the basis for a new algorithm, in place of the Hart, Nillson, and Raphael algorithm, for the heuristically guided search.

The basic steps of the algorithm are the following ones:

1. Given an AV problem $\bar{P}$, and its associated AV goal problem $\bar{F}$, an auxiliary AV goal problem $\bar{F}'$ is automatically constructed by eliminating some elements of the LC set L.

2. The computation of the estimate $\hat{h}(\bar{s})$ for the problem $\bar{P}$ is performed by solving the problem $\bar{P}'=(\bar{F}',\bar{s})$.

An evaluation of the complexity of this method, compared with the well known search strategies, will constitute the goal of a new research effort.

V. CONCLUSIONS

In this paper we have proposed a new description of SSPS, called syntactic description, which constitutes a framework useful to structure a rich content of informations about a given problem.

The semantic description has been shown to be equivalent to the syntactic description of SSPS.

A method has been briefly outlined, which is based on the semantic description, and which makes possible to extract, in an automatic way, i.e., by computation, the heuristic information useful to guide the search in the state space.

The future research work shall be addressed to the exploitation of the semantic description as a powerful basis on which to formulate procedures apted to solve the main goal of automatic problem-solving.

In particular the direction of heuristic guided search and learning will be investigated.

Acknowledgments

The authors express their gratitude to Dr.D.Mandrioli, for many stimulating discussions, and to all the researchers of the Milan Polytechnic Artificial Intelligence Project.

REFERENCES

Feigenbaum, E., and Feldman, J.(eds). Computers and Thought. Mc Graw-Hill Book Company, New York, 1963.

Minsky, M. (ed). Semantic Information Processing. The M.I.T. Press, Cambridge, Massachusetts, 1968.

Nilsson, N. Problem-Solving Methods in Artificial Intelligence. Mc Graw-Hill Book Company, New York, 1971.

Slagle, J. Artificial Intelligence: the Heuristic Programming Approach. Mc Graw-Hill Book Company, New York, 1971.

Amarel, S. On Representations of Problems on Reasoning about Actions. Michie,D. (ed.) Machine Intelligence 3, pp. 131-171. American Elsevier Publishing Company, Inc., New York, 1968.

Hart, P., Nilsson, N., and Raphael, B. A Formal Basis for the Heuristic Determination of Minimum Cost Paths. IEEE Trans. Sys.Sci. Cybernetics, vol. SSC-4, no.2, pp. 100-107, July 1968.

Mandrioli, D., Sangiovanni Vincentelli, A., and Somalvico, M. Towards a Theory of Problem-Solving. Marzollo, A. (ed). Topics on Artificial Intelligence. Springer-Verlag New York, Inc., New York, 1973.

Sangiovanni Vincentelli, A., and Somalvico, M. Theoretical Formalisation of the State-Space Method for Automatic Problem Solving. (In Italian). MP-AI Project. MEMO MP-AIM-6, October 1972.

Sangiovanni Vincentelli, A., and Somalvico, M. Theoretical Aspects of State-Space Approach to Problem-Solving. Proc. VII International Congress on Cybernetics, Namur, September 1973 (a).

Sangiovanni Vincentelli, A., and Somalvico M. Problem-Solving Methods in Computer-Aided Medical Diagnosis. Proc. XX International Scientific Conference on Electronics, Rome, March 1973 (b).

PERTURBATION THEORY AND THE STATEMENT OF INVERSE PROBLEMS

G.I.MARCHUK

Computer Center, Novosibirsk 630090, U.S.S.R.

Some aspects of the theory of inverse problems for linear and quasilinear equations of mathematical physics are investigated. The theory of perturbations is developed with respect to selected functionals of problems.

1. Conjugate Functions and the Notion of Importance

Let us consider the function $\varphi(x)$ which satisfies the equation

$$L\varphi(x) = q(x), \tag{1.1}$$

where L is some linear operator and $q(x)$ is the distribution of sources in the medium. Here x is the totality of all the variables of the problem (time and space coordinates, energy, velocity). Let us assume that the operator L and functions φ are real and that $\varphi \in \Phi$.

For the sake of definiteness we assume, for example, that the process under investigation is connected with diffusion or transfer of a substance though the conclusions of the theory are beyond the scope of this discussion.

Let us introduce Hilbert space functions with the scalar product

$$(g, h) = \int g(x)\, h(x)\, dx, \tag{1.2}$$

where integration is carried out over the whole domain D of the functions g and h.

The usual purpose of solving some physical problems is to obtain some quantity which is the functional of $\varphi(x)$. Any value linearly related to $\varphi(x)$ can be represented as such a scalar product. For example, if we are interested in the result of the measurement of some process in the medium with the charactecteristics of the device $\Sigma(x)$ the value is

$$J_{\Sigma} = \int \varphi(x)\, \Sigma(x)\, dx = (\varphi, \Sigma). \tag{1.3}$$

Thus we consider the physical quantities which can be represented as a linear functional of $\varphi(x)$

$$J_p[\varphi] = (\varphi, p),$$

where the quantity p is a characteristic of the physical process under consideration. Let us introduce along with the operator L its conjugate operator L^*, defined by the Lagrangian identity

(1.4) $$(g, Lh) = (h, L^* g)$$

for any functions g and h. Together with (1.1), which will be called the basic equation, we introduce, first formally, a conjugate inhomogeneous equation

(1.5) $$L^* \varphi^*_p = p(x),$$

where $p(x)$ is some arbitrary function and $\varphi^* \in \Phi^*$. Substituting solutions (1.1) and (1.5), φ and φ^*_p into (1.4) in place of the functions g and h, we obtain

(1.6) $$(\varphi^*_p, L\varphi) = (\varphi, L^* \varphi^*_p)$$

or, using equations (1.1) and (1.5),

(1.7) $$(\varphi^*_p, q) = (\varphi, p),$$

i.e.

$$J_q[\varphi^*_p] = J_p[\varphi].$$

Therefore, if we want to find the value of the functional $J_p[\varphi]$ we can get it in two ways : either by solving equation (1.1) and determining this value according to the formula

(1.8) $$J_p[\varphi] = (\varphi, p),$$

or by solving equation (1.5) and determining the same value according to the formula

(1.9) $$J_p[\varphi] = J_q[\varphi^*_p] = (\varphi^*_p, q).$$

Consequently, the function $\varphi^*_p(\varphi)$, satisfying equation (1.5) can be put to correspond with each linear functional $J_p[\varphi] = (\varphi, p)$ where the function $p(x)$, which characterizes the measuring device, should be used as the free term of this equation.

2. Perturbation Theory for Linear Functionals

If the properties of the medium with which the field interacts change i.e. if the operator of equation (1.1) becomes

$$L' = L + \delta L,$$

then both the field $\varphi(x)$ and the value of $J_p[\varphi]$ change :

$$\varphi(x) \to \varphi'(x), \quad J_p[\varphi] \to J'_p = J_p + \delta J_p .$$

Let us find the relation between the variation of the operator δL and that of the functional δJ_p . The perturbed system is described by the equation

$$L'\varphi' = (L + \delta L)\varphi' = q'. \tag{2.1}$$

The conjugate function of the unperturbed system corresponding to the functional J_p is described by the equation

$$L^* \varphi_p^* = p. \tag{2.2}$$

Multiplying scalarly equation (2.1) by φ^* , equation (2.2) by φ' and subtracting one from the other, using the determination of the conjugate operator of (1.4) we obtain, on the left,

$$(\varphi_p^* L'\varphi') - (\varphi', L^*\varphi_p^*) = (\varphi_p^*, \delta I \varphi') - (\varphi_p^*, \delta q), \tag{2.3}$$

and, on the right, in accordance with equation (1.7), we have

$$(\varphi_p^*, q) - (\varphi', p) = J_p[\varphi] - J_p[\varphi'] = -\delta J_p . \tag{2.4}$$

Equating (2.3) and (2.4) we obtain a general expression for the increment of the functional

$$\delta J_p = -(\varphi_p^*, \delta L \varphi') + (\varphi_p^*, \delta q). \tag{2.5}$$

Let us note an important peculiarity of the application of the formulas of the theory of perturbations, viz. since the formulas of the theory of perturbations are written with respect to the variation of the functional, and the allowable error is usually given in terms of some percentage, then it follows that it is not necessary to know the exact solution of the main and conjugate problems for the calculation of the variations mentioned above. It is sufficient to use their approximate solutions.

If the perturbation of the operator L (and, consequently, L^*) is small so that it does not distort the functions φ and φ_p^* much , then in (2.5), (2.6) we can substitute approximately $\varphi' \approx \varphi$, $\varphi^{*\prime} \approx \varphi^*$. In this case we obtain a formula for the theory of small perturbations* :

---

* Henceforth for simplicity formulas (2.5),(2.6) for the increment of functionals will be referred to as formulas of perturbation theory.

(2.6) $$\delta J_p = -(\varphi_p^*, \delta I\varphi) + (\varphi_p^*, \delta q).$$

Besides its direct application to evaluation of different effects and to the analysis of measurements, the formula, which we have obtained, can have another important application.

In theoretical treatments and practical calculations the method of substitution of a simplified system for a complex one is often used. Such a substitution should not bring about any changes in some main characteristics of the system which is evidently a necessary condition. As an example of such an approach to the differential equations one can cite the substitution of variable coefficients by constant ones. Among these methods is the method of effective boundary conditions which consists in a substitution of some simplified conditions for real ones. The conditions should be chosen so as to lead to the correct value of some selected functionals.

The formulas of perturbation theory obtained above allow us to formulate a rather general approach to certain types of problems. Let the system under study be characterized by the operator L , the functional $J_p[\varphi]$ being the most essential value in the problem . If the simple model sought for is characterized by the operator $L' = L + \delta L$, it is necessary that

(2.7) $$\delta J_p = -(\varphi_p^*, [L' - L]\varphi') = 0, \quad i.e. \quad (\varphi_p^*, L'\varphi') = (\varphi_p^*, L\varphi')$$

in order that J_p should not change in passing from the real system to the model.

If we are interested in some values J_{p_1} , J_{p_2} , etc., then we obtain several conditions of the type shown in equation (2.7) with solutions $\varphi_{p_1}^*$, $\varphi_{p_2}^*$ and so on.

Condition (2.7) does not identically determine the equivalent model sought for but is a necessary condition and, together with some other factors, can be helpful in finding the model.

3. Numerical Methods for Inverse Problems and the Planning of the Experiment

Suppose now that we have the set of functionals (measurements) δJ_i $(i = 1, 2, \dots, n)$ and denote it by ∂J_i . It is assumed that measurements may be of various types in character ; for example, measurements are made with one and the same device at different "points" of the domain of the solution definition, or with devices which re - solve different characteristics of the phenomenon under study. For the sake of simplicity we suppose that statistical errors in the measurements are removed and we deal already with a system of data preliminarily processed.

We set each functional J_p to correspond with the importance function for the unperturbed problem, i.e. the model in which the operator L and the domain of its definition are assumed known. Let us solve n different problems

(3.1) $$L^* \varphi_i^* = p_i, \qquad (i = 1, 2, \ldots, n),$$

find n functions φ_i^* and solve one main problem with the model (unperturbed) operator L with its conjugate L^*

(3.2) $$L \varphi = f.$$

Let $\varphi \in \Phi$ and $\varphi^* \in \Phi^*$ where Φ and Φ^* are the domains of the definition of the operators L and L^* correspondingly.

Then we construct n formulas for the small perturbations

(3.3) $$(\varphi_i^*, \delta L \varphi) = -\delta J_i, \qquad (i = 1, 2, \ldots, n),$$

where δL is the difference between the operator L' under study and the model one L. The structure of L is assumed to be known and have the form

(3.4) $$L = \sum_{k=1}^{m} \alpha_k A_k,$$

where the A_k are elimentary linear operators of differentiation or integration, for example, or a combination of both; and the $a_k(x)$ are coefficients to be sought, which are usually approximately known for the unperturbed (model) problem.

Now our task is to recover the coefficients α_k' and β_k' in the expression

(3.5) $$L' = \sum_{k=1}^{m} \alpha_k' A_k.$$

With (3.4) and (3.5) we have

(3.6) $$\delta L' = \sum_{k=1}^{m} \delta\alpha_k A_k ,$$

where

$$\delta\alpha_k = \alpha_k' - \alpha_k .$$

Substituting (3.6) to (3.3) under corresponding conditions we obtain

(3.7) $$\sum_{k=1}^{m} (\varphi_i^*, \delta\alpha_k A_k \varphi) = -\delta J_i, \quad (i = 1,2,\dots,n).$$

Our task now is the parameterization of variations $\delta\alpha_k$. First we consider the simplest case, where the α_k are constant. Under these conditions the system of equations (3.7) reduces to a problem of linear algebra

(3.8) $$\sum_{k=1}^{m} \delta\alpha_k (\varphi_i^*, A_k \varphi) = -\delta J_j, \quad (i = 1,2,\dots,n).$$

Here $(\varphi_i^*, A_k\varphi)$ are the matrix elements which can be calculated if φ, φ_i^* and A_k are known. Let $\underline{y}$ be the vector with the components $\delta\alpha_k$, $\underline{F}$ - the vector with the components δJ_i and $a_{ik} = (\varphi_i^*, A_k\varphi)$ - elements of the matrix Λ. Then we have

(3.9) $$\Lambda \underline{y} = \underline{F} .$$

If the number of functionals n is equal to the number of definable variations of coefficients α_k then system (3.9) allows us to find $\delta\alpha_k$. If $n > m$, then (3.9) is overdetermined and its "solution" is generally found by the method of least squares on the basis of the assumption that $\underline{y}$ minimizes the quadratic functional

(3.10) $$\| \Lambda \underline{y} - \underline{F} \|^2 = min .$$

The vector $\underline{y}$ minimizing this functional is sometimes called a quasi-solution to equation (3.9).

If $\delta\alpha_k(x)$ are functions then a solution to the inverse problem can be found by some parametrization methods described below. It is assumed that on the basis of a priori - statistical and correlation - analysis of the behaviour of the physical parameters, some complete orthogonal system of functions $u_{k\ell}(x)$ is found , such that with its help a sufficiently good approximation to the functions $\alpha_k(x)$ is possible for a small value, so that

$$\delta\alpha_k(x) = \sum_{\ell=1}^{m(k)} a_{k,\ell}\, u_{k,\ell}(x), \tag{3.11}$$

where the $a_{k\ell}$ are the coefficients to be defined.
Substituting expressions (3.11) to (3.7) we have

$$\sum_{k=1}^{m} \sum_{\ell=1}^{n(k)} a_{k,\ell}(\varphi_i^*, u_{k,\ell} A_k \varphi) = -\delta J_i, \tag{3.12}$$

$$(i = 1, 2, \dots, n).$$

Now we reorder the values $a_{k\ell}$, $b_{k\ell}$, denote them by y_ℓ, and define the matrix Λ such that the equation

$$\Lambda \underline{y} = \underline{F}$$

is equivalent to (3.12).
Then we again are led to the problem of linear algebra (3.12) which can be solved to obtain $a_{k\ell}$ and hence $\delta\alpha_k$.
We have considered only the case when a solution of the model problem is close to the real one, i.e. where we can substitute φ for φ' thus making use of the theory of small perturbations. If the unperturbed (model) state of the process substantially differs from the actual one, then the above algorithm can be regarded as the first approximation to the inverse problem. After variations $\delta\alpha_k$ are found we can correct the coefficients α_k and find

$$\alpha'_{k_1} = \alpha_k + \delta\alpha_k .$$

After that we must solve the "perturbed" problem

$$L'\varphi' = f \tag{3.13}$$

with the operator

$$L' = \sum_{k=1}^{m} \alpha'_k A_k$$

and consider a new approximation in the solution of the inverse problem instead of (3.7) making use of a more general formula of perturbations

$$\sum_{k=1}^{m} (\varphi_i^*, \delta\alpha_k \mathcal{A}_k \varphi') = -\delta J_i, \quad (i = 1, 2, \ldots, n) \tag{3.14}$$

and then repeat the cycle of calculations for specifying the variations $\delta\alpha_k$. We will call it the second approximation to the solution of the inverse problem. It is clear that the above procedure can be continued. The convergence of the methods of successive approximations can be proved when we have concrete information on the elementary operators of the problem $\mathcal{A}_k$ and the domain of the definition of the operators.

References

Marchuk, G.I., and Orlov, V.V. Sb."Neitronnaya Fizika",M., Gosatomizdat, 1961 (Russian).

Marchuk, G.I. Atomizdat , M., 1961 (Russian).

Marchuk, G.I. DAN SSSR, t.156, N°3, 1964 (Russian).

Marchuk, G.I. Zh."Kosmitcheskie issledovania", t.2, vyp.3, 1964 (Russian).

Marchuk, G.I. "Nauka", Novosibirsk, 1973 (Russian).

A MODEL FOR THE EVALUATION OF ALTERNATIVE POLICIES FOR ATMOSPHERIC POLLUTANT SOURCE EMISSIONS (MASC MODEL)

R.Aguilar, L.F.Escudero, J.F.G.de Cevallos, P.G.de Cos, F.Gómez-Pallete, G.Martínez Sánchez Madrid Scientific Center, IBM Spain

An ambitious long-term plan of action is being carried out to solve the atmospheric pollution problem facing the city of Bilbao (North of Spain) (1). In this work it has been necessary to develop certain specific lines of research. The approach and methodology of one of these lines constitute the subject matter of the present paper

1. OBJECTIVES

When faced with an atmospheric pollution problem, the following policy measures may be adopted:

- Analysis of the problem as it stands and outline of corrective alternatives to the pollution-causing activities in the event that they be necessary ones.
- On the basis of studies of the zone's economic, demographic, social, etc., development, projection of data concerning the processes involved in atmospheric pollution, and outline of alternatives previsional of future potentially-contaminating activities to obviate their negative impact on the environment.
- Outline of a permanent system for control of present and future pollution-causing activities alike in order to know at any given moment, both qualitatively and quantitatively, the current state of the situation, and thus be able to act accordingly on the basis of this knowledge.

The model, the approach and methodology of which is described further on, can be an effective tool in formulating different types of alternatives.

In establishing corrective alternatives for abatement of a pollution problem, use of the model provides necessary emission reductions in the areas responsible.

Determining to what extent each individual emission source making up an area subject to a single reduction level should correct its own emissions, is a phase of the problem which the model does not encompass. Governments can regard the information provided by the model as a reliable support on which to base their eventual decision making.

Today it is absolutely essential to take environmental protection criteria into account when drawing up development policies. Given the various aspects of the problem which the model treats, it can serve as an excellent working aid to planners

2. TERMINOLOGY AND CONCEPTS EMPLOYED

The emissions produced as a result of the different types of human activity are by their very nature dispersed into the atmosphere in a manner governed principally by meteorological factors. The resulting composition of the air or air quality, when deterioration exceeds certain limits, is harmful to the elements of the local ecosystem which interact with the atmosphere (human health, etc...). The model's use of data regarding these three concepts, requires that we define the space-time dimensions in which the data must be expressed.

2.1. EMISSIONS

Emissions are gauged by means of emission factors developed by the U.S.Environmental Protection Agency. These emission factors relate the quantity of the pollutant emitted to some index, the nature and units of measure of which depend on the type of activity involved, quantity of fuel burned, kilometers covered by the vehicle, etc. (5, 6, 10)

2.2. METEOROLOGY

Through a statistical analysis of the historico-meteorological data available we can define the meteorology-type present in each of the basic time periods chosen (for example: a month).

Given that the parameters utilized in the model are: wind direction, wind velocity and stability, the meteorology of each basic time period becomes structured in the form of a tridimensional probability matrix in which, therefore, each element represents the probability of a given meteorological situation (direction-velocity-stability) in a given period of time.

2.3. AIR QUALITY

The air quality criteria adopted by different countries are formulated using the parameters below: a). An absolute parameter: Maximum permissible concentration of the pollutant in the atmosphere. b). A relative parameter: Number of times within a given time period that said concentration may be exceeded.

With respect to the relative limit, the model deals with it using probability formulae: the maximum admissible probability that the concentration in a certain area exceed the absolute limit. In the references cited (9, 15, 22) may be found other normally employed criteria.

(In the tests conducted applying the model to the case of SO_2, 365μg SO_2/m^3 (2, 21) has been taken as the absolute limiting value, with a maximum admissible probability of 20 per cent).

2.4. SPACE-TIME DIMENSIONS

2.4.1. Space

Following the classic method of diffusion models, the geographic area under study is divided up using a grid pattern, with grid distances depending fundamentally on the basis of the geographic density of the polluting activities (8,18) (Tests have been carried out using two types of grids for: 1 km and 0.5 km to a side (2)).

A grid can be considered in two ways. When we speak of an emitter grid we are referring generically to the emissions produced by the polluting activities situated within it. Conversely, we regard a grid as a receptor when we are considering the polluting agents dispersed in its atmosphere.

Considering pollutant concentration in its various aspects, the following definitions may be established:

A polluted grid is understood to be one in which the probability of the concentration exceeding the absolute limit is greater than a certain value considered the maximum admissible one (relative limit).

Each polluted grid is influenced by the combined group of emitter-grids defined by the diffusion model. This group is designated the influence area.

On the basis of both concepts, the problem area is defined as the combined group of influence areas having at least one grid in common. Within the problema area itself the polluted area is taken to be the combined group of polluted grids.

2.4.2. Time

The discretization of the scale of times in the treatment of atmospheric pollution problems should be realized based on: 1) A thorough knowledge of the two fundamental factors which enter into play: meteorological matrix and emissions, and 2) The condition that these factors remain constant.

In the case of emissions, variation over time is closely linked to the type of emission source. There are polluting activities such as home heating systems having emissions fundamentally dependent upon temperature. Variation over time of this type of emission is then related to the meteorology.

Conversely, emissions produced by industrial activities may be considered in large degree independent from the meteorology.

Looking at both types of emissions together, the period t can be said to be established on the basis of meteorological criteria.

As far as establishing reduction alternatives is concerned, it must be remembered that polluting activities can be grouped into two general categories: 1) Ones whose operation is of an elastic nature, subject to variation in different degrees. 2) Those operating in such a way as to permit no variation.

Activities belonging to the first group are subject to a differential reduction policy extending over the year, in equalized periods nt, for which the reduction level remains constant. The value of n ($n \geq 1$) will decrease the more elastic the operation is.

Those operations which allow no variation should reduce a particular activity evenly throughout the period studied, in accordance with the highest reduction level.

Given that the proposed model is linear, both types of activities may be considered separately.

3. METHODOLOGY

In this section is found described the model which, using the emissions and meteorological matrix corresponding to a basic time period t for the study area, establishes emissions reduction alternatives on a grid level for the purpose of maintaining air quality within established standards, based on certain determined criteria. Description of this methodology has been divided into three sections:

3.1. DELIMITING THE PROBLEM

First, the probability distribution of the concentration of the polluting agent is estimated for each grid considered as receptor, based on: 1) the emissions of the influencing emitter grids and 2) the meteorological matrix.

Considering the meteorological matrix element by element, a set of pollutant concentrations for each grid is worked out. The probability of a particular concentration taking place is assumed to be equal to the probability of the corresponding meteorological condition. Concentration and probabilities values together make up the so-called probability distribution of the concentration.

Mean concentration, variance, maximum and minimum values, probability of the concentration exceeding the absolute limit being considered, etc... may be calculated for each probability distribution. Storing this information may be expedient in order to establish further reduction alternatives based on some other possible criteria.

In short, it is a question of determining the problem areas and their corresponding polluted areas, on the basis of results reached beforehand, relative to the probability of the concentration exceeding the absolute limit considered.

On the other hand, obtaining maps of isopleths of the different parameters, constitutes a supplementary aid toward a visualization of the dimensions of the problem. Nonetheless, we feel that, of the different possible maps, the one representing the lines of isofrequency with which the absolute concentration limit is exceeded offers a more reliable description of the situation. This is due to the fact that it is free of the defect of masking the hazard of large concentrations with situations of low concentration, as occurs in the representation of other parameters.

3.2. ESTIMATING EMISSIONS REDUCTION

The objective of this second stage consists of eliminating in each problem area its corresponding contaminated area. That is: to provide alternatives for emissions reduction so that in no grid are the established air quality standards exceeded.

These reduction alternatives may be estimated on the basis of the following criteria:

a) that the emission reduction to be imposed on each emitter grid be, on the one hand, the minimum one possible, and, on the other, proportional to its incidence on the polluted area.

b) that the probability of the concentration exceeding the absolute limit in each grid of the polluted area be minimum.

Both criteria necessitate a maximum emission reduction limit in each grid which is defined by the current emission and by socio-economic considerations.

These criteria may be applied alternatively. Should the second be adopted, previous establishment of a lower limit is necessary.

Both criteria may also be considered simultaneously thus creating through the "multi-criteria" technique contained in (3,4,19), a new criterion, optimization of which is obtained with that solution which, equilibrating the original objectives, is considered best, since there is no other which offers a greater optimization in each and every one of the proposed criteria.

In order to formulate reduction alternatives, it may be useful to consider the two following aspects: a) Type of period on the basis of which reduction alternatives are established, and b) Scale of reduction values.

In the present paper two kinds of formulation have been established: 1) The First formulation considers a scale of reduction values divided into discrete levels and a homogenized time period T for which the reduction level in each grid remains constant 2) The second formulation considers a basic time period and a continuous scale of reduction values.

3.3. SIMULATION OF THE CONTROLLED SITUATION

Finally, on the basis of the recommended reductions, the new emissions are calculated and the distribution of the pollutant concentration in each grid of the study area is simulated, from which we obtain a quantitative idea of the degree of correction which would be achieved.

4. PROCEDURES

In this section are found schematized the procedures for: 1) Simulation of concentrations and later definition of problem areas (Diffusion Model). 2) Evaluation of reduction alternatives.

4.1. SIMULATION OF POLLUTANT CONCENTRATION

4.1.1. Data

For simulation of the pollutant concentration the following data are employed for each basic time period t used: a) probability matrix of the meteorological situation b) elemental grids into which the study zone is divided. c) total emissions in each grid.

4.1.2. Diffusion model

Emissions for each grid (QA) are made up by 1) emissions produced by area sources (QA'), and 2) those due to point sources (QA'').

To determine the pollutant concentration in an urban zone, the following formula (8,11,17,18) presently subject to validation and corresponding adaptation have been selected:

- Area Sources:

$$C'_{imt}=\sqrt{\frac{2}{\pi}}*\frac{1}{U_m}*\frac{(\Delta X/2)^{(1-b_m)}}{(\Delta X)^2 * a_m(1-b_m)}*(QA'_{it}+\sum_{r=2}^{R}QA'_{rt}*((2r-1)^{(1-b_m)}-(2r-3)^{(1-b_m)})) \quad (1)$$

- Point Sources:

$$C''_{imt}=\sum_{r=1}^{R}\frac{\exp\left(-\frac{1}{2}\left(\frac{Y^2_{ri}}{\sigma Y^2_{rim}}+\frac{H^2_r}{\sigma Z^2_{rim}}\right)\right)}{\pi * \sigma Y_{rim} * \sigma Z_{rim} * U_m} * QA''_{rt} \quad (2)$$

being the total concentration:

$$C_{imt}=C'_{imt}+C''_{imt}=\sum_{r=1}^{R}K'_{irm}QA'_{rt}+\sum_{r=1}^{R}K''_{irm}QA''_{rt} \quad (3)$$

where C_{imt} represents the concentration ($\mu g/m^3$) of the pollutant which is present in the atmosphere in the grid i, in the basic period t for the atmospheric situation m. ΔX represents the side of the grid; a and b are the parameters which depend upon atmospheric stability (16,20); U is the wind velocity. Finally, σZ and σY are the standard deviations which depend on the distance between emitter and receptor grids as well as on the atmospheric stability; H is the point source height.

As reduction alternatives are considered as a whole for each grid, it is necessary to modify formulae (3) in the following manner:

$$C_{imt} = \sum_{r=1}^{R} K_{irm} \, QA_{rt} \tag{4}$$

where:

$$K_{irm} = K'_{irm} * \frac{QA'_{rt}}{QA_{rt}} + \frac{QA''_{rt}}{QA_{rt}} * K''_{irm} \tag{5}$$

Is is assumed that both QA'_{rt}/QA_{rt} and QA''_{rt}/QA_{rt} remain constant once an emission reduction plan is carried out.

4.1.3. Difussion model results

The above mentioned statistical parameters of each grid are calculated. On the basis of probability values, problema areas are defined. Finally the contribution W_{rti} (or the emitter grid r, to the concentration in the receptor grid i for the emission QA_{rt}) is:

$$W_{rti} = \sum_{m} K_{irm} * QA_{rt} * PME_{mt} \tag{6}$$

where K_{imr} is the contribution of the grid r to the unitary emission in the concentration in grid i for the meteorological situation m which causes the concentration C_{imt} to exceed the absolute limit LC; and PME_{mt} is the probability of the meteorological situation m in the basic period t.

4.1.4. Computer support

(a) The program designated APS supports the model described above. It is written in FORTRAN, operative in DOS, takes 60 K, and requires 2 m of CPU for a case of 504 grids and a probability matrix of 720 meteorological situations (3 types of stability * 16 wind directions * 15 types of wind velocity).

(b) The program STAMPEDE find the iso-lines of the parameter being considered. It is written in FORTRAN, operative in DOS and requires 64 K.

4.2. EVALUATION OF REDUCTION ALTERNATIVES

The objective of this phase consists of evaluating reduction alternatives to succeed in eliminating the polluted area of each one of the problem areas which have been defined by means of the formulation contained in the previous section.

4.2.1. Data and variables

The data and variables required to use those mixed integer linear programming models (7) are the following:

Subindexes

$t\epsilon T$: Basic study period. $r\epsilon R$: Emitter grid influencing the problem area. $i\epsilon I$: Receptor grid constituent of the polluted area where the concentration is to be reduced. $m\epsilon M$: Meteorological situation which in grid $\underline{i}$ for $t\epsilon T$ and given a set of emissions, causes the concentration C_{imt} to exceed absolute limit LC. $1\epsilon L$: Types of abatement levels.

Emission Inventory

QA_{rt}: Current emission (µg/seg) of the pollutant for the grid $\underline{r}$ in the basic period $\underline{t}$.

Abatement levels

N_1: Abatement level to be considered. In this study the following levels have been adopted: $N_1 = 0.05$, $N_2 = 0.10$, and $N_3 = 0.15$.

Pollutant concentration

CA_{imt}: The concentration (µg/m^3) of the pollutant which is present in the grid $\underline{i}$ in the meteorological situation $\underline{m}$ during the basic time period $\underline{t}$ for a current set of emissions QA_{rt}. LC: Absolute concentration limit.

Probability values

PME_{mt}: Probability of the meteorological situation. PM: Maximum admissible probability (relative limit) of the concentration (C_{it}) exceeding the absolute limit in a grid.

Weighted incidence

W_{rt}: Contribution of the grid $\underline{r}$ in the basic period $\underline{t}$ to the pollutant concentration in the sum of grids which make up the polluted areas, for the total number of meteorological situations which cause the concentration C_{imt} to exceed the absolute limit.

Emission variables

XQS_{rt}: New emission of pollutant for the grid $\underline{r}$ in period $\underline{t}$. XQR_{rt}: Pollutant emission to be reduced in the grid $\underline{r}$ in period $\underline{t}$. XPR_r: Percentage reduction of the pollutant over the present emission QA_{rt} in the grid $\underline{r}$ during the entire period studied T.

Concentration variables

XC_{imt}: New pollutant concentration in grid $\underline{i}$ in meteorological situation $\underline{m}$ in basic period $\underline{t}$, for the new set of emissions XQS_{rt}. XCB_{imt}: New concentration such that $XC_{imt} = XCB_{imt}$; or $XCB_{imt} = LC$, if $XC_{imt} \geq LC$. XCL_{imt}: New concentration in excess of LC, such that $XCL_{imt} = 0$ if $XC_{imt} \leq LC$; or $XCL_{imt} = (XC_{imt} - LC)$ if $XC_{imt} \geq LC$.

Decision variables

δ_{r1} may only equal 0 or 1. If equal to 1, then the emission in grid $\underline{r}$ is reduced in level $\underline{1}$. δ_{imt} may only equal 0 or 1. If equal to 1, then in period $\underline{t}$ for meteorological situation $\underline{m}$, concentration XC_{imt} exceeds the absolute limit LC.

4.2.2. Emission Reductions model

Two alternative types of formulation are possible as described in 3.2.:

A). Generalized formulation. The criteria to be considered simultaneously in compensated form (see section 4.2.3.) are the following:

(a) To minimize in weighted form (W_r) the total abatement or emission XQR_{rt} in each period t

$$\text{Min. } R = \sum_{t=1}^{T} \sum_{r=1}^{R} \frac{1}{W_{rt}} * XQR_{rt} \qquad (7)$$

(b) To minimize the sum of probabilities of the concentration exceeding absolute limit LC

$$\text{Min. } P = \sum_{t=1}^{T} \sum_{i=1}^{I} \sum_{m=1}^{M} PME_{mt} * \delta_{imt} \qquad (8)$$

Subject to the following conditions:

1). Condition of balance of emissions:

$$1 - \frac{1}{QA_{rt}} * XQR_{rt} = \frac{1}{QA_{rt}} * XQS_{rt} \quad \text{for } r\varepsilon R,\ t\varepsilon T \qquad (9)$$

2). Determination for each grid of the percentage emission reduction

$$XPR_r = \sum_{l=1}^{L} N_l * \delta_{rl} \quad \text{for } r\varepsilon R \qquad (10)$$

3). Requirement that each grid have at most one reduction level.

$$1 \geq \sum_{l=1}^{L} \delta_{rl} \quad \text{for } r\varepsilon R \qquad (11)$$

4). Condition of equal relative emission reduction for each grid r in all basic time periods.

$$XPR_r = \frac{1}{QA_{rt}} * XQR_{rt} \quad \text{for } r\varepsilon R \text{ and } t\varepsilon T \qquad (12)$$

5). Estimate for each grid i and basic period t of the concentration XC_{imt} in each meteorological situation m considered in this model.

$$XC_{imt} = \sum_{r=1}^{R} K_{irm} * XQS_{rt} \quad \text{for } i\varepsilon I,\ m\varepsilon M,\ t\varepsilon T \qquad (13)$$

Equation (13) is equivalent to the equations (4) and (5)

6). Condition that the variable δ_{imt} must adopt the value 1 if the concentration XC_{imt} exceeds the absolute limit LC.

$$XC_{imt} = XCB_{imt} + XCL_{imt} \quad (14); \qquad LC * \delta_{imt} \leq XCB_{imt} \leq LC \quad (15);$$

$$XCL_{imt} \leq (CA_{imt} - LC) * \delta_{imt} \quad (16); \qquad \text{for } i\varepsilon I,\ m\varepsilon M,\ t\varepsilon T.$$

7). Condition that the new set of emissions XQS_{rt} for $r\varepsilon R$, for each grid which makes up the polluted area, not cause a concentration such that its probability of exceeding the absolute limit LC, be greater than the permitted maximum PM

$$\sum_{m=1}^{M} PME_{mt} * \delta_{imt} \leq PM \quad \text{for } i\varepsilon I \text{ and } t\varepsilon T \qquad (17)$$

Making the appropriate substitutions in the preceding exposition, we are left with the following formulae in which appear variables XQR_{rt}, δ_{imt}, δ_{rl}, XCB_{imt} and XCL_{imt}.

$$\text{Min. } R = \sum_{t=1}^{T} \sum_{r=1}^{R} \frac{1}{W_{rt}} * XQR_{rt} \tag{7}$$

$$\text{Min. } P = \sum_{t=1}^{T} \sum_{i=1}^{I} \sum_{m=1}^{M} PME_{mt} * \delta_{imt} \tag{8}$$

Subject to:

1) $\frac{1}{QA_{r1}} * XQR_{r1} = \sum_{l=1}^{L} N_l * \delta_{rl} \quad r\epsilon R$ (18)

2) $1 \geq \sum_{l=1}^{L} \delta_{rl} \quad r\epsilon R$ (11)

3) $XQR_{rt} = \frac{QA_{rt}}{QA_{r1}} * XQR_{r1} \quad r\epsilon R, \ t\epsilon T, \ t\neq 1$ (19)

4) $XCB_{imt} + XCL_{imt} = CA_{imt} - \sum_{r=1}^{R} K_{irm} * XQR_{rt} \quad i\epsilon I, \ m\epsilon M, \ t\epsilon T$ (20)

5) $LC * \delta_{imt} \leq XCB_{imt} \leq LC$ (15)

$XCL_{imt} \leq (CA_{imt} - LC) * \delta_{imt}$ (16) $\quad i\epsilon I, \ m\epsilon M, \ t\epsilon T$

6) $\sum_{m=1}^{M} PME_{mt} * \delta_{imt} \leq PM \quad i\ \epsilon I, \ t\epsilon T$ (17)

B) Reduced Formulation:

(a) $\text{Min. } R = \sum_{r=1}^{R} \frac{1}{W_r} * XQR_r$ (7)

(b) $\text{Min. } P = \sum_{i=1}^{I} \sum_{m=1}^{M} PME_m * \delta_{im}$ (8)

Subject to the following conditions:

1) $XCB_{im} + XCL_{im} = CA_{im} - \sum_{r=1}^{R} K_{rmi} * XQR_r$ (21)

2) $LC * \delta_{im} \leq XCB_{im} \leq LC$ (22)

$XCL_{im} \leq (CA_{im} - LC) * \delta_{im}$ (23)

3) $\sum_{m=1}^{M} PME_m * \delta_{im} \leq PM$ (24)

the variable XQR_r being limited by

$XQR_r \leq RRM * QA_r$ (25)

provided that reduction levels have been eliminated and it is advisable that the quantity emission reduction over one be limited by a maximum admissible value (RRM).

4.2.3. Emission reductions model results

- Emission reduction values, expressed in percentages, in each grid, for the entire period studied, and in either discrete or continuous scale fashion.
- The corresponding new emission values.
- The new concentration distributions in each grid which had made up the polluted area.
- The probability of the new concentration exceeding the concentration limit.

4.2.4. Computer support

(a) Model-generating program. Based on the data contained in section 4.2.1. this program creates the equations and limits of the conditions and variables in the format required by the linear programming code employed. It is written in FORTRAN, operative in OS and takes 100 K.

Applying the generalized model in each problem area, the dimensions of same, on an average, are the following:

- Average total number of problem situations (from among all basic time periods, polluted grids and harmful meteorological situations): However, it depends upon the degree of pollution of the study area, the size of each grid, the absolute limit LC, the basic time periods for which we are trying to homogenize the emission reduction level, etc.
- Number of influencing grids in the problem area under consideration. In the cases studied the average is 50.
- Number of basic time periods the emission reduction level of which we want to homogenize. 3
- Reduction levels. 3

Thus the number of variables is: Decision variables: an average of 750; Continuous variables: an average of 1350; And the resulting number of conditions is an average of 2700.

(b) Program which on the basis of the criteria considered in the model, equations (7) and (8), creates, through the "multi-criteria" technique the new objective function. This program is written in FORTRAN, operative in OS and requires 200 K.

(c) Program MPSX/MIP (12,13) which obtains the optimal solution of the function created in program (b), as well as those alternative solutions whose functional value exceeds the optimum by no more than 10%.

(d) Program editing the results of the pollutant emission abatement. This program, via READCOMM (14), gathers the results of program (c). This program is written in FORTRAN, operative in OS, and requires 80 K.

5. CONCLUSIONS

- The model presented in this paper should be considered an effective tool in the establishment of bases for certain corrective alternatives for abatement of an atmospheric pollution problem. It should be regarded in addition as a truly valuable instrument for quantifying, within the development policies of a given area, ever more essential criteria concerning the protection of the environment.

- It should be pointed out that the fundamental statistical parameter considered in the formulation of the model is the maximum admissible probability of the concentration in a given grid exceeding the maximum permitted limit, as contrasted with models which consider quality standards on the basis of mean values. In this way we avoid masking the hazard of large concentrations with situations of low concentration.
- The criteria to be minimized are: a) Weighted emission reduction for each grid, according to the incidence of the latter on the pollutant concentration in the total number of grids comprising the polluted area, and/or b) Probability of the concentration exceeding the adopted limit.
- The possibility arises of considering these criteria simultaneously, optimizing them in uniform fashion.
- Lastly, the results provided by this model are formulated as emissions reduction percentages at grid level, for the entire period studied, without considering to what extent nor with that concrete technical alternatives each pollution source present in the grid should correct its respective emissions. Emission reduction values are expressed in either discrete or continuous scale fashion.

REFERENCES

(1) AGUILAR, R.; 1970, "Problemática de la Contaminación Atmosférica en la Zona del Gran Bilbao", IBM Dpto. Desarrollo Científico, Madrid, España.

(2) AGUILAR, R.; GOMEZ-PALLETE, F.; G.DE COS, P.; MARTINEZ SANCHEZ, G.; 1973 (próximo a aparecer) "Nervion River Valley Bilbao Air Pollution. Study: State of the art", IBM MASC 03.73, Madrid, Spain.

(3) BLENSON, S.M.; and KAPUR, K.C., 1973, "An algorithm for Solving Multicriterium Linear Programming Problems with examples", OPNS, RES.QUART., vol 24 n°1, pp 65-77

(4) BENAYOUN, R.; MOTGOLFIER, J.DE and TERGUY, J.; 1970 "Linear Programming with multiple objective functions: step method (STEM)", 7th Math. Program. Sympo. The Hague.

(5) EPA, 1972, "A guide for compiling a comprehensive emissions inventory" Work paper N.Y.

(6) EPA, 1972 "Compilation of Air Pollutant Emissions Factors"

(7) ESCUDERO, L.F.; 1973, "Operativa y aplicaciones de las técnicas de programación lineal: contínua, entera, mixta y bivalente", IBM-España, SR 10-8023, Madrid. España.

(8) GIFFORD, F.A.; and HANNA, S.R.; 1970, "Urban Air Pollution Modelling", International Air Pollution Conference of the International Union of Air

Pollution Prevention Associations

(9) GORR, W.L., and KORTANEK, K.O.; 1970, "Numerical aspect of pollution abatement problem: constrained generalized moment techniques", Institute of physical planning report, n°12, Carnegie - Mellon University.

(10) G.DE COS, P.; GOMEZ-PALLETE, F.; 1973, "Contaminación Atmosférica: Bases para la confección del Inventario de Emisiones de la zona del Gran Bilbao", Comunicación interna. IBM MASC, Madrid, España.

(11) HANNA, S.R.; 1972, "Description of ATDL Computers Model for Dispersion from Multiple Sources", Air Resources Atmospheric Turbulence and Diffusion Laboratory. Oak Ridge Tennessee.

(12) IBM MPSX (Mathematical Programming system) 1971, SH20-0968, N.Y.

(13) IBM MPSX/MIP (Mixed Integer Feature). 1971, SH20-0908, N.Y.

(14) IBM MPSX/READ COMMUNICATIONS FORMAT (READCOMM), 1971, SH20-0960, N.Y.

(15) KORTANEK, K.O. and GORR, W.L., 1971, "Cost benefit measures for Regional Air Pollution abatement models", I. of P.P. n°15, Carnegie, Mellon University.

(16) PASQUILL, F., 1962, "Atmospheric Diffusion". C.Van Nostrand Company Inc., Princeton, New Jersey.

(17) SHIEH, L.J., and HALPERN, P.K., 1971, "Numerical comparison of various models, Representations for a continuous area source", IBM PASC, G320-3293, Palo Alto, California.

(18) SHIEH, L.J.; HALPERN, P.K.; CLEMENS, B.A.; WANG, H.H. and ABRAHAN, F.F.; 1971 "The IBM Air Quality Diffusion Model with an application to New York City", IBM PASC, G320-3290, Palo Alto, California.

(19) SIMONNARD, M.; 1966, "Linear Programming", Prentice-Hall, Englewood Cliffs, New Jersey, pp. 92-112.

(20) SMITH, M.E.; (editor), 1968, "Recommended guide for the Prediction of the Dispersion of Airbone Effluents", ASME.

(21) STERN, A.C.; 1968, "Air Pollution Handbook", Academic Press, New York.

(22) TELLER, A., 1968, "The use of linear Programming to estimate the cost of some alternate Air Pollution abatement Policies", Proceedings IBM SCI, Comp. Symp. Water and Air Resource Management, IBM 320-1953, N.Y., pp. 345-354.

MATHEMATICAL MODELLING OF A NORDIC HYDROLOGICAL SYSTEM, AND THE USE OF A SIMPLIFIED RUN-OFF MODEL IN THE STOCHASTIC OPTIMAL CONTROL OF A HYDROELECTRICAL POWER SYSTEM

Magne Fjeld, The Engineering Research Foundation at the University of Trondheim, The Norwegian Institute of Technology (SINTEF), Division of Automatic Control, Trondheim, Norway.

Stein Lockwood Meyer*, University of Trondheim, The Norwegian Institute of Technology, Division of High Voltage Systems, Trondheim, Norway.

Sverre Aam*, The Norwegian Research Institute of Electricity Supply (EFI), Trondheim, Norway.

1. INTRODUCTION

In the planning of long-term hydroelectric power production, a kind of stochastic optimization where some particular assumptions on the boundary conditions are implicitely present, are widely used in Norway ("The water value method", based on the incremental cost principle).

Usually the calculation is also based on the assumption of no time-correlation in the stochastic part of the run-off, i.e. the white noise assumption. To get an idea of the effect of such a simplification, it is of great interest to investigate the importance of coloured noise in the run-off, i.e. the effect of dynamical states in the system which governs the run-off to the primary controlled hydroelectric water reservoirs to be controlled.

The first stage in such a project is the hydrological model-building. Such a model may have several purposes, as:

a. An aid in the simulation and better understanding of the dynamics of hydrological systems.

---

*) The main part of this work was done while the authors were with the Division of Automatic Control at The Norwegian Institute of Technology.

b. River flow prediction.

c. Simulations for sub-optimal hydroelectric power systems planning and production.

d. In the computation of stochastic optimal control laws of power production.

In the case a it is obviously preferable to have a model which is physically based, while this is not necessary for instance in the case d. In the latter case, a simple abstract model which posesses the main dynamics is appropriate, partly because of unavoidable uncertainty in the long range all the same, and partly because of the difficulties encountered when applying too complex models in optimization.

2. HYDROLOGICAL MODELLING

2.1. Process characteristics and the multilevel approach.

Three kinds of models of an IHD-representative basin are presented, where different degrees of complexity are suggested. All of them has a multilevel structure. The first level consists of lumped, interconnected nonlinear reservoirs, where the water contents are the dynamical state variables. The second level changes some parameters in the model when the states exceed certain definite values, and in dependence of some parameters governed on the third level. Finally, the third level governs some of the parameters according to the temperature history. This is necessary in Norway because of the alternating climatic conditions.

Consider a hydrological basin, as shown in figure 1. The hydraulic inputs/outputs are precipitation (v2, not shown), channel flow (q_s), groundwater flow (q_g) and evapotranspiration (q_e, not shown). The non-hydraulic inputs or disturbances as temperature, wind and sun radiation are also influencing the hydrological system to a greater or smaller extent.

It is difficult to make a reasonably simple and general model of such a distributed-parameter system like a hydrological basin.

A widely used approach in flow systems, for instance in chemical engineering, is to apply physical lumping of the system. Hence, we subdivide the basin into partial basins where the water storage parts of the model are considered as stirred tanks. In this way, the sub-basins can more easily be adapted to general, physically based, mathematical models. It is assumed that the lumping is done such that an acceptable accuracy in the description is obtained for the application in question.

A typical partial basin is shown in figure 2. The components v_i, which together with q_s and q_g are considered to be the main inputs/outputs (inflows/outflows) of the system, are measured. $q_{u(out)}$ is the total evapotranspiration [m³/day]. The vector $\underline{y}$ is the measurement vector ($\underline{y} = \underline{y}(\underline{x})$), while $\underline{q}_{(out)}$ is the outputs (outflows) from the model.

Observe that the flows $q_{s(in)}$ and $q_{g(in)}$ in general may consist of several contributions. Firstly, we assume that the partial basin is sufficiently homogeneous such that mean values characterizing the disturbances, the surface and the soil (precipitation, evapotranspiration, temperature etc.) are good approximations. Secondly, we assume that the basin is an uncontrolled, natural basin with soil, i.e. urban basins, glaciers and areas with naked mountains only are not considered. This forms the basis of the physical lumping in the model-building. The idea is of course not new in hydrological model building; physical approximation and representation of underground reservoirs by tank models have been used with success [2].

The crust of frozen earth and the snow during the winter season complicate a Nordic model, since the temperature and its history (the temperature is in fact a state variable in a possibly enlarged model of nature in this respect) is of importance for the discharge from the basin. Another problem is how the infiltration progresses, because infiltration is not measured systematically by the hydrologists.

Considering the time aspect, we are interested in a model encompassing the most important long-term properties, since its potential use is for economical dispatch of hydroelectric power at long sight. However, it ought to have a certain degree of accuracy with respect to estimated run-off, such that prediction errors important to the economical dispatch are reasonably well minimized. Expressions like this, and "degree of accuracy" will be given special attention elsewhere [5].

It is seen that the nature may be considered to function like a multilevel system. The complete structure is illustrated in figure 3.

In this paper the 1st level will be represented by a dynamical water balance system, which is assumed to be nonlinear and lumped.

Its simplified mathematical representation in continuous form is the vector differential equation

$$\dot{\underline{x}} = \underline{f}(\underline{x},\ v2(k), \overline{v3},\ \underline{q}_{(in)},\ \underline{p}_1(k),\ \underline{p}_2, \underline{\alpha}) \tag{1}$$

and

$$\underline{q}_{(out)} = \underline{q}(\underline{x},\ \underline{q}_{(in)},\ \underline{p}_1(k),\ \underline{p}_2,\ \underline{\alpha}) \tag{2}$$

$$\underline{y} = \underline{g}(\underline{x},\ \underline{q}_{(in)},\ \underline{p}_1(k),\ \underline{p}_2,\ \underline{\alpha}) \tag{3}$$

Here $\overline{v3}$ is the mean evaporation during the spring and the summer, v2 is precipitation, $\underline{q}_{(in)}$ is the inflow vector and $\underline{p}_1(k)$, $\underline{p}_2$ are parameter vectors steered from the higher levels of the model. $\underline{p}_1$ is piecewise constant in time, and is changed discretely in time with fixed intervals. $\underline{\alpha}$ is the unknown parameter vector (to be determined), and finally, $\underline{x}$ is the state vector, comprising the volumes of water in the tanks of the model. $\underline{q}_{(out)}$ is the outflow vector, being a direct function of the parameters, inflow and states, and $\underline{y}$ is the measurement vector.

The second level consists of a system governing state-dependent parameters $\underline{p}_2$,

$$\underline{p}_2 = \underline{p}_2(\underline{x},\ \underline{p}_1(k)) \tag{4}$$

On the third level, the "seasons" are used as "states", and these are governed by the temperature (vl) history, the latter being an input to the model. On this level, certain temperature-dependent parameters $\underline{p}_1$ are directly given by the season vector $\underline{p}_0$,

$$\underline{p}_1(k) = \underline{p}_1(\underline{p}_0(k)) \tag{5}$$

whereas the transitions of $\underline{p}_0$ are given by a Huffman table, which formally may be written as

$$\underline{p}_0(k+1) = \underline{h}(vl(k),\ \underline{p}_0(k)) \tag{6}$$

The components of $\underline{p}_1$ and $\underline{p}_2$ are of "on-off" type (zero and one). A diagram illustrating the possible transitions of "seasons" is given in figure 4. The Huffman table approximates the dynamics and hysteresis of the seasonal transitions. The components of $\underline{p}_0$ are the "season", a counting parameter to registrate the TMEAN-days period and the integrated temperature (in order to calculate its mean v1MEAN over TMEAN days).

2.2. Parameter observability.

All parameters of a practical hydrological model cannot be determined from simple observations and selective measurements of specific physical parameters. It is also clear that since a hydrological model is a simplified one of a distributed process, even exact knowledge of physical parameters is less valuable, since such parameters in greater or smaller extent will loose their physical interpretation in the approximate model. Hence, many parameters of the model have to be adjusted on the basis of measured input/output time series for the basin. The output measurements will normally be relatively few in number compared to the number of unknown parameters, and the question of state and parameter observability [10] of nonlinear models comes heavily into the problem of sensible model building. This question has been neglected in hydrological model building.

Of course a yes/no answer to the observability question is valuable. However, for practical design of a model, information about how observable the model is, is equally important. Information about this may for instance be obtained from the covariance of the parameter estimation error of an estimation algorithm [4], [5]. This problem will not be treated in this paper.

2.3. Model A.

For the first level, this version is shown in figure 5. (Level 1A.) State variables and parameters can as a rule be given a hydrological explanation, but this will not be done in detail here. However, in brief we have as states:

- x1: Land-surface water storage (water, ice, snow),
- x3: Reservoir storage (lakes), referred to the discharge threshold level
- x4: Accessible soil moisture
- x5/x6: The part of the groundwater volume which does not/does interact with the reservoir storage.

The parameters K_i ($i = 1,2,\ldots$) multiplied by the volumes x_i contribute to the rate of change of the volumes. Hence, a K_i is in principle the inverse of a time-constant. These parameters depend on a number of physical parameters like area, crust in the soil, the specific yield of the soil, the specific hydraulic conductivity, hydraulic inclination, depth to bedrock and the roughness and vegetation of the surface. The dimensionless parameters G_i $(i = 1,2,\ldots,7)$ are difficult to determine a priori, but they are mainly dependent on area. The parameters A_i can be determined directly from a topographical map, since they depend on area only. The Q_i-parameters are dimensionless distribution parameters.

As is clear from figure 2, the measurements in this system are the groundwater level, water stage in the reservoir and the downstream flow rate from the reservoir. However, the latter is partly related to the water stage. The model on level 1A is thus given by 5 non-linear differential equations, 3 output flows given as functions of 5 states and 3 inputs, and finally 3 measurements.

On the second level (Level 2A) the value of the parameter vector $\underline{p}_2^T = (B1,B3,B4,B6)$ is dependent of the state vector $\underline{x}$ and the parameter vector $\underline{p}_1^T = (F1,F2,F3)$. The components of $\underline{p}_2$ change their values when the components of $\underline{x}$ exceed certain treshold values, the "D"-parameters.

On the third level (Level 3A), possible transition of the "season" is done every TMEAN days. We found that the representation of eq. (6) by a Huffman table was more convenient for the problem at hand than a cumbersome formulation with discrete-time equations containing logical expressions. The motivation for this level of the model, is the inertia in the temperature-dependent "parameters". Rapid temperature variations affect the hydrological system very little: The specific heat, melting and evaporation heat of water are large, and snow is a good insulator, too. This also means that the value and the duration of a positive temperature gradient must be larger to get the system switch from "winter" to "spring", than those required for a switch from "autumn" to winter". These phenomena are represented by hysteresis functions. The evapotranspiration is larger in the "spring" than in the "autumn", because of the increasing temperature and since larger areas are covered by water in the spring.

In this way, level 3A represents approximately the complex dynamics of freezing and melting in the nature. A first order differential equation describes approximately the melting (decay of x1).

Parameter observability of model A.

If the Schoenwandt criterion for local observability [10] is used, observability can easily be tested for the model, since the model is piecewise analytic in the states. A test can be made for each of the situations occurring with respect to reservoir levels versus the threshold values. It is then not surprising that the model A is not observable. There are 14 completely unknown parameters and 5 state variables to be estimated. In addition, it is to be noted that we have assumed that all the parameters on the 3rd level can be fairly well rated, and that the unknown "reference value" H1 (which is that part of the groundwater reservoir assumed not to influence the discharge from it, see figure 5) can be rated a priori.

The conclusion is that the model has to be simplified in order to get a model of a complexity which matches the amount of information got in this basin.

It may also be observed that model 1A is simpler than the now well-known Stanford Watershed Model [2].

2.4. Model B.

For this version, the levels 2B and 3B are the same as 2A and 3A respectively.

The 1st level, level 1B, is shown in figure 6, and is a simplified version of level 1A. The parameters and states of this model can however to a less extent than for model A be given a physical interpretation, apart from the fact that $\underline{x}$ still contains the "available" water resource in the basin. In particular, it is to be noted that the infiltration is not described by a differential equation in model B. G5 (= 1 - G6) encomprises in one constant the specific hydraulic conductivity, surface roughness and hydraulic inclination. Assume now that x1 can be estimated from measurements of v2, or by a measurement y4 using snow pillows. Assume also that as many of the parameters as possible are rated a priori with good accuracy, this includes all parameters on level 2-3. It then turns out that the following states and parameters must be estimated:

x2, x3, K4 (or K5), G3 (= 1 - G4), G5 (= 1 - G6) and G8.

If v2 $\neq$ 0 or x1 $\neq$ 0 one can prove by applying the Schoenwandt observability criterion that model B is locally observable in any state, provided the winter season is not present. This also applies if AL2 = 0 such that $\underline{y}^T = (y_1, y_2)$. During the winter, it turns out that G5 is not observable.

Such peculiarities of a hydrological model must be taken into account if a sequential state/parameter estimator is constructed, since non-observable parameters within certain time intervals should not be adjusted. This will not cause any trouble to us, since batch estimation is used, such that the best constant-valued set of parameters is found.

2.5. Model C.

In order to compare model B with a simpler version with respect to the 3rd level, model C contains Level 1B and Level 2A. On the third level, the Huffman table is not included, and "seasons" are made directly dependent on v1MEAN.

Under the same conditions as put on model B, this model is observable.

2.6. Adaption of the parameters.

In order to get some feeling of the problems encountered in this first investigation, a simple batch estimation of the parameters and states was tried. Although it is obvious that some of the parameters depend on the climatic conditions in a much more subtle way than in the models here, it is of interest to get an idea of how well such lumped models could be fitted to the measurement data. Since model A is not observable, the unknown parameters and states of the models B and C were adapted to measurements from a part of the IHD-representative basin "Sagelva". This part of the basin, which is illustrated in figure 7, is a small basin, but unfortunately not very homogeneous.

The well-known principle of many parameter estimation schemes is shown in figure 8, where $\underline{\alpha}$ represents the four unknown parameters (of model B) to be estimated. As adjustment strategy a simple hill-climbing method has been applied ("one-at-a-time") over a data interval of 2 years with very changing climatic conditions. (In a later work [5], a SIMPLEX search method included in a batch estimation program for the UNIVAC 1108 [4] was used, being considerably more efficient.) The loss functional to be minimized for optimal parameter values was taken as

$$S = \int_{t_1}^{t_2} (|y_{1m}(t) - y_1(t)| + 8\cdot|y_{2m}(t) - y_2(t)|)dt \qquad (7)$$

Results from a "ballistic" simulation forcing the model B with the input data over 1 year, are shown in figure 9. $\overline{T}$ is the mean temperature during 15 days, and vl is precipitation per day. xi, i = 1,2,3, are simulated water storages in the basin, respectively land-surface water storage, groundwater storage and reservoir storage. yl is simulated groundwater level, while y2 is simulated reservoir water storage level. y_mi, i = 1,2, are the corresponding measured levels.

With the parameters obtained from the estimation, so-called recession ("dry weather"-) curves were simulated. These are shown in figure 10. Here q_s is surface discharge from the groundwater storage. They are both simulated according to the temperature history shown. In addition parts of recession curves being characteristic of each season are plotted: q_{ss} denotes pure summer surface discharge, q_{sa} pure autumn surface discharge, and q_{sw} correspondingly for the winter season.

Similarly, estimation and simulations were performed for model C, but the results were less reliable than for model B under unnormal winter conditions.

The conclusion is that for a Nordic hydrological model it seems necessary with some kind of sequential control of temperature-dependent parameters, which also in an approximate way takes care of the dynamics of melting and freezing under different conditions. It seems worth while to make further investigations on the basis of a model having a structure like model B.

3. STOCHASTIC OPTIMIZATION OF HYDROELECTRIC POWER DISPATCH

3.1. System description.

In the long term planning for the economical dispatch of hydroelectric power, the optimization interval over which the given performance functional is to be minimized (or maximized), usually is in the range of a few months to about one year. Because of uncertainty in the future run-off into the reservoirs, a reasonable goal is to minimize the expected value of the functional. Hence, we will have to consider a system model where the environmental model representing the run-off contains stochastic state variables. See figure 11, where we have

$\underline{a}$. a mathematical process model for the production system, with control vector $\underline{u}$ and states (volumes) $\underline{x}_1$,

$\underline{b}$. a lumped state variable model for the environment (state vector $\underline{x}_2$), yielding the run-off $\underline{r}(\underline{x}_2)$ to the reservoirs. The input to this model is an expected mean function v_o plus a white noise sequence Δv with a given distribution (the precipitation $v = v_o + \Delta v$).

In addition, there are given data for the power demand, which possibly also may be decomposed like the precipitation, in a mean value function plus a stochastic term.

In Norway it is usual to divide the optimization interval into sub-intervals of one week, and use the so-called "water value method" based on the incremental cost principle. (A description of the basic principle may be found in [9].) An analysis of this approach will show that the run-off is considered as pure stochastic (white noise) around a deterministic function of time. Considering for instance figure 10, it is observed - especially during the winter season - that such an approximation is less accurate relative to the fineness of the time discretization the smaller this discretization interval is. There is considerable dynamics in the run-off, which may be expressed by the autocorrelation function (in the linear case), or more generally, by a set of 1st order differential equations.

The dynamics will show up in the evolution of the probability distribution, as sketched in figure 12, which shows the "stationary" probability distribution of Δr as a function of time. In the linear, Gaussian case, the evolution of the probability density is uniquely given by the differential equation for the covariance $E\{\Delta r^2(t)\}$.

To be more specific, the complete system may be formulated as

$$\dot{\underline{x}}_1(t) = \underline{f}_1(\underline{x}_1(t), \underline{r}(\underline{x}_2(t)), \underline{u}(t), t) \qquad (8)$$

$$\dot{\underline{x}}_2(t) = \underline{f}_2(\underline{x}_2(t), \underline{v}(t), t) \qquad (9)$$

$$\underline{x}(t) \in X \quad (\underline{x}_1(t) \in X_1), \quad \underline{u}(t) \in U.$$

3.2. Discussion of the run-off model.

For long-term optimization problems of the kind discussed here it is obvious that uncertainty is very pronounced, as observed from figure 12. There seems to be no practical reason - at least for reasonably homogeneous or small basins - to work with higher order run-off models.

An abstract, 1st order linear model with a time-variable parameter ("time-constant") established, say, on the basis of initial condition responses ("recession curves") of a more complex model like the responses of figure 10, has the form

$$\dot{x}_2(t) = -a(t)x_2(t) + v(t) \qquad (10)$$

where

$$v(t) = v_0(t) + \Delta v(t)$$

We may then assume a linear relationship between the environmental state x_2 of eq. (10) and the run-off r,

$$r(t) = k \cdot x_2(t) = r_0(t) + \Delta r(t) \qquad (11)$$

Substituting into eq. (10), we have

$$\dot{r}(t) = -a(t)r(t) + k(v_0(t) + \Delta v(t)) \qquad (12)$$

The recession function is given by the unforced solution of eq. (12),

$$r(t) = r(0) \cdot e^{-\int_0^t a(\theta)d\theta} \qquad (13)$$

By letting a(t) be a function of time, it is possible to take into account the expected main seasonal changes in the climatic conditions. A sensible approximation is to apply three different values for a, these values respectively referring to the winter season, the snow-melting period and the period without snow, snow-melting and frost. The time constant $\frac{1}{a}$ is dependent on the basin, and is typically between 10 and 90 days, having its largest value during the winter.

During the snowmelting period, the water from the melted snow will usually be a dominating part of the run-off. A main part of this flow will be discharged into the reservoirs from the surface.

In this work, no attempt is done to make use of an optimal adaption of a(t) to the behaviour of the basin in question.

It is quite obvious that inertia in the run-off dynamics is of greater and greater importance the smaller the ratio between reservoir volume and integrated run-off to the reservoir through one year is. For instance, if a reservoir can accumulate on an average the run-off through 2-3 years (without discharge from the reservoir), it is obvious that a dynamical run-off model, characterized by a time-constant of about a month, will have almost no effect on the economical dispatch of such a system.

3.3. The optimization problem.

A dynamical description of the stochastic part of the run-off implies two essential distinctions for the economical dispatch problem, compared to a run-off which is not correlated in time.

a. Instead of using the "stationary" distribution of the run-off and possibly consider it as white noise, the dynamical evolution of the run-off and its probability density from a given initial condition, is taken care of (possibly with a given uncertainty in the initial condition).

b. Since we work with the expected evolution of the environmental states, these functions and their associated density functions are per definition given for the whole optimization interval. As is well known, this will in a control problem result in a realizable "feedforward" coupling from the environmental states to the control vector. Further, there will be a coupling from the reservoir volumes to the control vector, which is the "feedback part" of the control law. (Of course, in a nonlinear problem, these parts cannot be separated, but the principle is still there.) See figure 13.

To apply solution by Stochastic Dynamic Programming (S.D.P.), the system equations are used in their time-discrete form. With a discretization interval T, we have for a single reservoir,

$$x_1((k+1)T) = x_1(kT) - u(kT) + kx_2(kT) \tag{14}$$

and for the environmental model

$$x_2((k+1)T) = e^{-aT} x_2(kT) + \int_{kT}^{(k+1)T} e^{-a\left((k+1)T-\tau\right)}(v_0(\tau)+\Delta v(\tau)) \tag{15}$$

If v(t) is considered constant within the interval $\{kT, (k+1)T\}$, and Δv is taken as a discrete-time white noise sequence, the latter equation simplifies to

$$x_2((k+1)T) = e^{-aT}x_2(kT) + \frac{1}{a}(1-e^{-aT})(v_0(kT) + \Delta v(kT)) \tag{16}$$

To simplify the notation, we will in the sequel use $x_i(k)$ for $x_i(kT)$ etc.

The objective function for the optimal control of the system is as follows. In Norway it is commonly assumed that the marginal incomes/expenditures dependent on the dispatch are a given function

$PF(u_p(k) - u(k))$, where PF is price per energy unit (öre/kWh). $u_p(k)$ is power as ordered by contract from customers within the optimization interval, and $u(k)$ is the actual power production. (GWh/month.) This function is often given as a staircase function like the one in figure 14. There is however uncertainty in the future power prices, so it might have been sensible to take this uncertainty into consideration. In S.D.P. this can be done without any difficulties, but with an increase in computation time. In the example here, however, the smooth curve as shown on figure 14 has been used without uncertainty on it.

The expenditure within an interval $[k, k+1]$ is

$$W_k = \int_0^{u_p(k)-u(k)} PF(\mu)\, d\mu \tag{17}$$

The optimal criterion is to minimize the expected expenditures during the optimization interval $(0,N)$,

$$E\{J\} = E\{\sum_{k=0}^{N-1} W_k(u_p(k) - u(k))\} \tag{18}$$

As data, the functions $u_p(\cdot)$ and $v_o(\cdot)$ and the probability density distribution $p(\Delta v)$ of Δv are given.

Since the main purpose here is to obtain a feeling of the importance of dynamical modelling of the environment of a hydroelectric power system for the economical dispatch, straightforward S.D.P. [1] is applied without any subtleties. The basis of the method can be studied in the textbook of Aoki [1]. An advantage in such applications as this using D.P., is that the state space is constrained because of maximum and minimum reservoir volumes. Also, maximum/minimum values for the run-off states may be rated fairly well. Complicated optimization criteria imply no difficulties. The most serious draw-backs are the well-known dimensionality problem and long computation time. The storage requirements for reasonably low-order systems (max. 4-5) may be solved by applying a mixture of different kinds of extensions of ordinary D.P. techniques [7], [8].

3.4. Example.

Computation of optimal controls for the first month in an optimization interval of five months in a certain year has been done using data for a small power station in the middle of Norway, named "Julskaret". The data of the production system are:

Power station.

Maximum storage capacity: 60 mill. m^3
Mean height difference between power station and the reservoir: 100 m
Mean energy conversion: 4.17 mill. $m^3 \to 1$ GWh
Machine installation: 8 MW.

This gives the constraints

$$0 \leq x_1(k) \leq 14.4 \quad \text{(GWh)}$$
$$0 \leq u_1(k) \leq 5.6 \quad \text{(GWh/month)}$$

$u_p(k)$ is given in the following table (dim u_p = GWh/month):

Month:	1	2	3	4	5
u_p:	2.7	1.9	4.6	4.4	3.9

The run-off system.

The total precipitation basin for the station is $A = 149.5$ Km^2 = 149.5×10^6 m^2. The time-constant for the run-off is estimated to $T_1 = \frac{1}{a} = 1.2$ months on the basis of a recession curve. For simplicity, a^{-1} is assumed constant. We assume $r = k \cdot x_2 = x_2$. The run-off equation with dim$[x_2] = m^3$, dim $[v]$ = m/month, is

$$x_2(k+1) = e^{-\frac{T}{T_1}} x_2(k) + \frac{AT_1}{4.17 \cdot 10^6}(1 - e^{-\frac{T}{T_1}})(v_o(k) + \Delta v(k))$$

or

$$x_2(k+1) = 0.434\, x_2(k) + 24.8(v_o(k) + \Delta v(k))$$

which is assumed valid throughout the optimization interval. Realistic values of $x_2(k)$ are assumed to be within $0 \leq x_2(k) \leq 10$. The density function $p(\Delta v)$ is estimated on the basis of precipitation through 40 years. The data are not given here, but to get an impression of the spread, the variance $\sigma^2_{\Delta v}$ is given in the following table, where also $v_o(k)$ is tabulated:

Month k	1	2	3	4	5
$10^3 \cdot v_o(k)$	43	39	41	34	37
$\sigma_{\Delta v}(k)^2$	90	94	87	57	61

Performance criterion.

For the objective function the smooth curve $PF(u_p(k) - u(k))$ in figure 14 is used.

The results would be rather uninteresting in practice if the terminal state $x_1(N)$ is not considered in the optimization problem, since this would imply a policy which aims at emptying the reservoir towards the end of the optimization interval. Many kinds of criteria taking the expected final state into account could be thought of. For instance, an analysis of the principle of the procedure used in [9], shows that within the assumption of linearity in the process equations, the policy is to aim at reproducing the reservoir volume after one year [3]. A reasonable policy might be to let the expected final state $x_1(N)$ have a sensible value based on experience for that month of the season. A more direct, and in fact an equivalent approach, is to include a weighting on the final state in J, with such a weighting that the expected final state has a reasonable value. Hence, we use as an optimal criterion

$$E[J'] = E\{J + dx_1(N)\} \tag{19}$$

where J is given by eq. (17) - (18).

Results.

It is interesting to find the variation in the optimal power production $u_{opt}(0)$ of the first month as a function of the initial condition $x_2(0)$ in the run-off model. The results are shown in figure 15 for three different initial storages $x_1(0)$ in the power station reservoir and $\underline{d = 3}$. As expected, the initial state $x_2(0)$ has a considerable effect on the optimal policy. The expected final state $E\{x_1(N)\}$ (applying the expected run-off and picking the control from the computed tables of optimum stochastic controls) is 7.4 GWh at $d = 3$, and 8.2 GWh at $d = 6$. The two different values of d gave no difference in the optimum control for the first stage. However, at $d = 0$, $u_{opt}(0) = 3.2$ at $x_1(0) = 100\%$ (14.4 GWh). The control policy for the first stage is rather insensitive to the weighting factor on $x_1(N)$, as long as the expected final state has a reasonable value for the month in question. This is mainly an effect of the uncertainty of the future, and also indicates that it should not be necessary to use larger optimization intervals than, say, half a year, in order to compute the optimal control for the first month.

An interesting comparison is to compute the optimum control if Δr is pure stochastic (white) with approximately the same probability density as that one which can be estimated from the run-off observations. It is not surprising that the computed value in this case, u_{opt} = 4 GWh/month, at $x_1(0)$ = 100% corresponds to a value (see figure 15) which is close to the mean in the run-off for that month.

Of course, the numerical values obtained here should not be used in a general discussion of the goodness of approximation by using a non-dynamic run-off description in the computation of the economical dispatch for any hydroelectric power system. However, the example clearly shows that the problem should be given attention.

4. CONCLUSIONS

Results on simple batch parameter estimation of a hydrological system have been presented in the first part. The number and kind of measurements justify the synthesis of a rather crude model only. This conclusion has been drawn on the basis of observability analysis. Hence, it is not surprising that the goodness of fit will vary somewhat dependent on the season, and that the simple model has deficiencies like inaccurate reservoir level during the winter and the spring, and too low groundwater level during the late autumn. However, it should be kept in mind that the errors in the fitting will distribute on each variable according to the weighting factors in the loss functional [5].

In the last section, with respect to the application of a hydrological model in the stochastic optimization of a hydrological power system, it has been demonstrated that the use of a dynamical run-off model may be necessary in the computation of the optimal control. Although it is open for discussion how complex such a model should be, it is likely that significant improvements in the control policy can be attained by representing the most important dynamics of the environmental system in a simple first-order, stochastic model with time-varying parameters.

[1] Aoki, M.: Optimization of Stochastic Systems; topics in discrete-time systems. Academic Press, 1967.

[2] Crawford, N.H., Linsley, R.K.: "Digital Simulation in Hydrology: Watershed Model IV". Technical Report No. 39, Department of Civil Engineering, Stanford University, California, USA.

[3] Fjeld, M.: Dynamic Programming and Its Application to the Dispatch Problem in Hydroelectric Power Systems. Report 72-90-S, The Engineering Research Foundation at the University of Trondheim, The Norwegian Institute of Technology (SINTEF), Division of Automatic Control, Trondheim, Norway. (In Norwegian.)

[4] Hertzberg, T.: MODTLP, a General Digital Computer Program for Fitting of General, Nonlinear, Dynamic Models to Experimental Data. The Norwegian Institute of Technology, The Chemical Engineering Laboratory, N-7034, Trondheim, Norway. (In Norwegian.)

[5] Holmelid, A.E.: State and Parameter Estimation of Hydrological Systems. Thesis 1973. The Norwegian Institute of Technology, Division of Automatic Control, N-7034, Trondheim, Norway. (In Norwegian.)

[6] Jacobson, D.H., Mayne, D.Q.: Differential Dynamic Programming. American Elsevier Publ. Co., 1970.

[7] Larson, R.E.: A Survey of Dynamic Programming Computational Procedures. IEEE Trans. on Autom. Control, AC-12, No. 6, pp. 767-774 (December 1967).

[8] Larson, R.E.: State Increment Dynamic Programming. Elsevier 1968.

[9] Lindquist, J.: Operation of a Hydrothermal Electric System: A Multistage Decision Process. AIEE Journal, April 1962, pp. 1-7.

[10] Schoenwandt, U.: On observability of Nonlinear Systems. Preprints of the IFAC Symposium on Identification and Process Parameter Estimation, Prague, June 1970.

ACKNOWLEDGEMENT

The authors wish to express their appreciation to the staff at the Division of Hydraulic Engineering at the Norwegian Institute of Technology for giving data from the IHD representative basin "Sagelva" at our disposal, and their helpfulness in various other questions concerning hydrology.

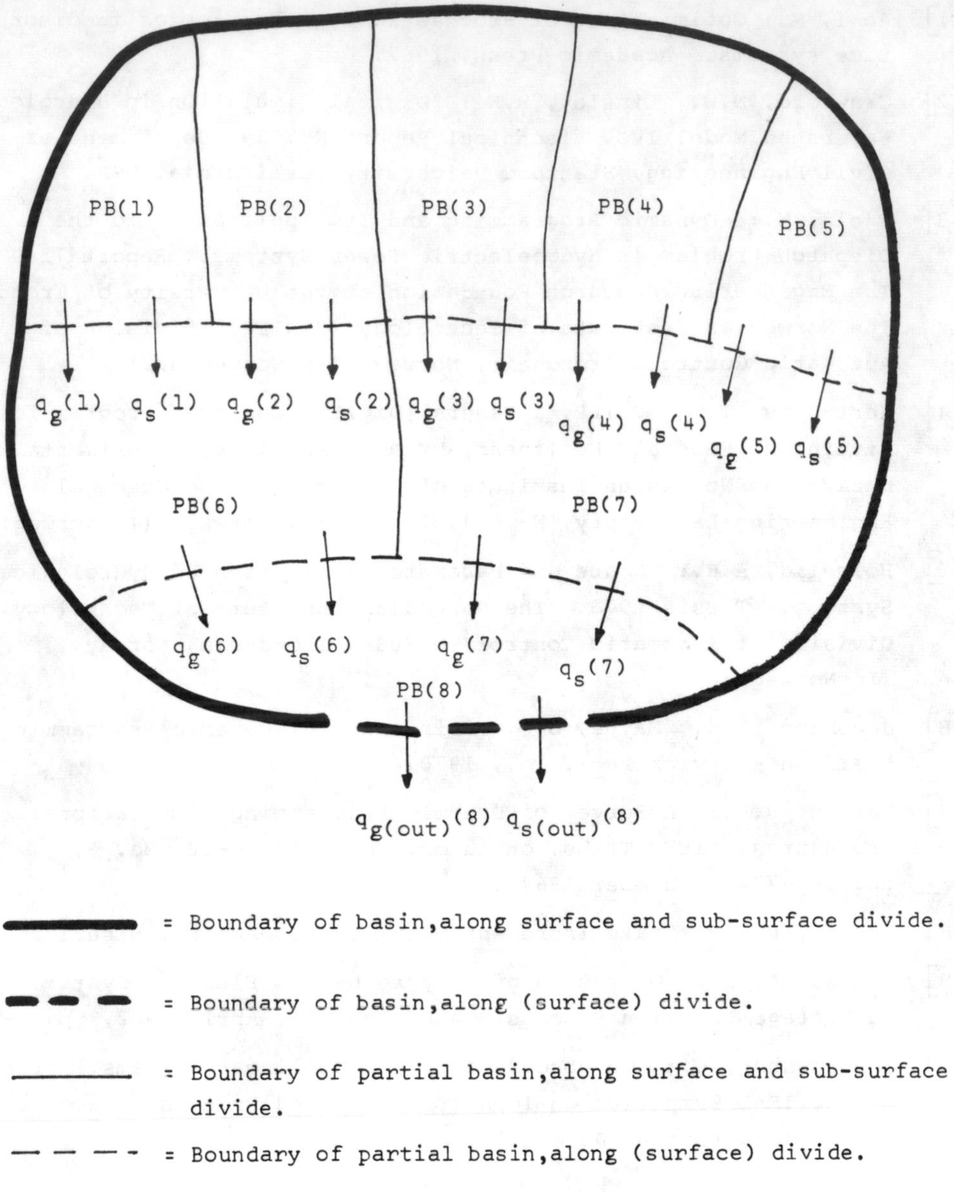
PB(1)
PB(2)
PB(3)
PB(4)
PB(5)
q_g(1) q_s(1) q_g(2) q_s(2) q_g(3) q_s(3) q_g(4) q_s(4) q_g(5) q_s(5)
PB(6)
PB(7)
q_g(6) q_s(6) q_g(7) q_s(7)
PB(8)
q_g(out)(8) q_s(out)(8)
= Boundary of basin,along surface and sub-surface divide.
= Boundary of basin,along (surface) divide.
= Boundary of partial basin,along surface and sub-surface divide.
= Boundary of partial basin,along (surface) divide.

Fig. 1. A large (hydrological) basin.

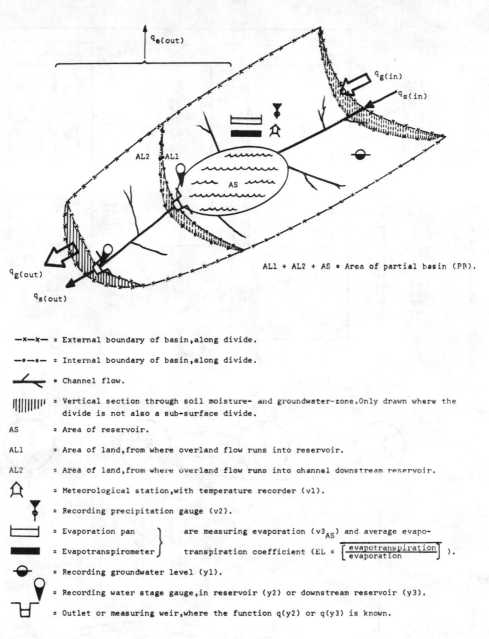

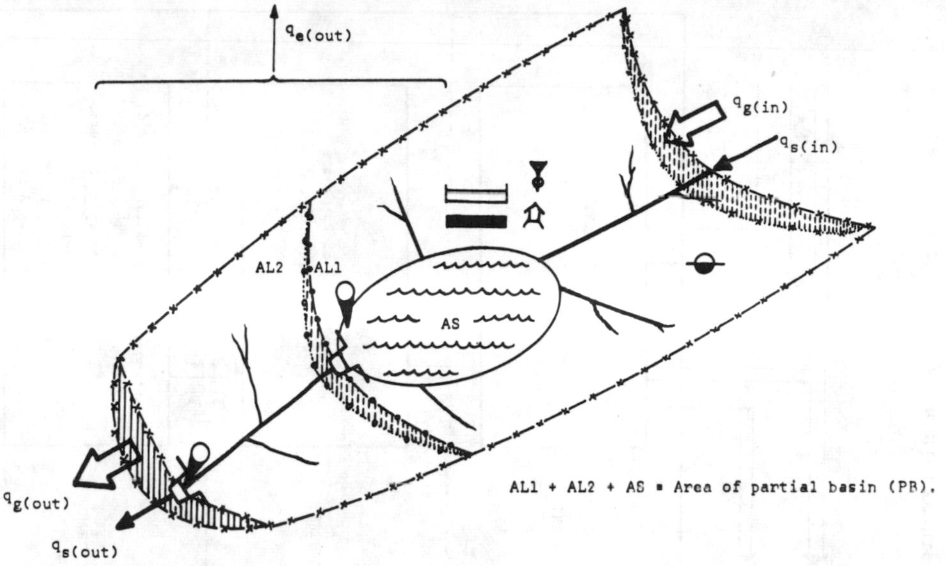

—x—x— = External boundary of basin, along divide.

—•—•— = Internal boundary of basin, along divide.

= Channel flow.

= Vertical section through soil moisture- and groundwater-zone. Only drawn where the divide is not also a sub-surface divide.

AS = Area of reservoir.

AL1 = Area of land, from where overland flow runs into reservoir.

AL2 = Area of land, from where overland flow runs into channel downstream reservoir.

= Meteorological station, with temperature recorder (v1).

= Recording precipitation gauge (v2).

= Evaporation pan } are measuring evaporation ($v3_{AS}$) and average evapo-

= Evapotranspirometer } transpiration coefficient ($EL = \left[\frac{\text{evapotranspiration}}{\text{evaporation}}\right]$).

= Recording groundwater level (y1).

= Recording water stage gauge, in reservoir (y2) or downstream reservoir (y3).

= Outlet or measuring weir, where the function q(y2) or q(y3) is known.

Fig. 2. A typical partial basin.

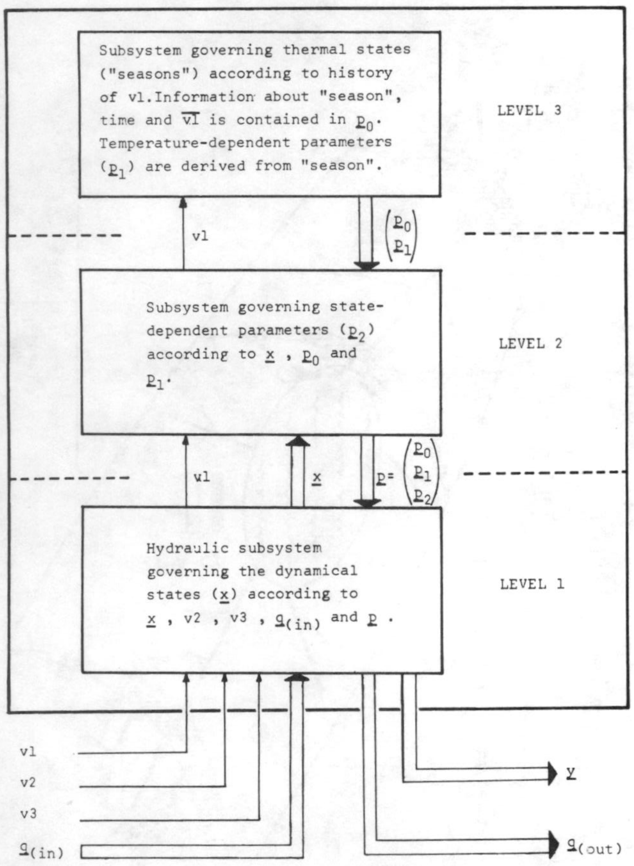

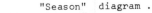

Fig. 3. The basin sketched as a hierarchical system.

Subsystem governing thermal states ("seasons") according to history of vl.Information about "season", time and $\overline{VI}$ is contained in $\underline{p}_0$. Temperature-dependent parameters $(\underline{p}_1)$ are derived from "season".

LEVEL 3

Subsystem governing state-dependent parameters $(\underline{p}_2)$ according to $\underline{x}$, $\underline{p}_0$ and $\underline{p}_1$.

LEVEL 2

Hydraulic subsystem governing the dynamical states $(\underline{x})$ according to $\underline{x}$, $v2$, $v3$, $q_{(in)}$ and $\underline{p}$.

LEVEL 1

v1

v2

v3

$q_{(in)}$

v1

$\underline{x}$

$\underline{x}$

$\underline{p} = \begin{pmatrix} \underline{p}_0 \\ \underline{p}_1 \\ \underline{p}_2 \end{pmatrix}$

$\begin{pmatrix} \underline{p}_0 \\ \underline{p}_1 \end{pmatrix}$

$\underline{x}$

$q_{(out)}$

Fig. 4. "Season" diagram.

SPR 2

SPR 1

WIN 2

WIN 1

SUM 1

SUM 2

AUT 1

AUT 2

Synchronism:

TMEAN days between each change of "season".

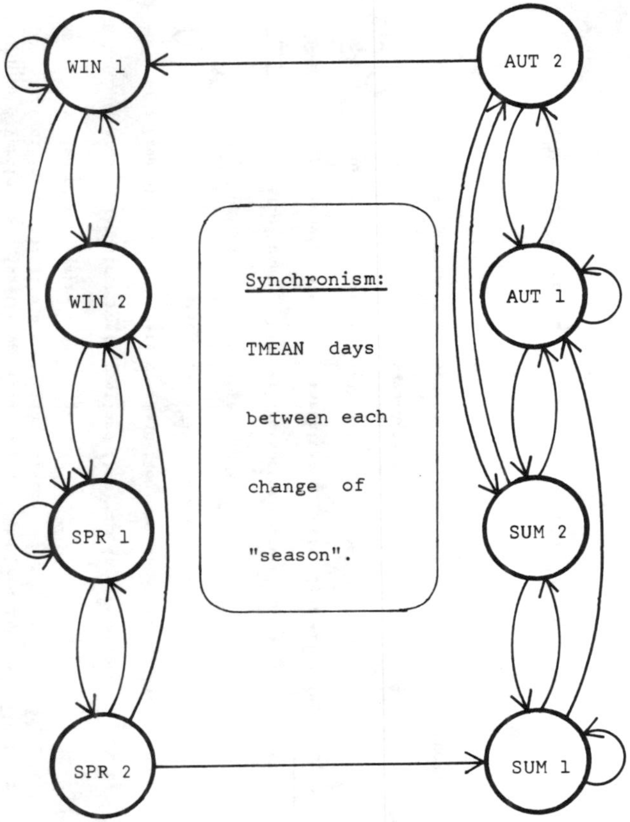

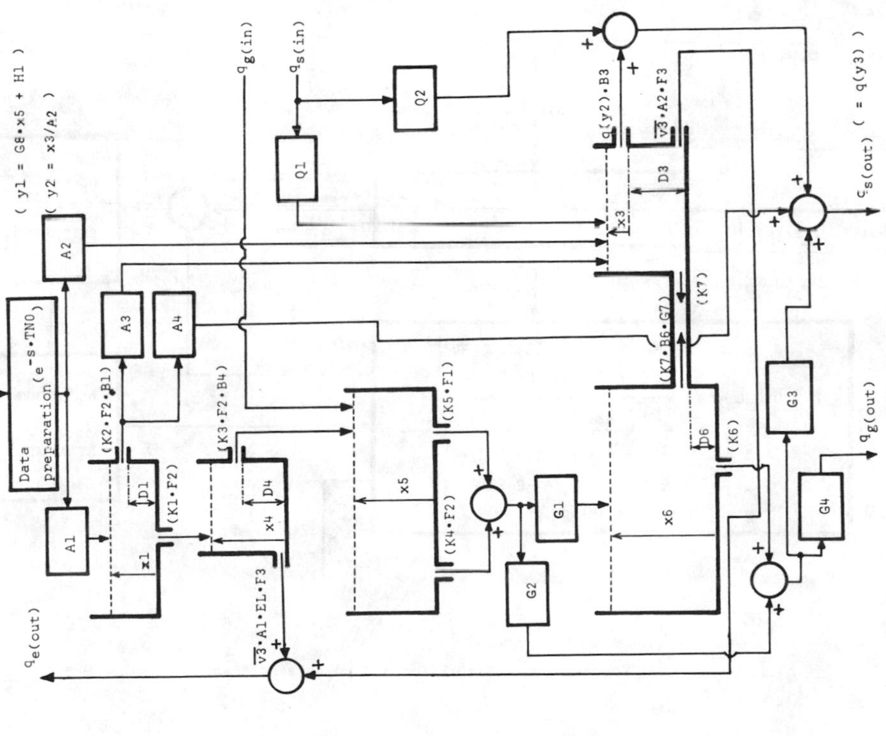

Fig. 5. LEVEL 1A.

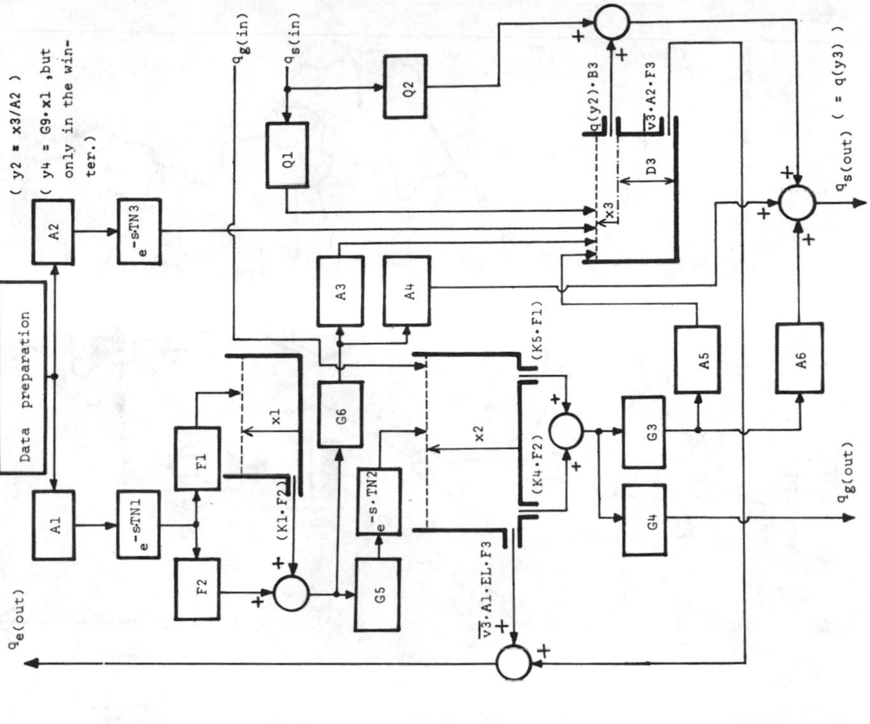

Fig. 6. LEVEL 1B.

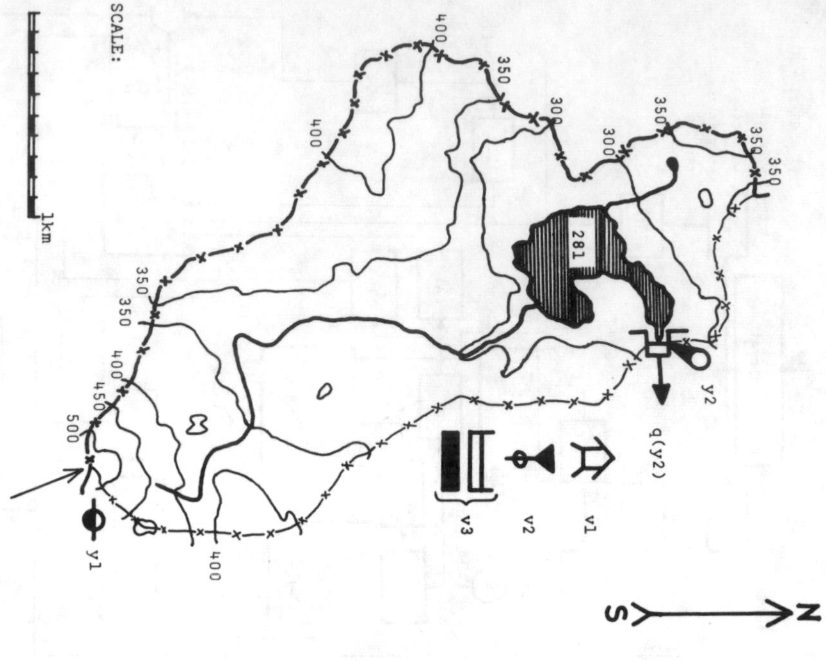

Fig. 7.   A part of Sagelva IHD Representative Basin,Norway.

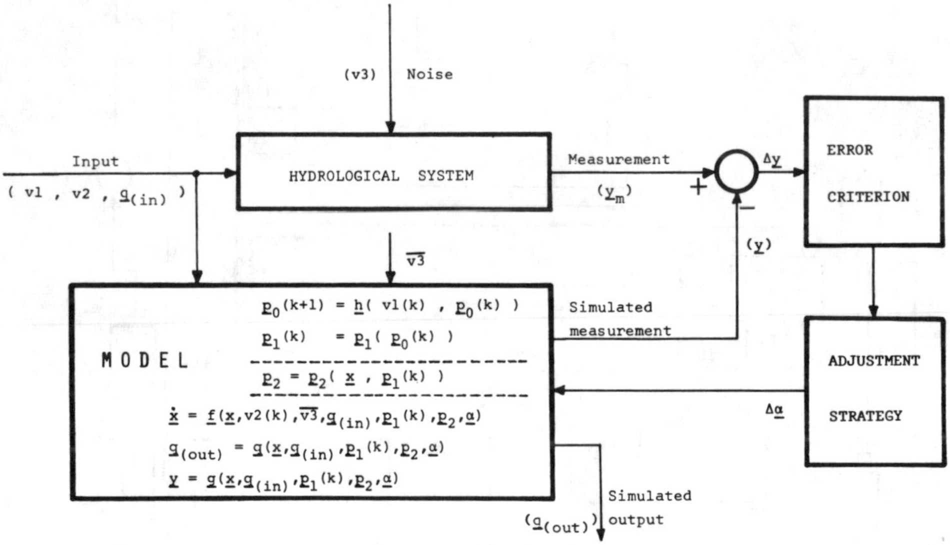

Fig. 8.   Simulation and adjustment plan.

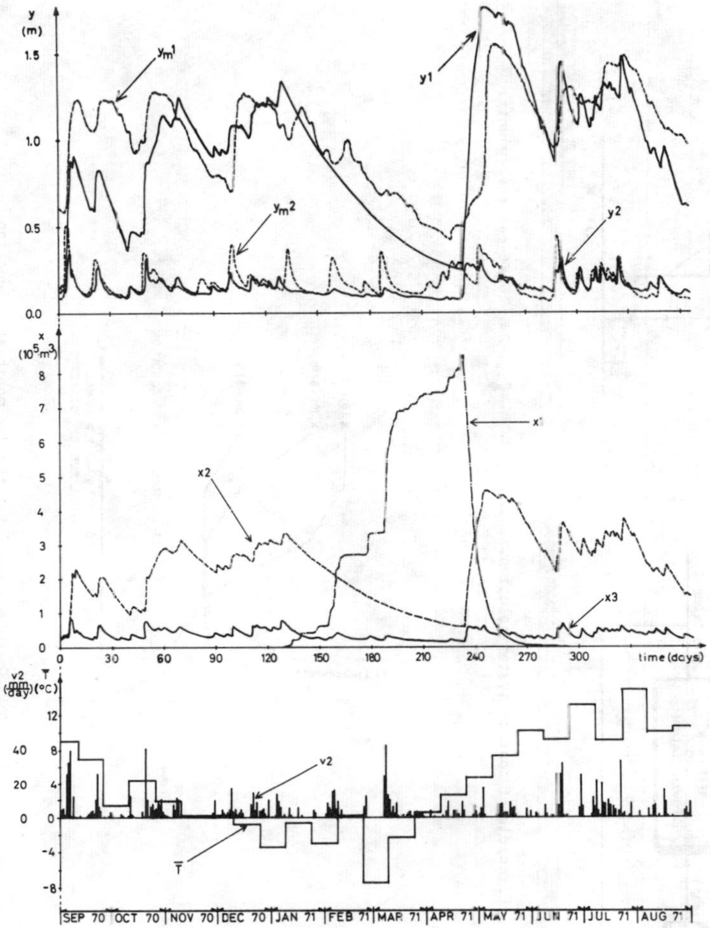

Fig. 9. Simulation and measurement.

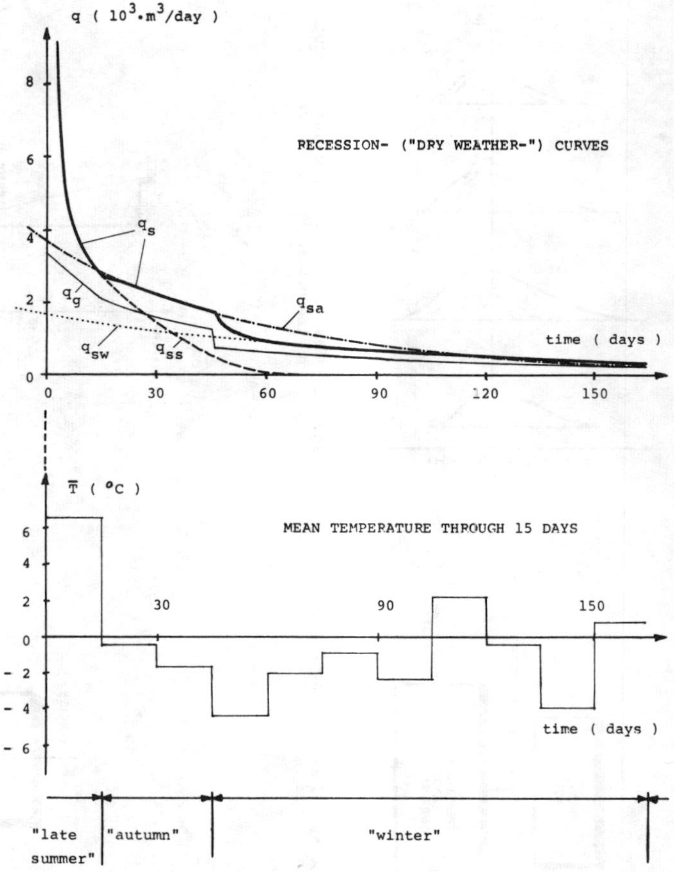

Fig. 10. Simulation with precipitation = constant = 0.

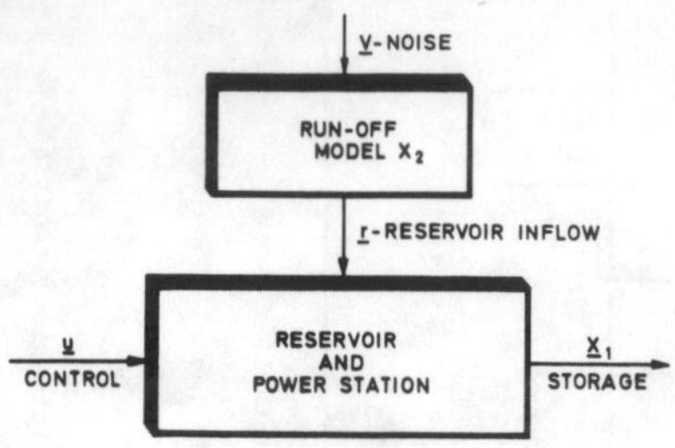

Fig. 11. Process and environmental model.

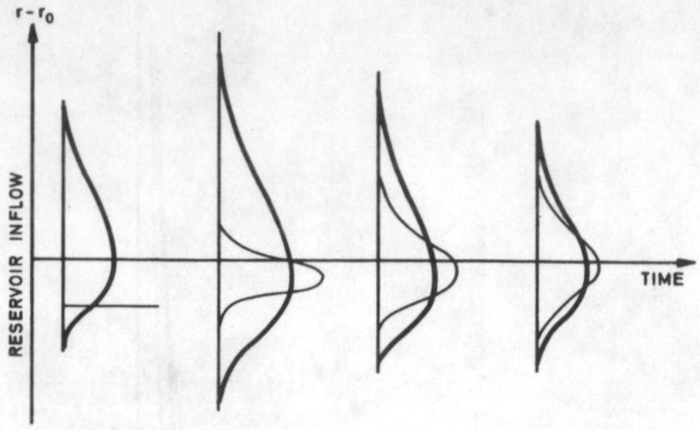

Fig. 12. "Stationary" and conditional evolution of the probability density.

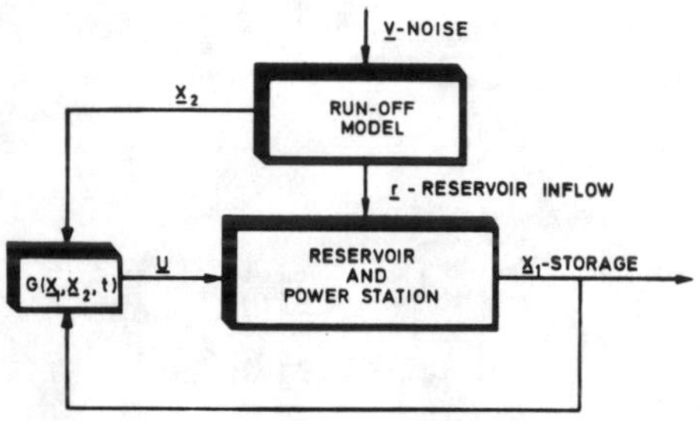

Fig. 13. Principle of control system solution.

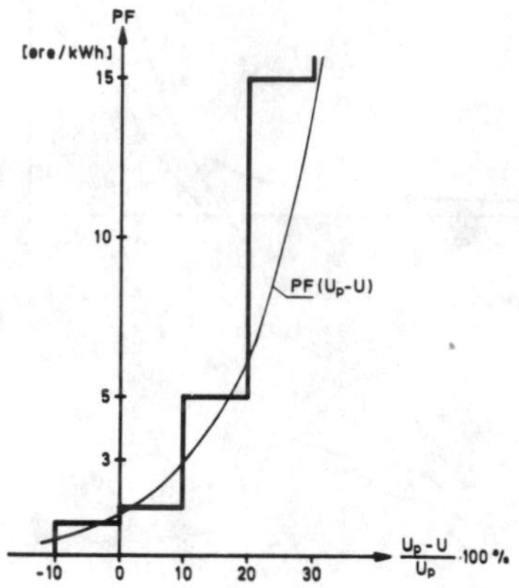

Fig. 14. Cost per energy unit.

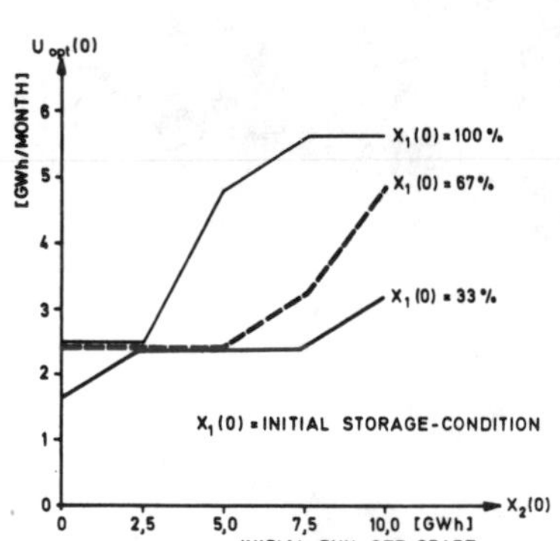

Fig. 15. Optimum control for the 1st month as a function of the initial values in the states.

A TWO-DIMENSIONAL MODEL FOR THE LAGOON OF VENICE

C. Chignoli, R. Rabagliati
IBM Venice Scientific Center, Italy

CONTENTS:
- The Navier-Stokes equations and the mathematical formulation of the problem.
- The two-dimensional model and the validity of the assumption of the non linear shallow water theory.
- The numerical methods, stability and accuracy.
  - The P. Lax method.
  - Some results obtained with the P. Lax method.
  - The staggered method (W. Hansen).
  - About some simulations of the propagation of the tidal wave inside the Venetian Lagoon.

The Navier-Stokes equations and the mathematical formulation of the problem.

The basic approaches to the study of the dynamics of a basin are essentially two, with either a Lagrangian or an Eulerian formulation of the problem. With the Lagrangian approach the trajectories of single particles of the fluid are followed; with the Eulerian approach it is the change in time of the variables which is studied in a fixed point.

In our case we used the Eulerian approach with the Navier-Stokes equations, which define the fluid behavior within a space domain, which we shall refer to as Ω. This domain Ω is defined by the free surface of the fluid

$$Z_1 = \eta\ (x,y,t) \tag{1}$$

and by the bed of the basin.

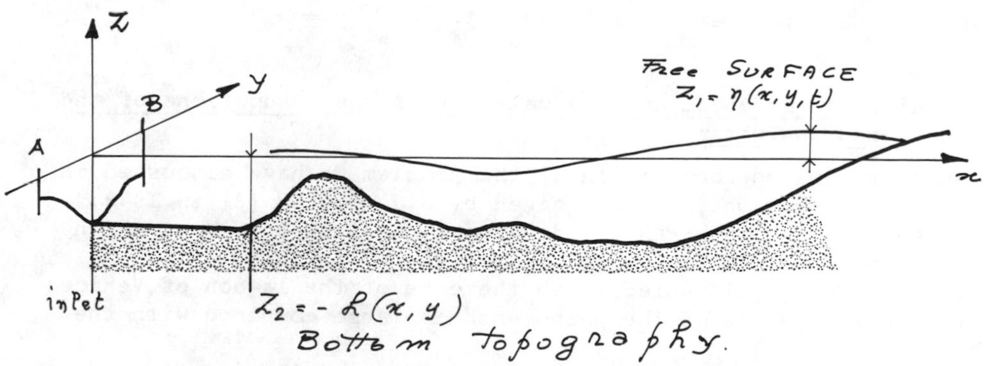

Fig.1 Example of the geometry of the propagation problem. The z-axis is parallel to the gravity field and opposite in direction.

The bottom erosion is not taken into account in this work; therefore the equation which defines the bed of the basin is not time dependent and we can write this function as follows

$$Z_2 = - h(x,y) \tag{2}$$

If we use the following notation

- u, v, w — the three orthogonal components of the velocity in each point of Ω
- p — the pressure field
- g — the acceleration due to gravity
- ρ — the density of the fluid
- ν — the kinematic coefficient of viscosity,

the Navier-Stokes equations are written in the following form:

$$\begin{aligned} \frac{\partial u}{\partial t} + u\frac{\partial u}{\partial x} + v\frac{\partial u}{\partial y} + w\frac{\partial u}{\partial z} + \frac{1}{\rho}\frac{\partial P}{\partial x} &= \nu \nabla^2 u \\ \frac{\partial v}{\partial t} + u\frac{\partial v}{\partial x} + v\frac{\partial v}{\partial y} + w\frac{\partial v}{\partial z} + \frac{1}{\rho}\frac{\partial P}{\partial y} &= \nu \nabla^2 v \\ \frac{\partial w}{\partial t} + u\frac{\partial w}{\partial x} + v\frac{\partial w}{\partial y} + w\frac{\partial w}{\partial z} + \frac{1}{\rho}\frac{\partial P}{\partial z} &= \nu \nabla^2 w \end{aligned} \tag{3}$$

Furthermore we must write the conservation law for the mass:

$$\frac{\partial u}{\partial x} + \frac{\partial v}{\partial y} + \frac{\partial w}{\partial z} = 0 \tag{4}$$

The following boundary conditions, respectively on the free surface and on the bottom, must be included:

$$u\, h_x + v\, h_y + w = 0 \qquad \text{on} \quad Z_2 = - h(x,y) \tag{5}$$

$$\eta_t + u\,\eta_x + v\,\eta_y = w \quad \text{and} \quad p(x,y,\eta,t) = p_a \quad \text{on} \quad Z_1 = \eta(x,y,t) \tag{6}$$

Another boundary condition will be introduced in order to define the time evolution of the pressure field at the entrance of the basin (A-B in fig. 1a).

This kind of assumptions for the boundary not the best one, but this method gives acceptable results for our problem. For a detailed analysis of this problem see (1) and (2).

The two-dimensional model and the validity of the assumptions of the non linear shallow water theory.

An approximated formulation of the problem we have discussed in the previous paragraph, has been given by J.J. STOKER (3) in 1948. In 1958 CARRIER and GRENSPAN (4) developed the Stoker theory, which now is known as non linear shallow water theory (NLSWT).

The NLSWT is applicable, as in the case of the lagoon of Venice, when the lateral scale of the phenomenon is large compared with the depth of the basin.

The first order NLSWT may be written in the following form:

$$\frac{\partial U}{\partial t} + U\frac{\partial U}{\partial x} + V\frac{\partial U}{\partial y} + g\frac{\partial \eta}{\partial x} + g\frac{U(U^2+V^2)^{1/2}}{c^2(h+\eta)} = 0$$

$$\frac{\partial V}{\partial t} + U\frac{\partial V}{\partial x} + V\frac{\partial V}{\partial y} + g\frac{\partial \eta}{\partial y} + \frac{V(U^2+V^2)^{1/2}}{c^2(h+\eta)} = 0 \qquad (7)$$

$$\frac{\partial \eta}{\partial t} + \frac{\partial}{\partial x}\{(h+\eta)U\} + \frac{\partial}{\partial y}\{(h+\eta)V\} = 0$$

The equations (7) will be solved in a domain ω whose boundary (γ) is the orthogonal projection on the {x,y}-plane of the intersection of the two surfaces Z_1 and Z_2.

γ is an open line, on which has been imposed the vanishing of the normal component of the velocity to the line. On the open boundaries (see for example A-B in fig;1) the time evolution of the water level is given.

As we have briefly discussed the NLSWT is not always applicable to the study of the propagation of the tidal waves, namely when the lateral scale of the phenomenon is comparable with the depth.

For the lagoon of Venice the three inlets are the only regions where the assumptions of the NLSWT are not fully verified.

Moreover the NLSWT does not take into account the existence of multiple-connected regions defined by land emergent at low tide. In spite of the absence of an appropriate theory, the finite difference formulation of the problem allows to introduce new boundaries whose shapes depend on the depth of the water in the basin.

Let us now discuss the problem of the reduction of the number of spatial coordinates.

If the vertical acceleration and velocity fields are not taken into account the last Navier-Stokes equation may be written in the form

$$\frac{\partial p}{\partial z} = -\rho g \qquad (8)$$

Equation (8) leads to a linear dependence of the pressure p from the water height (z). By integrating equation (8) between the limits Z_1 and Z_2 we obtain the value of the pressure p. The assumption defined by equation (8) is usually known as the hydrostatic hypothesis.

The numerical methods; stability and accuracy.

In this section we consider the solution of equation (7) using finite difference methods. Two explicit methods are used with two different kinds of network.

.1) The P. LAX method [5].

This numerical method considers the solution of equation (7) on an orthogonal network. Such a kind of network is called "dense" because on each node of the network all the variables of the problem are computed: u, v, and η (fig.2).

We must notice that u and v are the mean velocities in the x and y directions. The time derivaties are computed using two time levels (fig.3) connected by the following relationship

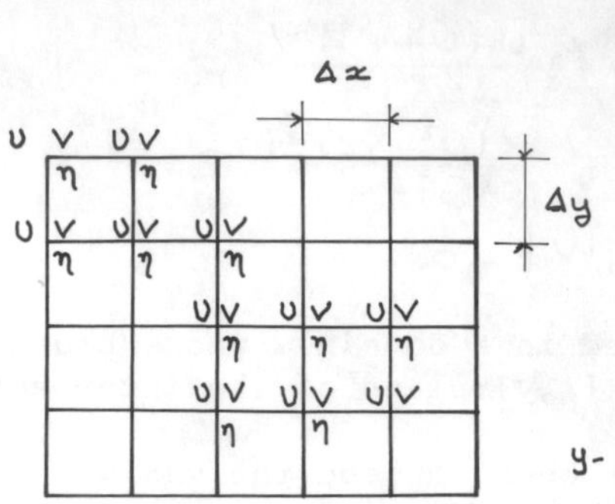

Fig.2 Example of 'dense' network. u, v, η are computed at the same time level.

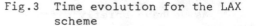

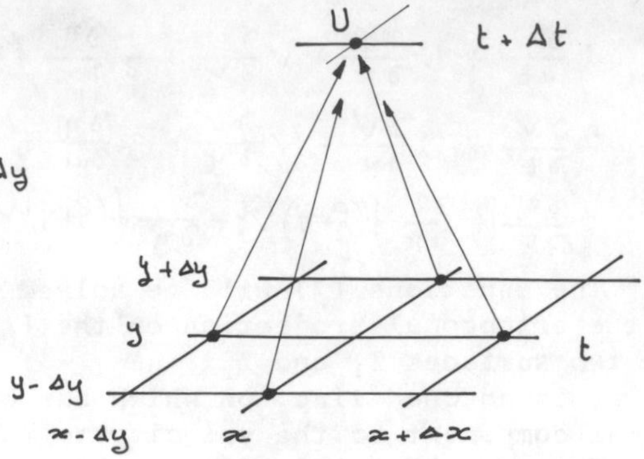

Fig.3 Time evolution for the LAX scheme

$$U(t+\Delta t, x, y) \cong \bar{U}(t, x, y) + \Delta t \frac{\partial U}{\partial t} \tag{9}$$

where

$$\bar{U}(t,x,y) = \frac{1}{2}\left[U(t, x+\Delta x, y) + U(t, x-\Delta x, y)\right]\frac{\Delta y^2}{\Delta^2} + \frac{1}{2}\left[U(t, x, y+\Delta y) + U(t, x, y-\Delta y)\right]\frac{\Delta x^2}{\Delta^2} \tag{10}$$

and

$$\Delta^2 = \Delta x^2 + \Delta y^2$$

In order to display the terms which are responsible for the artificial viscosity of the LAX scheme we substitute expression (10) into (9). Finite difference equation (9) may then be written in the following way

$$U(t+\Delta t, x, y) = U(t,x,y) + \Delta t \frac{\partial U}{\partial t} + \frac{\Delta x^2}{2}\left(\frac{\partial^2 U}{\partial x^2}\right)\frac{\Delta y^2}{\Delta^2} + \frac{\Delta y^2}{2}\left(\frac{\partial^2 U}{\partial y^2}\right)\frac{\Delta x^2}{\Delta^2} \tag{11}$$

where the last two terms are responsible for the strong artificial viscosity of this method. For this scheme (11) we must work with the maximum time step (the Courant-Friederich-Lewey criterion must be satisfied) and with the minimum mesh size to obtain a minimum for the artificial viscosity.

The numerical scheme may be obtained writing equation (7) in the finite difference form, using the expression (10) to evaluate the time derivative and a centered formula to compute the spatial derivative. We obtain in this way only a first order scheme because of the averaging introduced with the equation (10).

The last part of this section is devoted to demonstrate the order of accuracy of the Lax scheme.

Let us consider a one-dimensional problem written in symbolic form:

$$L\,\psi = f \tag{12}$$

where f is a known function and L is a linear differential operator.

For the present work we consider the following operator:

$$L \equiv \frac{\partial}{\partial t} + \frac{\partial}{\partial x} \tag{13}$$

the corresponding finite difference equation for the Lax scheme may be written in the following form:

$$\frac{\psi_i^{n+1} - \frac{1}{2}\left(\psi_{i+1}^n + \psi_{i-1}^n\right)}{\Delta t} - \frac{\psi_{i+1}^n - \psi_{i-1}^n}{2\,\Delta x} = f_i^n \tag{14}$$

where $t = n\,\Delta t$ and $x = i\,\Delta x$.

We substitute now the proper Taylor expression about the point $\{x,t\}$ for the various ψ_i^{n+1} or $\psi_{i\pm1}^n$. The equation (14), as it is shown in (6) may be written as follows:

$$\left(L_{\Delta x}[\psi]_{\Delta x}\right)_i^n = \left\{\psi_t - \psi_x - \frac{\Delta x^2}{2\,\Delta t}\psi_{xx} + \frac{\Delta t}{2}\psi_{tt}\right\}_{\substack{t=n\Delta t\\ x=i\Delta x}} + O\left(\Delta t^2, \Delta x^2, \frac{\Delta x^4}{\Delta t}\right) \tag{15}$$

As the equation (15) shows, the finite difference scheme approximates the differential equation (12) or the equation

$$\psi_t - \psi_x - k\,\Delta x\,\psi_{xx} \ .$$

depending on the relationship between the time step and the mesh size. If we choose a quadratic dependence between Δt and Δx

$$\Delta t = k\,\Delta x^2 \tag{16}$$

the finite difference scheme approximates to the first order the differential equation (12).

$$\left(L_{\Delta x}[\psi]_{\Delta x}\right)_i^n = \left(\frac{\partial\psi}{\partial t} - \frac{\partial\psi}{\partial x}\right)_{\substack{t=n\Delta t\\ x=i\Delta x}} + O\left(\Delta x\right) \tag{17}$$

.2) <u>Some results obtained with the LAX method</u>.

As far as the stability of the Lax scheme is concerned and its dependence on the input wave frequency four simulations were carried out. For these simulations we used a network with 1250m mesh size. Figure 4 displays the four simulations, which have as input a sinusoidal wave with a period of

12 H	Fig. 4a
6 H	Fig. 4b
3 H	Fig. 4c
1 H	Fig. 4d

As we can see in fig.4 no instability effects may be found for the different input waves. However the artificial viscosity of the method which is partially responsible for this strong stability, introduces an excessive attenuation of the waves inside the lagoon. For example with a semidiurnal tidal wave (a sinusoidal wave with a period of 12H)

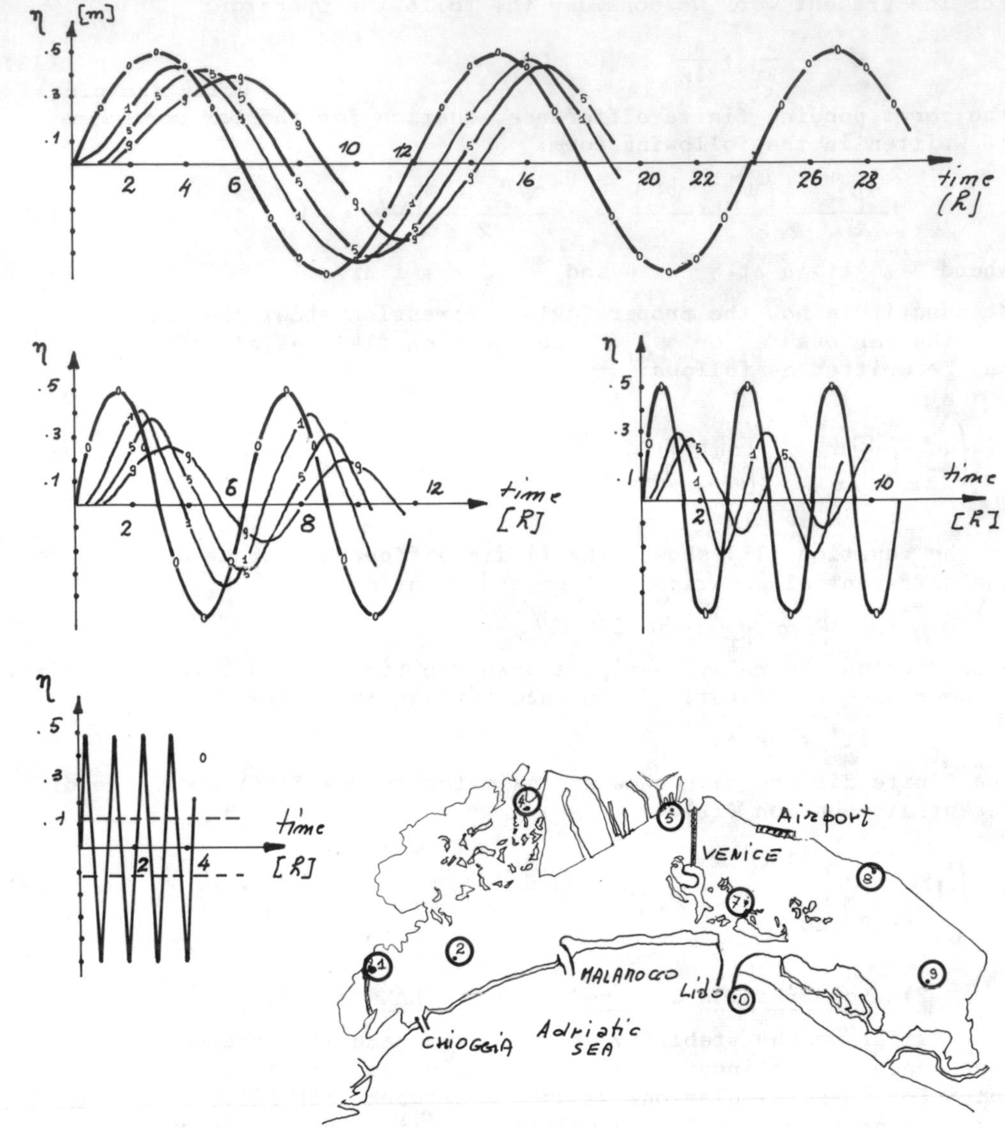

Fig. 4 Time evolution of the water level in different points of the lagoon.

we can see in some parts of the lagoon a small amplification of the tidal wave. The simulation with a semidiurnal tide (fig. 4a) does not show such kind of amplification. Only a reduction of the mesh size may avoid this effect. At present a network with a mesh size of 300 m is almost completely set down.

.3) The staggered method (W. HANSEN).

This numerical scheme due to W. HANSEN may be described with four time levels. As it is shown in fig.5, starting from the continuity equation we can compute the new water level $\eta^{t+\Delta t}$ using the old water level $\eta^{t-\Delta t}$ and the velocities u^t, v^t. Now with the motion equations the new velocities $v^{t+2\Delta t}$, may be obtained using the old velocities u^t, v^t and the water level $\eta^{t+\Delta t}$. For this scheme the following centered formulas to evaluate the time and spatial derivatives will be used:

$$\frac{\partial U}{\partial t} = \frac{U_{i,J}^{n+1} - U_{i,J}^{n-1}}{2\Delta t} + O(\Delta t^3) \quad ; \quad \frac{\partial U}{\partial x} = \frac{U_{i+1,J}^{n} - U_{i-1,J}^{n}}{2\Delta x} + O(\Delta x^3) \tag{18}$$

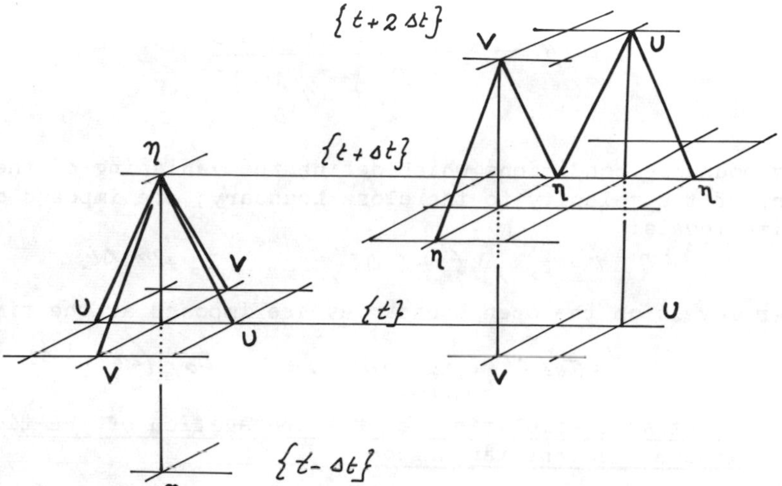

Fig. 5 Time evolution for the staggered scheme.

Moreover for this scheme a set of equations, equivalent to equation (7), have been used:

$$\begin{aligned} &\frac{\partial U}{\partial x} + g(h+\eta)\frac{\partial \eta}{\partial x} + g\,\frac{U\{U^2+V^2\}^{1/2}}{C^2(h+\eta)^2} = 0 \\ &\frac{\partial V}{\partial y} + g(h+\eta)\frac{\partial \eta}{\partial y} + g\,\frac{V\{U^2+V^2\}^{1/2}}{C^2(h+\eta)^2} = 0 \\ &\frac{\partial \eta}{\partial t} + \frac{\partial U}{\partial x} + \frac{\partial V}{\partial y} = 0 \end{aligned} \tag{19}$$

In this set of equations (19) u, v are the x and y mass flux. The advective terms are not taken into account because of their small influence, especially with the present network.

Moreover the elimination of these advective terms allows the construction of a second order scheme (cfr. (6), chapter III, part 1). However we must take care in the computation of the last term which takes into account the loss of energy due to the friction (7). This term must be evaluated during the second step as described at the beginning of this paragraph, but at the time $t+\Delta t$ only the water levels are known. For that reason we must write the finite difference equation in the following form:

$$U_{ij}^{n+2} = \ldots\ldots \quad g\,\frac{U_{ij}^{n+2}\left\{\left(U_{ij}^{n}\right)^2+\left(V_{ij}^{n}\right)^2\right\}^{1/2}}{C^2\left(h_{ij}+\eta_{ij}^{n+1}\right)^2} \tag{20}$$

The boundary conditions which define the vanishing of the normal component of the velocity to the close boundary, are imposed only at these time levels:

$$t = t_0\,,\quad t_0+2\Delta t\,,\quad \ldots\quad t_0+2n\,\Delta t \tag{21}$$

The water levels on the open boundaries are imposed at the time levels:

$$t = t_0+\Delta t\,,\quad t_0+3\Delta t \quad \ldots\quad t_0+(2n+1)\,\Delta t \tag{22}$$

.4) <u>About some simulations of the propagation of the tidal wave inside the Venetian Lagoon</u>.

For this method too the stability analysis in the frequency domain was performed. The scheme is stable even if the period of the input wave is of 1 hour. The simulation of the semi-diurnal tide showed at the point 6 (fig. 4c) a gain factor (ratio between the tide inside and outside the lagoon) of 1.2 with a good agreement with the experimental data.

A few remarks about the convergence of the numerical scheme.

Four 9-day simulations have been carried out with a 1250m-size mesh network and with time step Δt 10,20,30, 60 seconds.

In order to illustrate the rate of convergence of the numerical solution $\eta(\Delta t,x,y)$ to the true solution $\eta(x,y,t)$ we have plotted in fig. 6 the function $F(t,\Delta t)$ which is a sort of measure of the space averaged deviation of $\eta(\Delta t_i,x,y,t)$ from $\eta(10,x,y,t)$.

$$F(t,\Delta t) = \iint_D \left|\,\eta(10,x,y,t)-\eta(\Delta t_i,x,y,t)\right|\,dx\,dy \tag{23}$$

D is the whole surface of the lagoon
$\eta(10,x,y,t)$ is the computed water level with a time step of 10 secs
$\eta(\Delta t,x,y,t)$ is the computed water level

The graph in fig.7 clearly shows the following properties:

a) the magnitude of the errors is dependent on the rate of variation of the water level but it seems to be limited
b) the maximum mean error is of 3 mm

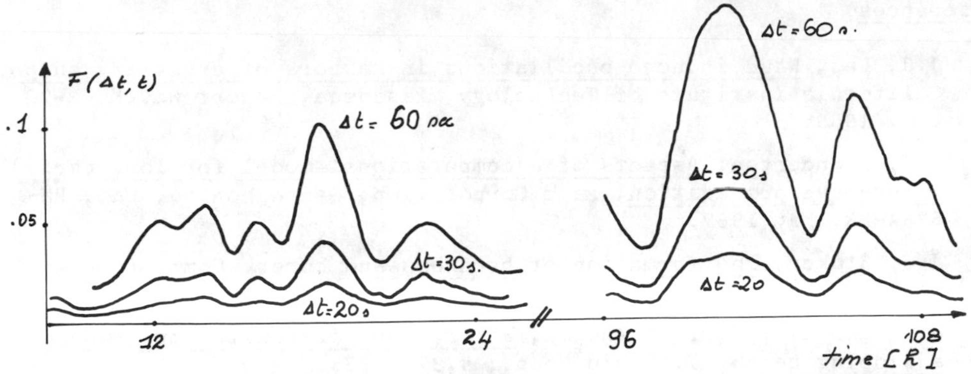

Fig. 6 An example of the convergence of the numerical solution.

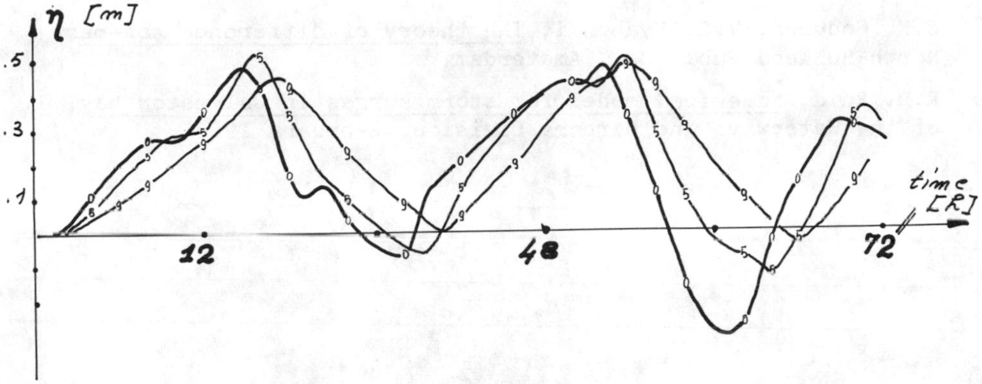

Fig. 7 A 9-day simulation with the 1250m-size mesh network.

An example of simulation is shown in fig.7 ($\Delta x = \Delta y = 1250$m , $\Delta t = 60$ sec).

The roughness coefficient c was set to 60 over the whole lagoon.

The simulation shows a good qualitative agreement with the experimental data. For instance the oscillation in the northern part of the lagoon (lines 8, 9) is strongly damped as in the field.

From the inspection of the right part of fig.8 we can evaluate the phase of the tidal wave at different points of the lagoon: the agreement with the experimental data is satisfactory as the relative error never exceeds 20 percent.

Another simulation was carried out with a network of 1600 nodes (mesh size = 625 m). The main results of it were an encouraging decrease of the relative error in the phase evaluation and a better description of the tidal wave in the northern part of the lagoon.

This fully justifies our efforts to build a new network (300 m, for instance) and to analyze a sufficient amount of mareographic data in order to obtain a reliable calibration of the model.

References

1. J.J. Lee, Wave induced oscillations in harbors of arbitrary shape, California Institute of Technology, Pasadena, Report No. KH-R-20, Dec. 1969.

2. J.J. Leendertse, Aspects of a computational model for long-period water-wave propagation, Rand Corporation, Santa Monica, Ca., RM-5294-PR, May 1967.

3. J.J. Stoker, The formation of breakers and bores, Comm. Pure Appl. Math., 1,1, 1948.

4. G.F. Carrier, H.P. Greenspan, Water waves of finite amplitude on a sloping beach, J. Fluid Mech., 4,97, 1958.

5. P. Lax, Weak solutions of non linear hyperbolic equations and their numerical computation, Comm. Pure Appl. Math., 7,159,193, 1954.

6. S.K. Godunof, V.S. Ryabenki, The theory of difference schemes, North-Holland Publ. Co., Amsterdam.

7. R.D. Reid, Numerical model for storm surges in Galveston bay, J. of the Waterways and Harbors Division, February 1968.

SEA LEVEL PREDICTION MODELS FOR VENICE

by A. Artegiani,* A. Giommoni,* A. Goldmann,^ P. Sguazzero,^ A. Tomasin*

* CNR Laboratory for the Study of the Dynamics of Large Masses, Venice
^ IBM Scientific Center, Venice

In the northern side of the Adriatic sea (Fig. 1) the water level, under certain weather conditions, can rise beyond the ordinary value of the high tide and cause severe floods in the coastal areas. The effects of these phenomena are particulary famous for Venice, but they threaten a much larger area, which certainly includes the whole lagoon where Venice is located.

Fig. 1 - Central Mediterranean Area Mareographic (.) and meteorologic (o) stations, whose data were used in the present work.

The possibility of forecasting the floods was limited, until recently, to the experience gained by the local offices in the past, and enabled to a prediction made two or three hours in advance.

But an objective forecast with a longer time lag is necessary, and the need will increase when sluices for the lagoon, which have already been planned, will be built.
From a practical point of view, we can observe that winter time is the period in which the danger for high water in Venice becomes most threatening. It is in this period that cyclones coming from the British Isles or from the Western Mediterranean (Fig. 2), move towards Italy.

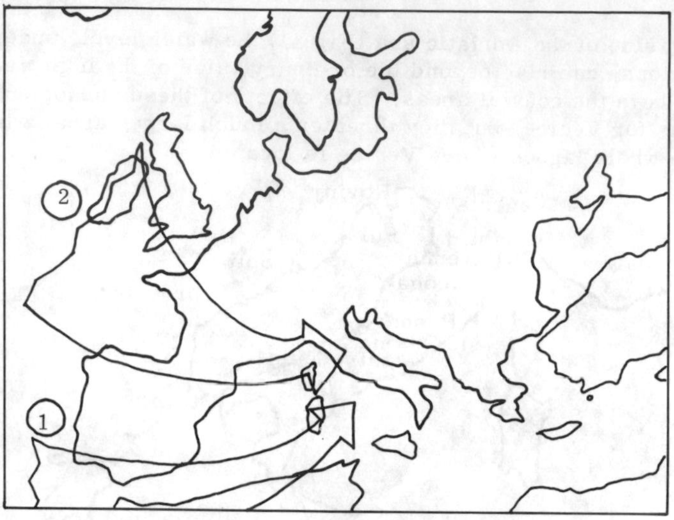

- Fig. 2 Preferential tracks of Atlantic cyclones entering the Mediterranean

These atmospheric disturbances generate winds blowing from the South or the South-East which pile up water in the northern part of the Adriatic Sea.
An experienced weatherman, just looking at the meteorological maps, can guess whether dangerous storm surges will occur or not, but if we want to give a numerical forecast of the sea level with the explicit indication of the time of the event, a more careful analysis is necessary.
The observed level at any point in the Adriatic can be regarded, in a first approximation, as the sum of two components which we assume to be indipendent:

a) the astronomical tide due to the attractions of the sun and the moon on the waters of the sea;
b) the meteorological tide due to the atmospheric forces, namely the tangential wind stress and the normal barometric pressure.

As the astronomical tide can be predicted with high accuracy with standard techniques we shall speak from now on only about meteorological tide.

Given the geometric characteristics of the Adriatic, the surges in the sea can be described by the hydrodynamic equations in one-dimensional form: the continuity equation which follows from the principle of conservation of mass

$$b(x)\frac{\partial \eta}{\partial t} + \frac{\partial Q}{\partial x} = 0$$

the motion equation which simply expresses the balance between the terms of inertia and surface slope and the atmospheric pressure gradient, the wind stress, the bottom stress

$$\frac{\partial Q}{\partial t} + g S(x)\frac{\partial \eta}{\partial x} = S(x)\left[-\frac{1}{\rho_w}\frac{\partial P_a}{\partial x} + \frac{1}{h(x)}(T_s - T_b)\right]$$

where

x	is the basin axis
$Q(x,t)$	the discharge
$\eta(x,t)$	the displacement of water from equilibrium surface
$S(x)$	the cross section area
$b(x)$	the basin width
$h(x)$	the basin depth from equilibrium surface
P_a	the atmospheric pressure
T_s	the surface wind stress
ρ_w	the density of the water
g	the acceleration of gravity

The bottom stress T_b is usually simulated by a quadratic or linear function of the mean velocity of the fluid: given the use we will make of this kind of analysis the choice of a frictional term proportional to the discharge is satisfactory. It is easy to see that in this case the solution of the hydrodynamic equations (under proper initial and boundary conditions) takes the form of a series of fundamental oscillations (modes): each of them can be obtained as a convolution integral of a suitable forcing function (depending on the meteorological variables) and a response function of a sinusoidal-damped type. The behaviour of the Adriatic Sea seems sufficiently close to this simplified scheme; it's known from many years that the basin oscillates on characteristic frequencies, when it has been excited by atmospheric disturbances. The first eigenfrequencies correspond to the dominant oscillations or seiches, whose shapes, derived from the previous theoretical considerations, are shown in figures 3 and 4.
The first shows the uninodal seiche, with the node at Otranto, the mouth of the basin, and with a period of about 22 hrs. The second figure shows the binodal seiche which as a period of about 10 hrs.

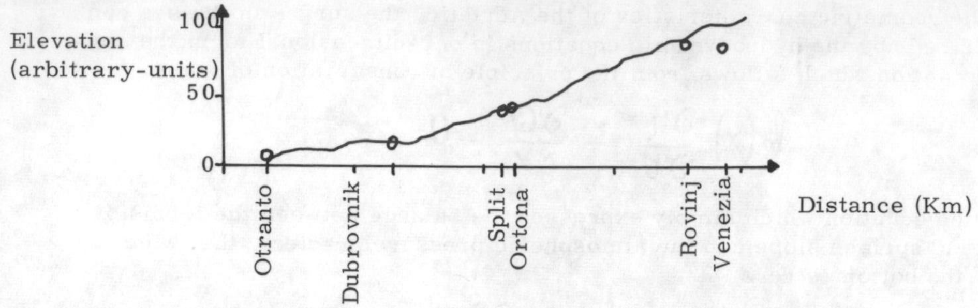

- Fig. 3 Theoretical profile of the first seiche along the Adriatic. Experimental estimations are also given

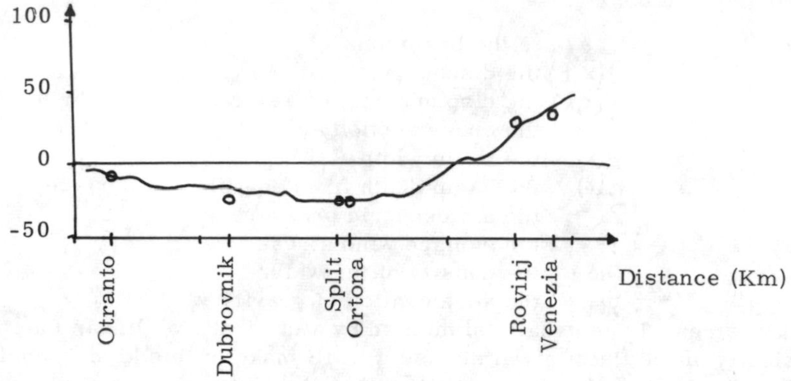

- Fig. 4 Theoretical profile of the second seiche along the Adriatic Experimental estimations are also given

Spectral techniques were used in order to obtain an experimental evaluation of the periods and put in evidence the importance of the seiches.
The spectrum of many months of the recorded sea level at Venice, 1966, from which the astronomical tide has been subtracted is shown in Fig. 5.
Apart from a very low frequency band which is due to long period meteorological fluctuations, the spectrum contains two distinct energy bands (B, C) at the seiches periods of 21.7 and 10.8 hrs; the periods computed with the simple analytical model previously presented are of 21.69 and 9.92 hrs, and the qualitative agreement is good indeed.

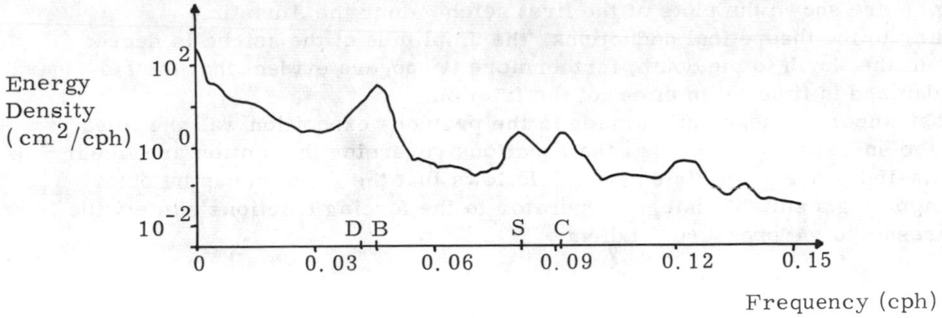

- Fig. 5 Spectrum of the meteorological tide in Venice, 1966

A more direct analysis can be carried out by filtering the time series of the sea level of several harbours in the Adriatic Sea by suitable band-pass filters centered on the eigenfrequencies and by comparing the outputs.

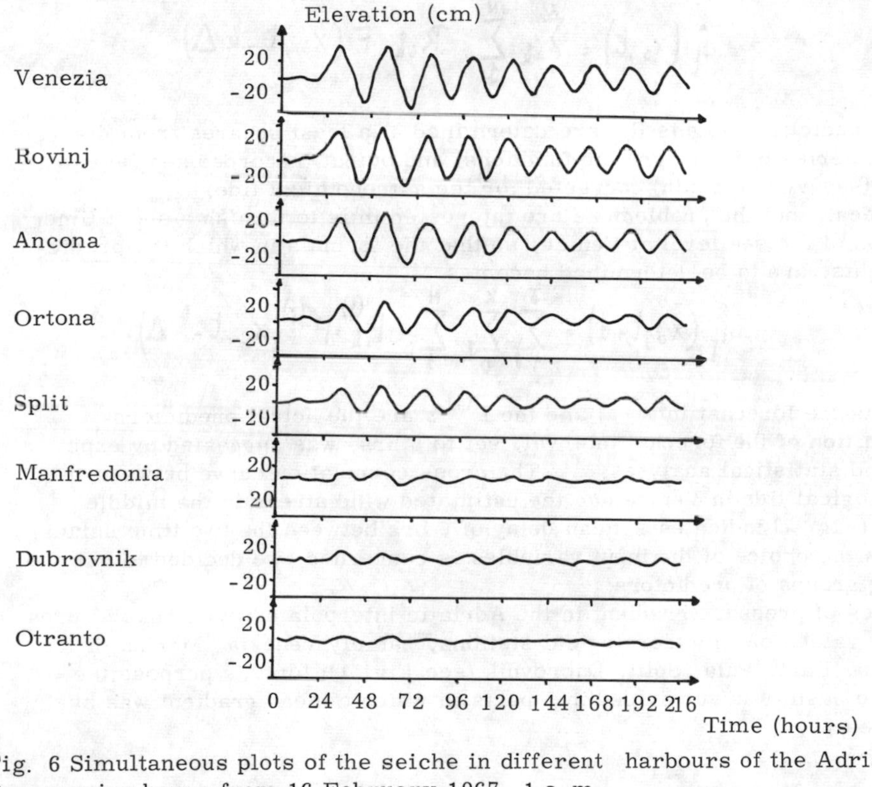

- Fig. 6 Simultaneous plots of the seiche in different harbours of the Adriatic. Progressive hours from 16 February 1967, 1 a. m.

Ing Fig. 6 are shown the plots of the first seiche along the Adriatic. According to the theoretical deductions, the amplitude of the seiche is decreasing from the North to the South; furthermore it appears evident that the seiche is damped in time as an effect of the friction.
The most important assumption made in the previous exposition was the linearity of the surge; this means that the equations governing the motion are linear with time-indipendent coefficients. It follows that the solution can be obtained by applying a suitable integral operator to the forcing functions namely the wind stress and the pressure gradient:

$$\eta(x_0, t) = \int_0^{\infty}\int_0^{l} R(x_0, x, T)\, \mathfrak{F}(x, t-T)\, dx\, dT$$

The prediction problem for the sea level at the point x_0 (say Venice) is to find the function R (x_0, x, T) (the response function) wich relates to $\eta(x_0, T)$ its principal causes $\mathfrak{F}(x, T)$
We approach the problem of determining the response function of the Adriatic, with statistical methods. Essentially the following steps are involved:

a) the integral operator is discretized

$$\hat{\eta}(x_0, t) = \sum_{0}^{K}{}_{k} \sum_{1}^{N}{}_{i} R_{ik}\, F(x_i, t-k\Delta)$$

b) the prediction weights R_{ik} are determined à la least squares from discrete time series of input (driving functions) and output (recorded sea level at x_0, (say Venice again) corrected for the astronomical tide).

Let us recall that the problem we are interested in is to furnish a short time prediction of the sea level at Venice, so that the formula in which the prediction weights, are to be determined becomes

$$\hat{\hat{\eta}}(x_0, t+d) = \sum_{1}^{J}{}_{j} \sum_{0}^{K}{}_{k} \sum_{1}^{N}{}_{i} P_{ik}^{(j)} F^{(j)}(x_i, t-k\Delta)$$

where d is the forecast interval and the $F^{(j)}$'s are the actual predictors.
The definition of the forecast interval, set to 6 hrs, was suggested by experience and statistical analyses. The cross correlation curve between the meteorological tide in Venice and the estimated wind stress in the middle Adriatic (Fig. 7) indicates a mean delay of 6 hrs between the two time series.
As far as the choice of the input variables is concerned, we decided to use the following groups of predictors:

1) values of pressure gradient in the Adriatic interpolated from coastal pressure values of 7 meteorological stations, namely Venezia, Marina di Ravenna, Bari, Pula, Split, Dubrovnik (see Fig. 1); for this purpose the whole basin was subdivided in 5 parts in which a mean gradient was assumed;

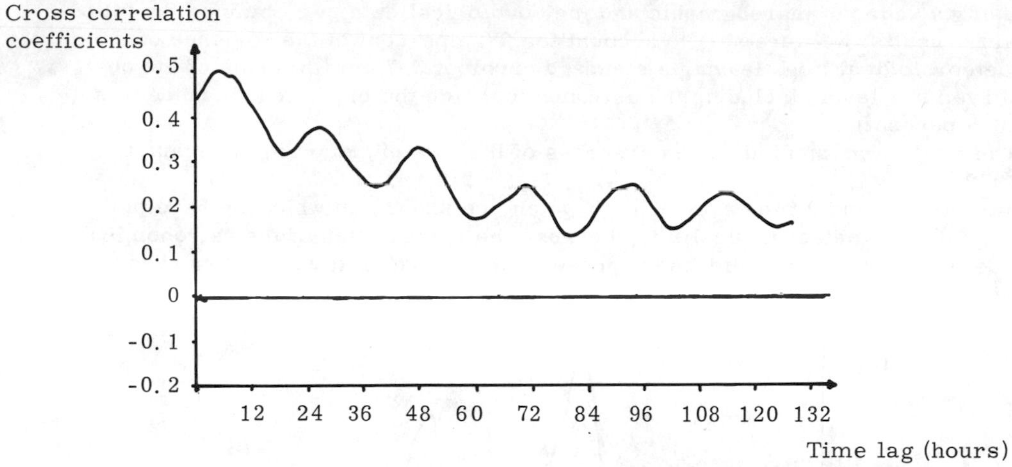

- Fig. 7 Cross correlation functions for the wind stress in the Adriatic and the meteorological tide in Venice (1966)

2) values of $|\nabla P|\nabla P$ for the same zones; they are representative of the surface wind stress by accepting the geostrophic hypothesis and a quadratic relationship between stress and wind.

It can be shown that for the meteorological tide the autocovariance remains large for a considerable amount of time. In other words the meteorological tide is highly predictable from its own past.
Se we decided to introduce as an additional input in the set of predictors

3) the past of the sea level itself

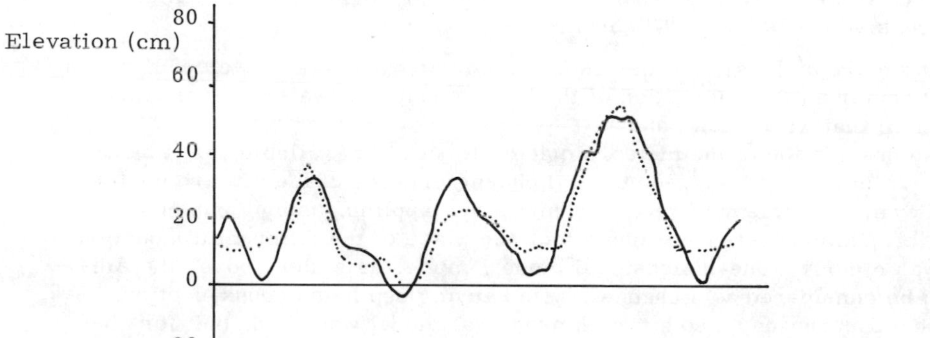

- Fig. 8 Observed (——) and predicted (- - -) meteorological tide in Venice. Progressive hours from 1 April 1971, 7p.m.

Using 4 years of mareographic and meteorological data, we obtained the follo wing results: we were able to account for 87. 5 percent of the variance of the meteorological tide, leaving a standard error of 5. 7 cm (in terms of the observed sea level, including the astronomical tide the predicted variance was 96. 5 percent).
The model was applied during all cases of high water, starting from autumn 1970.
In figures 8 and 9 two cases of high water are shown, in which we have plotted only the meteorological tide, because the computation of the astronomical tide is standard: the solid line represents the surge as it can be observed, the dashed line the surge forecasted 6 hrs in advance using the statistical model.

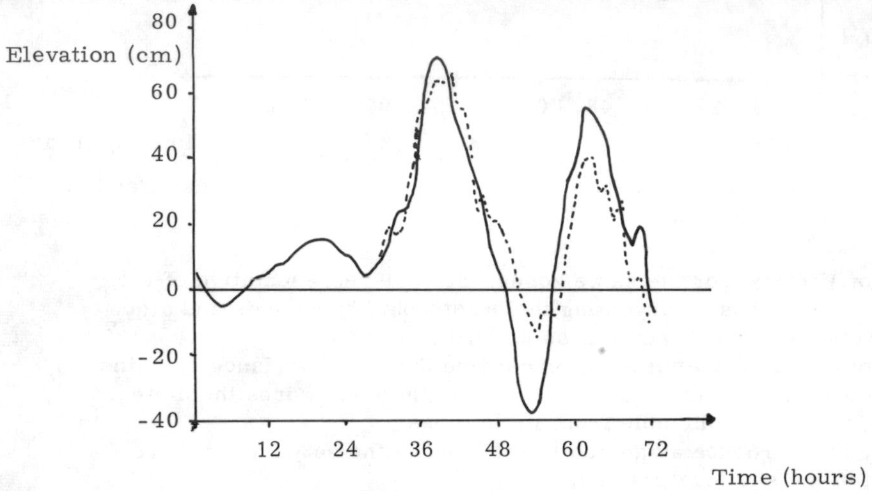

-Fig. 9 Observed (——) and predicted (---) meteorological tide in Venice. Progressive hours from 12 February 1972, 4a. m.

Following more or less the same line, another approach was developed, which directly performs the integration of the basic equations (water conservation and motion) that were shown above.
Since the description of the meteorological "forces" is available, as it has been shown above, and the geometrical characteristics of the sea are obviously known, a simple numerical technique was applied, using a computer. An explicit finite-difference scheme took the place of the differential equations, and it was clearly a one-dimensional model, since, as stated above, the Adriatic can be considered as a channel, with varying depth and cross-section. (For special purposes, also a two-dimensional model was used, but for the practical task of surge prediction the former one was sufficient).
When using this method, one can integrate the equations and deduce the levels and currents only as long as the forcing functions are known. To predict future conditions, one has either to use the weather forecastings or just to keep the latest weather reports as valid in the following hours.

This surprising "wind freezing" turns out to be useful when forecasting about six hours in advance for the reason that this lag corresponds to the typical response time (in the obvious sense), so that the changes in the meteorological field that will occur during this time will not appreciably affect the results.

This method also was tested by a "field use", and again the results were satisfactory, as one cas see from Fig. 10, whose accuracy is typical.

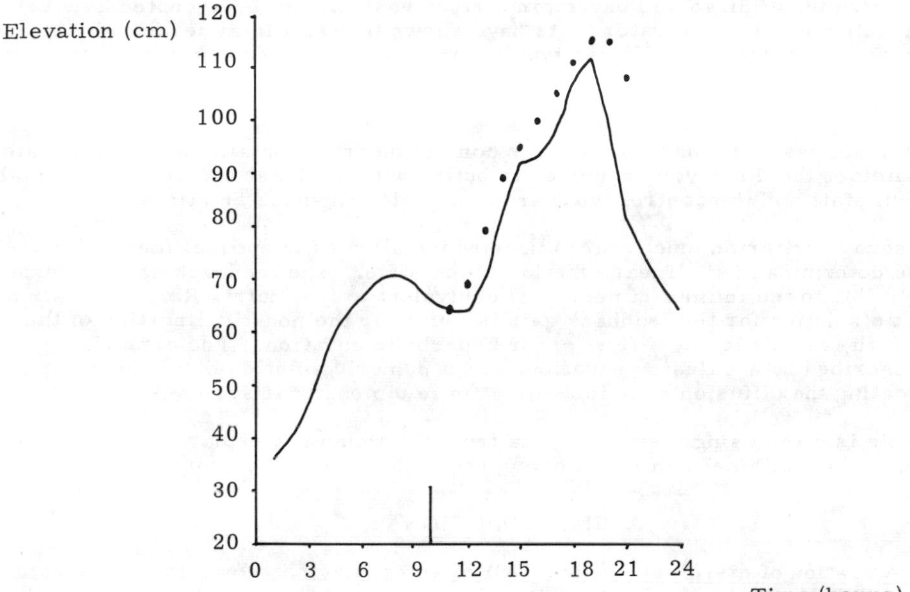

- Fig. 10 An example of prediction of floods in Venice by hydrodynamical numerical models. Using the meteorological reports of the time marked on the horizontal axis, a forecast was issued (square points) announcing the flood, as soon as the reports were made available (more than an hour later).
  Both the solid line (observed values) and the prediction refer to the 'total' tide, without astronomical subtraction. The time scale begins at 9. 0, November 9, 1971

The conclusion that one can draw from our experiment is that the forecast of the floods in the Adriatic can be made, with a good accuracy, up to six hours in advance. A longer time lag for prediction will require a deeper involvment in the meteorological forecasting.

We are indebted to the Ufficio Idrografico del Magistrato alle Acque di Venezia, the Servizio Meteorologico dell'Aeronautica Militare Italiana, the UNESCO and many of our colleagues for their help in this research.

OPTIMAL ESTUARY AERATION: AN APPLICATION OF DISTRIBUTED PARAMETER CONTROL THEORY

Wayne Hullett*
The Singer Company
Librascope Systems Division

ABSTRACT

The concentration of dissolved oxygen in a river has come to be accepted as a criterion of water quality. In-water tests have shown that artificial aeration by means of in-stream mechanical or diffuser type aerators can be an economically attractive supplement or alternative to advanced wastewater treatment as a means of improving water quality.

This paper applies distributed parameter control theory to obtain the aeration rate that maximizes the dissolved oxygen distribution with the least control effort. Both the system state and the control input are distributed in space and time.

A mean square criterion functional is used which allows the optimal feedback control to be determined as a linear function of the state. The feedback gain is found as the solution to the infinite dimensional equivalent to the matrix Riccati equation. An analytic solution for the feedback gain is found for the non-tidal portion of the river, which is modelled by a first order hyperbolic equation. The estuarine portion is described by a diffusion equation, and a numeric solution obtained by approximating the diffusion equation with a finite dimensional system.

An example is given using historical data from the Delaware estuary, and the dollar cost of the optimal control is compared with other ad hoc control strategies.

1. INTRODUCTION

The concentration of dissolved oxygen (DO) in an estuary has come to be accepted as a criterion of water quality. It has been suggested (Whipple, 1970) that the use of artificial in-stream aeration can be an economically attractive supplement or alternate to advanced wastewater treatment.

The problem to be addressed here is the determination of the aeration rate that achieves the best increase in the DO level with the least control effort. This is approached as a distributed parameter control problem, where both the system state and the control are distributed in space and time. A solution in the form of a feedback control is sought.

2. WATER QUALITY MODEL

The dissolved oxygen level in an estuary is decreased by the oxidizing action of an effluent, described by its biochemical oxygen demand (BOD). The DO is increased by atmospheric aeration and by artificial in-stream aeration. Other factors affecting DO but not included here are benthal demand, algal respiration, photosynthesis and the nitrogen cycle. In order to illustrate the action of the optimal control, the following partial differential equation (Harleman, 1971) is used to describe

---

*The research described in this paper was conducted under the supervision of Professor A. V. Balakrishnan while the author was a Research Assistant at the University of California, Los Angeles.

the distribution of dissolved oxygen:

$$\frac{\partial C}{\partial x} = \frac{1}{A}\frac{\partial}{\partial x}\left(AE\frac{\partial C}{\partial x}\right) - v\frac{\partial C}{\partial x} + K_3 (C_s - C) + U - K_1 L \qquad (1)$$

with initial condition

$$C(x, 0) = C_o(x)$$

and boundary conditions

$$C(0, t) = f(t)$$
$$C(x_f, t) = g(t)$$

where

$A(x, t)$ = estuary cross section area (sq. ft.)
$C(x, t)$ = dissolved oxygen concentration (mg/ℓ)
$C_s(x, t)$ = DO saturation value (mg/ℓ)
$E(x, t)$ = tidal dispersion (sq. ft. / day)
K_1 = BOD oxidation rate (1/day)
K_3 = atmospheric aeration rate (1/day)
$L(x, t)$ = biochemical oxygen demand (mg/ℓ)
$U(x, t)$ = aeration rate (mg/ℓ/day).

The BOD concentration as a function of the effluent input obeys a similar equation:

$$\frac{\partial L}{\partial t} = \frac{1}{A}\frac{\partial}{\partial x}\left(AE\frac{\partial L}{\partial x}\right) - v\frac{\partial L}{\partial x} - K_1 L + J$$

where

$J(x, t)$ = effluent discharge rate (mg/ℓ /day

Many criterion functionals are relevant: the least squares has the advantage that the optimal control is a feedback control. The functional to be minimized is therefore:

$$S(U) = \int_0^T \int_0^{x_f} Q(x)\,[C - C_s]\,dx\,dt + \lambda \int_0^T \int_a^b R(x)\,U^2\,dx\,dt$$

where the control interval (a, b) is contained in the estuary length $(0, x_f)$, and T is the time duration of the control.

Minimizing this functional over $U(x, t)$ will tend to drive the DO level toward the saturation value, while penalizing large control inputs. The functions $Q(x)$ (non-negative) and $R(x)$ (positive definite) are appropriate weighting functions.

3. OPTIMIZATION PROBLEM

Let

$$C(x, t) \in L_2 (0, x_f; 0, T) = H_1$$

$$U(x, t) \in L_2 (a, b; 0, T) = H_2$$

and define the mapping

$$B: H_2 \rightarrow H_1$$

by

$$BU = \begin{cases} U(x, t); & x \in (a, b) \\ 0; & \text{otherwise} \end{cases}$$

The model equation (1) can be written in the form

$$\frac{\partial C}{\partial t} = FC + BU + V$$

where F is the operator defined by

$$FC = \frac{1}{A} \frac{\partial}{\partial x} (AE \frac{\partial C}{\partial x}) - v \frac{\partial C}{\partial x} - K_3 C$$

with domain $D(F) = \{C \in H_1 \mid FC \in H_1; \quad C(0, t) = f(t); \quad C(x_f, t) = g(t)\}$

and

$$V = K_3 C_s - K_1 L.$$

The criterion functional is the sum of inner products in H_1 and H_2

$$S(U) = \langle (C-C_s), Q(C-C_s) \rangle_1 + \lambda \langle U, RU \rangle_2$$

The feedback control that minimizes S(U) (Lions, 1968) is

$$U = -\frac{1}{\lambda} R^{-1} B^* (MC + G)$$

where M is a self adjoint linear operator and G(x, t) is a function that satisfies

$$\frac{\partial M}{\partial t} = -F^* M - FM - \frac{1}{\lambda} MBR^{-1} B^* M + Q$$

and

$$\frac{\partial G}{\partial t} = -F^* G + \frac{1}{\lambda} MBR^{-1} B^* G - MV + QC_s. \qquad (2)$$

The equation for M is the infinite dimensional analog of the matrix Riccati equation where equality $M_1 = M_2$ is to be interpreted in the sense

$$M_1 C = M_2 C \quad \text{for all } C \in D(F)$$

The final conditions are

$$MC(x,t)\Big|_{t=T} = G(x,T) = 0$$

and boundary conditions are

$$MC(x,t)\Big|_{x=0} = MC(x,t)\Big|_{x=x_f} = G(0,t) = G(x_f, t) = 0.$$

Since final conditions are given for the equations for M and G, these must be solved backwards in time, so that knowledge of all future values of the system parameters A, E, v and C_s and the system input $V = K_3C_S - K_1L$ for $t \in (0,T)$ is required. This is not as bad as it seems, however, and for the systems under consideration, is actually a benefit, because it permits the use of all available information in computing the optimal control. In general, one is able to predict several days in advance from meteorlogic conditions and producţion schedules the temperature and BOD discharges upon which C_s and L depend. Since this information is available, it is reasonable that a control based on more information would be better than one computed without this extra knowledge.

One can envision a system, then, that would compute an optimal control (M and G) for a period of, say, 10 days, for which the temperatures and discharges could be reasonably predicted. When more data becomes available, after 5 days for instance, a new M and G could be computed for the next 10 days, or for the remaining 5 days if it is desired to stop control (aeration) at that time. One then has, at each time, a control system that is optimal based on all information available at that time.

4. ANALYTIC SOLUTION

The differential equations for M and G are extremely difficult to solve, and usually only a numeric solution is possible. In the special case of a stream (no tidal action) with a constant cross-sectional area and velocity, and assuming the molecular dispersion to be negligible compared with the bulk transport due to the stream velocity, the differential operator F becomes

$$FC = -v\frac{\partial C}{\partial x} - K_3C$$

In this simple case, an analytic solution for the feedback gain and forcing function can be found (Koppel and Shih, 1968). Assuming a solution of the form

$$MC = P(x,t)\,C(x,t),$$

and for definiteness letting

$$Q(x) = 1; \quad R(x) = 1$$
$$a = 0; \quad b = x_f$$
$$T = x_f/v$$

The Riccati equation becomes

$$\frac{\partial P}{\partial t} = -v \frac{\partial P}{\partial x} + 2K_3 P + \frac{1}{\lambda} P^2 - 1$$

with

$$P(x, T) = P(x_f, t) = 0,$$

which has the solution

$$P(x,t) = \begin{cases} \lambda [\alpha \tanh(-\alpha t + \beta) - K_3]; & x < vt \\ \lambda [\alpha \tanh(-\frac{\alpha}{v} x + \gamma) - K_3]; & x > vt \end{cases}$$

where

$$\alpha = \sqrt{K_3^2 + \frac{1}{\lambda}}$$

$$\beta = \alpha T + \tanh^{-1}\left(\frac{K_3}{\alpha}\right)$$

$$\gamma = \frac{\alpha}{v} x_f + \tanh^{-1}\left(\frac{K_3}{\alpha}\right).$$

Substituting this solution into (2) and solving (noting that the subsidiary conditions are $G(x, T) = G(x_f t) = 0$) yields

$$G(x,t) = \begin{cases} \operatorname{sech}(-\alpha t + \beta) \displaystyle\int_t^T \Big\{ \sqrt{\lambda} \sinh(\alpha [T - \sigma]) V[x - v(t-\sigma), \sigma] \\ \qquad - \cosh(-\alpha\sigma + \beta) C_s [x - v(t-\sigma), \sigma] \Big\} d\sigma; & x < vt \\ \frac{1}{v} \operatorname{sech}(-\frac{\alpha}{v} x + \gamma) \displaystyle\int_x^{x_f} \Big\{ \sqrt{\lambda} \sinh[\frac{\alpha}{v}(x_f - \sigma)] V[\sigma, t - \frac{x-\sigma}{v}] \\ \qquad - \cosh(-\frac{\alpha}{v}\sigma + \gamma) C_s [\sigma, t - \frac{x-\sigma}{v}] \Big\} d\sigma; & x > vt \end{cases}$$

These solutions for $P(x, t)$ and $G(x, t)$ completely define the optimal control as a feedback function of the current DO distribution:

$$U(x,t) = -\frac{1}{\lambda} (P(x,t)\, C(x,t) + G(x,t))$$

5. NUMERIC APPROACH

In order to obtain results for the general estuarine case, the diffusion equation was approximated by a finite system of difference equations (Sage, 1968) and the discrete optimal feedback control found in the usual manner.

This approach was applied to an example using historic data from the Delaware Estuary for the year 1964. The estuary was partitioned into 23 sections of 20,000 feet each and the control was constrained in sections 6 through 15, representing about 38 miles of estuary. The weighting function Q(x) in the criterion functional was adjusted to penalize DO deviations from the saturation value over the same interval.

The optimal control for a 15 day period is shown for several sections in Figure 1, and the resulting DO levels in Figure 2 can be compared with the DO levels without control in Figure 3.

6. COST OF AERATION

The energy required for aeration is given by

$$\text{Energy} = \int_0^T \int_a^b \frac{U(x,t)\,A(x,t)}{\eta(x,t)}\,dx\,dt$$

in horsepower hours, where

$$\eta(x,t) = \text{oxygen transfer efficiency}\left(\frac{\text{Lb } O_2}{\text{Hp-Hr}}\right).$$

Measured values (Whipple, 1970) of $\eta(x,t)$ for standard conditions of temperature and pressure in anaerobic water vary from 0.68 to 1.36 $\frac{\text{Lb } O_2}{\text{Hp-Hr}}$ depending on the depth of the aerator. For the depths under consideration, an average value of 1.0 was used. This value can be converted to test conditions by the relation

$$\eta_T = \frac{\frac{P}{29.92}(C_s)_T - C\,\phi^{T_c-20}}{(C_s)_{20}}$$

where

P = pressure (inches of mercury)
T_c = temperature (Deg. C)
ϕ = empirical constant (1.025)
$(C_s)_{20}$ = DO saturation value at standard conditions (9.02 mg/ℓ)
$(C_s)_T$ = DO saturation value at test conditions
$C(x,t)$ = DO concentration at aerator

The cost of the optimal control example for 15 days aeration is compared with two other control schemes in Table 1.

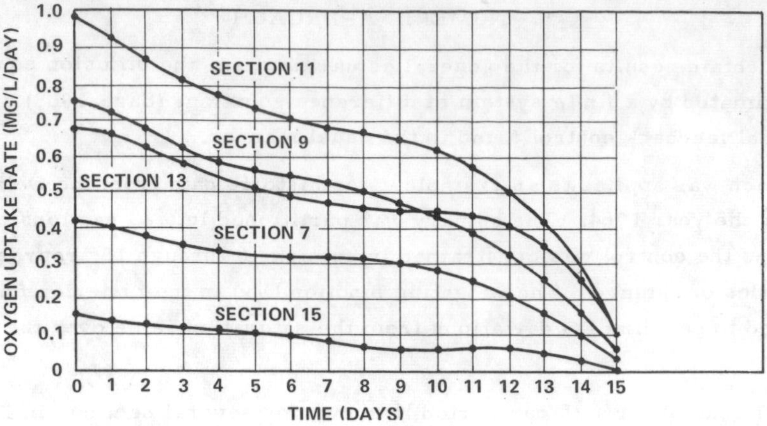

FIGURE 1. OPTIMAL CONTROL (λ=10)

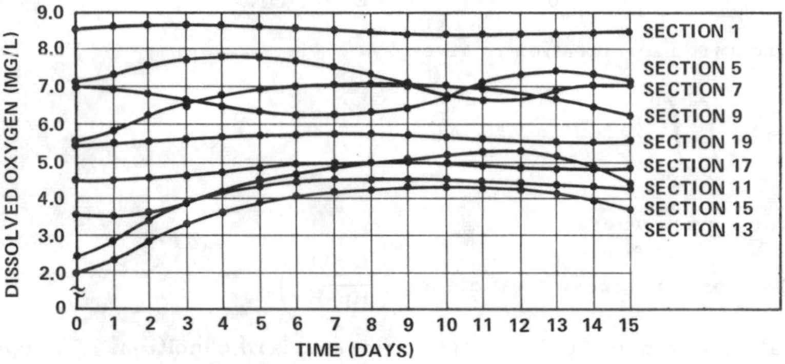

FIGURE 2. RESPONSE TO OPTIMAL CONTROL (λ=10)

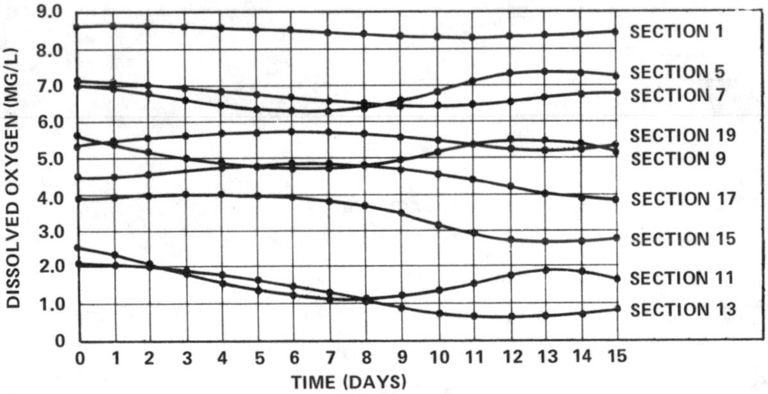

FIGURE 3. UNCONTROLLED DO

TABLE 1
COMPARISON OF CONTROL STRATEGIES

	Criterion Functional	Average DO Level (mg/ℓ)	Energy Cost (Dollars)	Savings (Dollars)
Optimal Control	0.115	5.5	89,327	--
Control 1	0.168	5.5	145,351	56,024
Control 2	0.156	5.5	105,046	15,719

Control 1 is an aeration rate that is constant over the same distance and duration as the optimal control. The constant rate is 0.5 mg/ℓ /day. Control 2 is an aeration rate that is proportional to the initial DO deficit in each section and is constant over the fifteen day duration. The proportionality factor is 0.116.

The criteria used for comparing the effectiveness of the control strategies are the value of the criterion functional and the average DO concentration over the controlled sections and the time interval. The cost for electrical energy was estimated at $0.006 per kilowatt hour, which is an approximate rate for large industrial users.

It is seen for this example that the savings in energy costs obtained by using the optimal control over the other controls is about $1000 per day. Of course there may exist other control strategies for which the savings would be less, but the two chosen for comparison are reasonable. In this example the optimal control approach is certainly a feasible alternative.

7. CONCLUSIONS

A potential savings in energy costs of estuary aeration using an optimal control approach has been demonstrated. The control of the aeration input rate can be implemented as a feedback control, which has well known advantages in terms of the sensitivity to incomplete knowledge of or variations in system parameters or inputs.

Future plans include comparisons with other sub-optimal control schemes, other numeric approaches and application of this approach to the control of other estuary variables such as the location and timing of BOD discharges.

8. REFERENCES

Harleman, D. R. F. "One Dimensional Models", Estuarine Modeling, an Assessment, U.S. Government Printing Office, Washington, D.C., 1971, Water Pollution Control Research Series, 16070 DZU 02/71.

Koppel, L. B. and Shih, Y. P. "Optimal Control of a Class of Distributed Parameter Systems with Distributed Controls," I & EC Fundamentals, Vol. 7, No. 3, pp. 414-422, 1968.

Lions, J. L. Controle Optimal De Systems Gourvenes Par Des Equations Aux Derivees Partielles, Dunod and Gauthier-Villars, Paris, 1968.

Sage, A. P. Optimum Systems Control, Prentice Hall, Englewood Cliffs, New Jersey, 1968.

Whipple, W., et al "Oxygen Regeneration of Polluted Rivers," Environmental Protection Agency, Washington, D.C., 16080 DUP 12/70, December 1970.

INTERACTIVE SIMULATION PROGRAM FOR WATER FLOOD ROUTING SYSTEMS

F. Greco and L. Panattoni
IBM Pisa Scientific Center - Italy

Introduction

A computing system is here presented for the study of water flow in open channels, with special regard to flood routing in rivers. This system has been built up during the study for the construction of the mathematical model of the Arno river basin. It allows us to reproduce the flood wave propagation in natural or artificial watercourses. The system has been built in an interactive mode in order to facilitate the setting up of the mathematical model and the simulation of the phenomenon under different conditions.

In particular we shall try to emphasize how the use of an interactive system and of a video unit has made it very easy to use the model for any purpose.

The flood routing problem lies in studying the propagation of a flood wave along the river, determining the variation of the wave's shape and height in its motion towards the sea. This is achieved by determining the evolution of the known water levels and discharges at the initial time (Initial Conditions), according to the variation in time of the water levels and/or discharges at the starting section, and the inflow due to the tributaries (Boundary Conditions).

The solution of this problem could then allow us to foresee the water levels and discharges in any section of the river at any time, assuming that the river bed geometry and its resistance to the water flow are known.

The Mathematical Model

The solution of the flood routing problem in rivers needs the construction of a mathematical model, that is a set of elements and relationships, mathematical or logical, between them.

In the case of flood routing the mathematical relationships are the well known equations (fig. 1) established by Barré de Saint Venant in 1871 [De Saint Venant], which express mass and momentum conservation respectively. The validity of these equations [Yevdjevich] implies substantially that the components of the water velocity and acceleration, other than the longitudinal one, are negligible, i. e. the motion of the water can be considered as unidimensional.

These equations together with the levels z and discharges Q, which are the unknowns of the problem, contain geometrical quantities, (like wet area A, superficial width B, hydraulic radius R), and other quantities depending on the nature of the river bed, like the passive resistance J of the bed to the water flow. Furthermore q is the lateral inflow or outflow by unit of lenght and g the acceleration of the gravity.

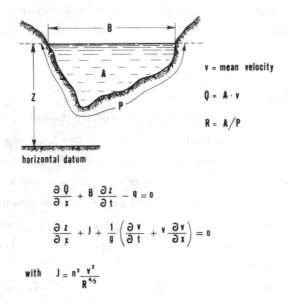

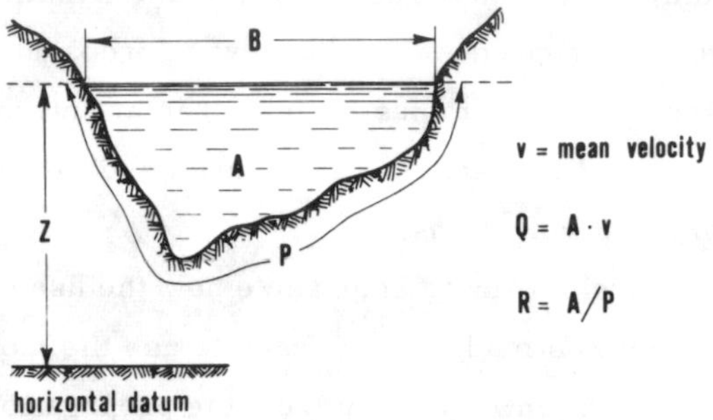

$$\frac{\partial Q}{\partial x} + B\frac{\partial z}{\partial t} - q = 0$$

$$\frac{\partial z}{\partial x} + J + \frac{1}{g}\left(\frac{\partial v}{\partial t} + v\frac{\partial v}{\partial x}\right) = 0$$

with $J = n^2 \frac{v^2}{R^{4/3}}$

Fig. 1

The description of the geometrical shape of the river bed can be obtained by means of several cross sections, the number of which depends on the regularity of the watercourse and on the accuracy we want to achieve.

In the case of our model of the river Arno, [Gallati and others] , for example, the mean distance between two consecutive sections is about 400 meters. Each section is defined by several points of its boundary.

Many formulas are available in order to estimate the passive resistance; in our model we have adopted the Manning formula containing the coefficient n which depends on the nature of the river bed, and which, obviously, cannot be measured directly. Furthermore in watercourses there are other factors which cause energy losses, like, for example, bridges, bends, sudden broadenings or narrowings and so on. All these factors have not been taken into account separately, but they have been included in the above mentioned Manning formula.

The coefficient n then must be deduced by choosing the value which gives us the best fit of the known levels or discharges with the computed ones [Gallati and others] . In doing so, however, besides the friction losses, all the other approximations made in the geometrical description and mathematical formulation of the phenomenon also affect the value of this parameter; then it, like all the other parameters which influence the phenomenon, completely looses its original meaning and becomes simply a fitting parameter.

In order to solve the Saint Venant equations an original finite difference implicit scheme has been adopted [Greco and Panattoni] . We did not use an explicit scheme because of its constraints on the size of the temporal step, which preliminary investigations showed to be too little (a few seconds) for our problems. The use of an implicit scheme, although a little more complicated, allowed us to use temporal steps of one hour, or more, which are more suitable for flood routing problems. However it is noteworthy that, if we use the Newton method to solve the nonlinear algebraic system, resulting from an implicit scheme, the matrix of the coefficients is always very simple [Greco and Panattoni] and this makes the solution of the system very easy.

Using this scheme, then, we are able to compute from the known state of the river, that is, from the level and discharge values in any section, at a given moment T, the same quantities at next time $T+\Delta T$. Starting then from the knowledge of the river at the initial time, we can determine the levels and discharges in any section at any time. In those sections, for which recorded data are available, we can then compare the measured and computed hydrographs and get some idea of the model's efficiency.

With the help of these comparaisons we can, as we said before, set up the model varying the parameters in order to obtain the best fit between the two hydrographs. And it is specially in this connection that the use of an interactive computing system and of a video unit is very helpful. In fact the contemporaneous influence of the various parameters and their interdipendence are not known a priori, and by means of an interactive system we can vary the values of the parameters in an extremely simple way and by the use of a video unit we can immediately display the complete results making the evaluation of the influence of the various parameters on them very easy.

The same advantages can be obtained also during the use of the model for the simulation of hypothetical floods, for the management of hydraulic works, either existing or in project, for the real time control of the river, and so on.

The Computation System

The figs. 2.1 and 2.2 show the structure of the Flood Routing Interactive Simulation System (FRISS).

The first step involves the initialization of the System, and, among other things, it assigns the default values to all the variables. The user, then, can comunicate, by means of a communications terminal, with the system defining :

Fig. 2.1

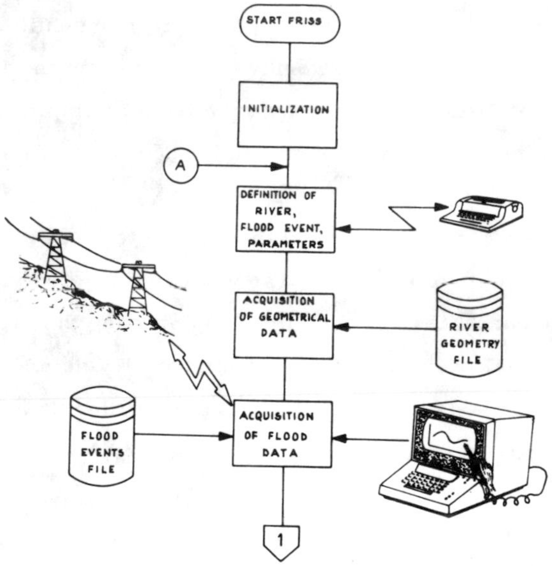

Fig. 2.2

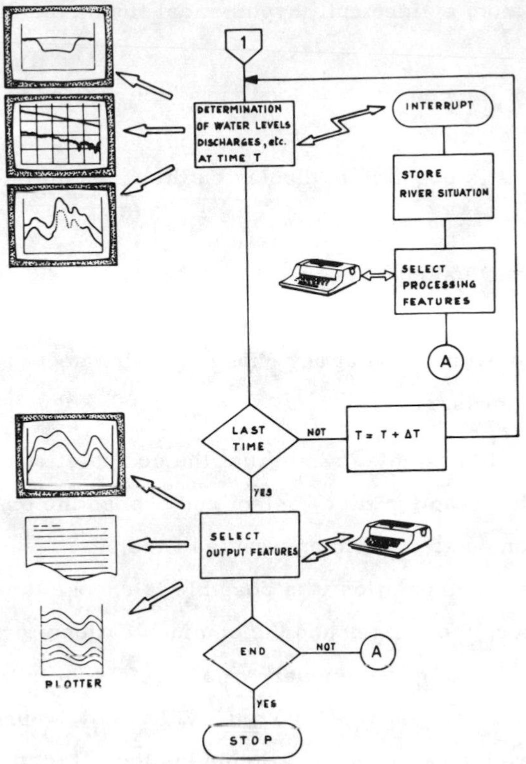

- the watercourse reach, natural or artificial, by means of its starting and ending sections,
- the geometry of the reach, either through the cross sections which describe it, or through its slope and shape,
- the hydraulic roughness of the reach (i. e. roughness formula and coefficient),
- the flood event, either past or hypothetical,
- the boundary conditions,
- the lateral inflow or outflow,
- the numerical solution algorithm,
- the size of the time step,
- the accuracy of the iterative process,
- the use of the Jones correction formula for the stage-discharge relationship.

When the problem is completely defined, the program acquires the geometrical data from a river geometry file and then acquires flood data from the past flood events file or from a video unit (hypothetical flood), or directly from a real time system.

Fig. 2.2 shows the computation loop, in which the water levels and discharges are computed at all times.

At any time T it is possible to display on the video unit some pictures, as, for instance :

- the water profile along the reach,
- the shape of any chosen cross-section and the water level at that time T,
- the computed hydrograph at any chosen gauging section of the reach compared with the measured one.

Furthermore it is possible to interrupt the computation, and, after storing the situation at the interruption time, select and change the processing features and to restart the computation from any wanted time.

At the end of the computation it is possible to choose the outputs: using the video unit, the plotter, the printer and the communications terminal.

Some results, relating to the river Arno, are shown in the Figs. 3 to 8.

Fig. 3 which has been obtained from the video unit, represents a longitudinal profile of a river reach about eighteen kilometers long starting from the section 54 km from the mouth. The lower line is the river bottom and the upper one is the

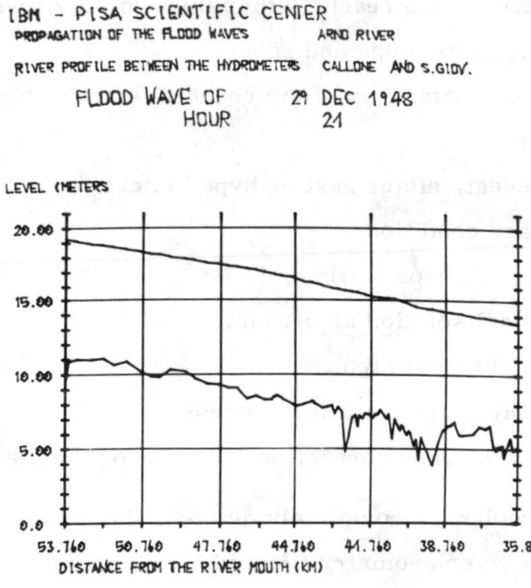

Fig. 3

Fig. 4

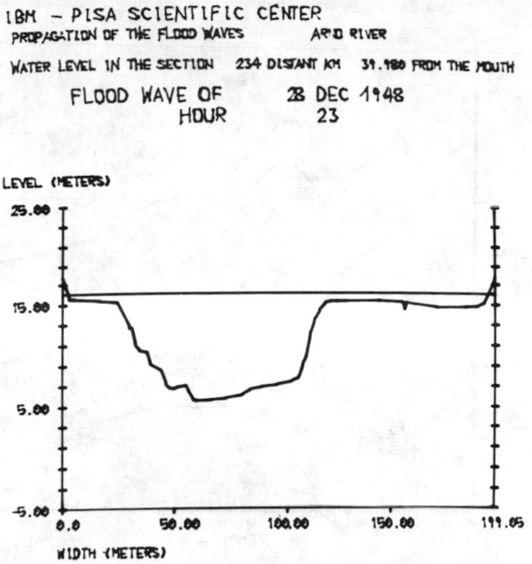

water level as computed by the model for a given time of the flood of December 1948.

Fig. 4 represents a specific cross section, about 40 km from the mouth, with the computed water level at a given time for the same flood.

Finally in Fig. 5, besides the measured flood wave in the upstream section of the considered river reach, both the measured (the dotted line) and computed

Fig. 5

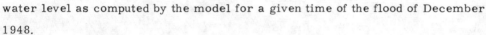

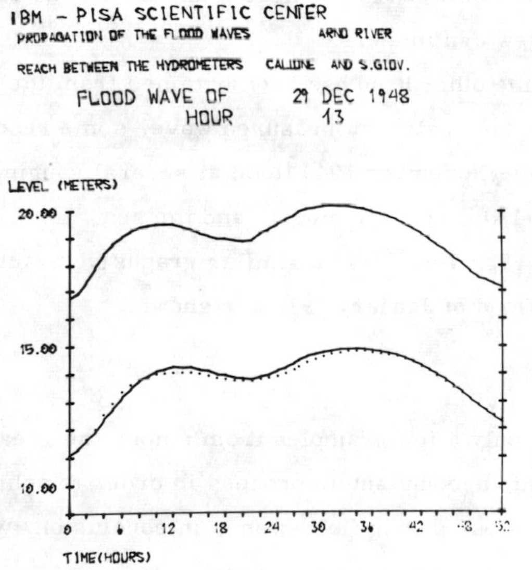

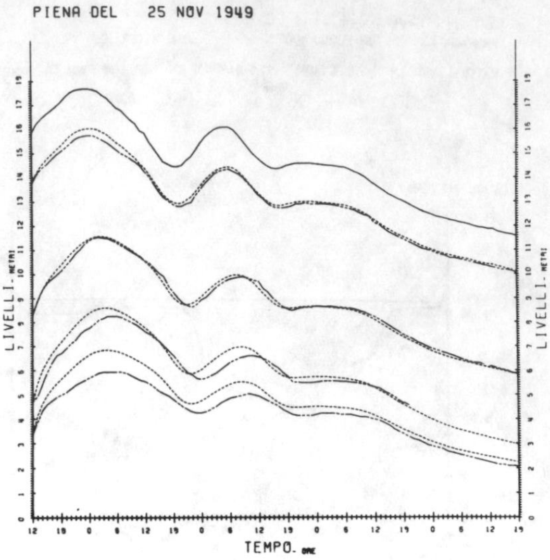

Fig. 6

(the full line) waves in the downstream section are drawn. It is clear that for the flood of December 1948 the reconstruction of the actual phenomenon is quite good and the differences between the computed and measured values of the water levels never exceed a few centimeters.

Fig. 6 on the other hand has been obtained from the plotter unit. On it you can see, besides the upstream measured wave, some reconstructions of the hydrographs of the November 1949 flood at several gauging stations between the first one, still 54 km from the mouth, and the sea.

Finally in Fig. 7 and Fig. 8 similar graphs relevant to the flood of January 1948 and to the flood of January 1949 are shown.

Conclusion

These are only a few examples from among the great number of events we needed to take into account and to process in order to achieve an accurate knowledge of the river behaviour during floods; and, in conclusion, we want to emphasize again how the use of an interactive system and a video unit have made it possible to set

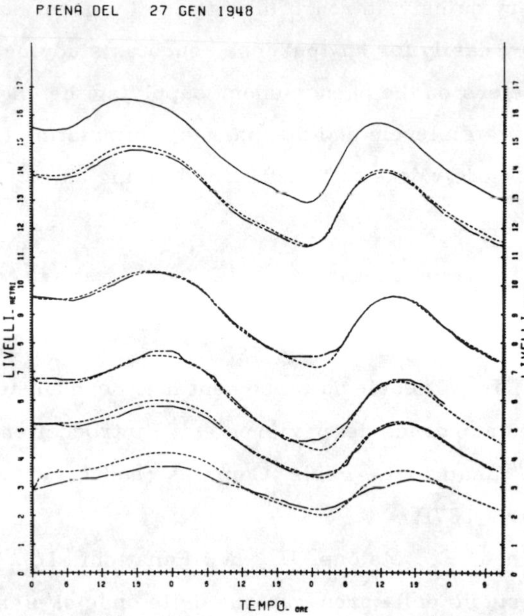
PIENA DEL 27 GEN 1948
LIVELLI. METRI
TEMPO. ORE

Fig. 7

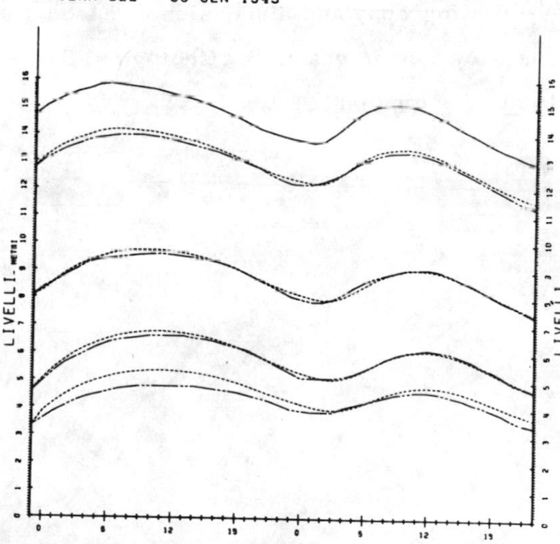
PIENA DEL 03 GEN 1949
LIVELLI. METRI
TEMPO. ORE

Fig. 8

up our model very quickly, through the fitting of a great number of past events, and to use it very easily for any purpose, such as studying the concurrence of different parameters on the phenomenon, supplying the engineering design with the necessary data (water levels and discharges), simulating the behaviour of engineering hydraulic works to be built, and giving objective and timely data for flood control.

References

De Saint Venant, B. "Théorie du movement non-permanent des eaux avec application aux crues des rivières et à l'introduction des marées dans leur lit" Acad. sci. Paris Comptes rendus, V. 73, p. 148-154, 237-240 Paris 1871.

Gallati, M., Greco, F., Maione, U., and Panattoni, L. "Modello matematico per lo studio della propagazione delle onde di piena nei corsi d'acqua naturali" XIII Convegno di Idraulica e Costruzioni Idrauliche, Milano, Sept. 1972.

Greco, F., and Panattoni, L. "An implicit method to solve Saint Venant equations". To be published.

Yevdjevich, V. "Bibliography and discussion of Flood-Routing methods and unsteady flow in channels" Geological Survey Water-Supply, Paper 1690. Washington 1964.

AN AUTOMATIC RIVER PLANNING OPERATING SYSTEM (ARPOS)

Enrico Martino - Bruno Simeone - Tommaso Toffoli

IAC, Consiglio Nazionale delle Ricerche, Rome, Italy

Summary

An analysis of the general structure of LP models of water resource systems and of the operations required in their construction suggests the opportunity of designing a special purpose interactive system to assist the engineer in developing such models.

As a working example of the ideas discussed, a simple operating system has been implemented.

1. Introduction

The optimal planning and management of complex water-resource systems requires the development of adequate models. Linear-programming (LP) models are almost invariably used as a starting point, even though they may be complemented, at a later stage, by more refined descriptions that use simulation or non-linear techniques (Buras and Herman, 1968, Maass and Hufschmidt, 1962).

ARPOS is a special-purpose interactive system (Galligani, 1971) aimed at assisting the analyst--typically a systems-oriented hydraulic engineer--in the construction of LP models of a water-resource system.

In designing ARPOS, the development stage of a model has been kept in mind. In fact, models evolve together with the analyst's appreciation and understanding of the system's determining features. The model current version's inadequacies are a source of feedback onto earlier stages of the construction process. Structural and numeric data are repeatedly reevaluated until a satisfactory behavior is achieved.

On the other hand, since river models tend to be quite large (thousands of rows in the LP matrix), every iteration of the revision process may require a large amount of error-prone work.

For those reasons, aside from providing a general framework where constructive activities and feedback loops are conveniently placed, ARPOS aimed at providing fast and reliable response to user feedback. This was achieved after an accurate analysis of the operations involved in the modeling process. In brief:

(a) The whole process can be broken down into a sequence of well-delimited operations.

(b) Many of the operations are well-defined and can be easily automatized. For instance, given a network representing the basin topology and the location of relevant activities (fig.1), the equations expressing continuity conditions at each node as linear constraints can be generated automatically in a straightforward way.

(c) Other operations, like, for instance, the removal of infeasibilities, are required at certain well-defined stages of the process but are less amenable to a formal description and must rely ultimately on the analyst's judgement.

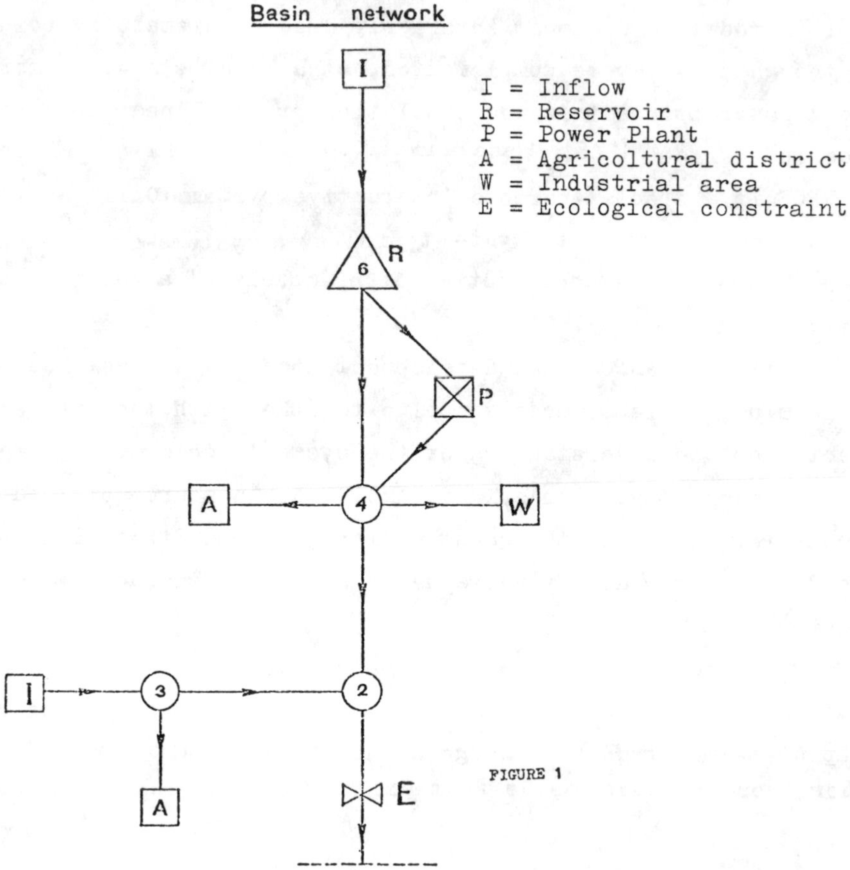

FIGURE 1

A more specific analysis will be carried out in Section 2. Section 4 summarizes the results of the synthesis process, in the form of an actual operating system, while section 3 illustrates some of the techniques on which such synthesis is based.

2. The ARPOS structure.

ARPOS takes into account the fact that certain features are almost invariably present in LP basin-planning models. These are usually multi-period, which often causes the model to require thousands of rows in the LP matrix. On the other hand, the model structure can be described more conveniently by considerably fewer symbolic constraints in which the time parameter appears only as a dummy index(+); to this, a list must be added containing, for all periods, only numeric data. Such a time-independent symbolic-constraint description represents a convenient break-point in the modeling process: it has a familiar aspect for the hydraulic engineer, it simplifies the task of revising the model, and at the same time can be used as output or input of automatic procedures.

Basic to the ARPOS structure is also the concept of standard constraints, defined as those that represent continuity conditions at individual nodes, or flow bounds at individual nodes or branches. Standard constraints (a) make up a large part (90% or more) of most known models, (b) are maintained in successive versions of the same model, and (c) can be automatically written down in symbolic form once the basin network is given. It is possible to express as standard constraints:

- physical constraints: continuity equations at each node;
- technological constraints: upper bounds due to the maximum capacity of channels, reservoirs, and power plants;

(+) - For instance, continuity at node 6 (fig.1) can be expressed by the equation $I_{6,t} + R_{6,t} - R_{6,t+1} = X_{6,t} + P_{6,t}$ (I inflow, R current reservoir storage, X outflow, P flow through the power plant turbines, all these variables beeing referred to node 6 at the period t).

- economic constraints: lower bounds due to minimum local water demand for agriculture, industry, or power generation;

- ecological constraints: lower bounds locally imposed on flows for pollutant dilution, upper bounds for flood control; etc.

The structure of the standard submodel (i.e., that consisting of standard constraints) makes it possible to test its feasibility by means of a special-purpose algorithm, namely, a variation of the well-known Hoffman-Fulkerson algorithm.

Other constraints, which we may call peculiar (for instance, global economic constraints) vary from model to model and must be supplied directly by the analyst.

Basically, ARPOS supports the following logic pattern of operations (fig.2):

Figure 2

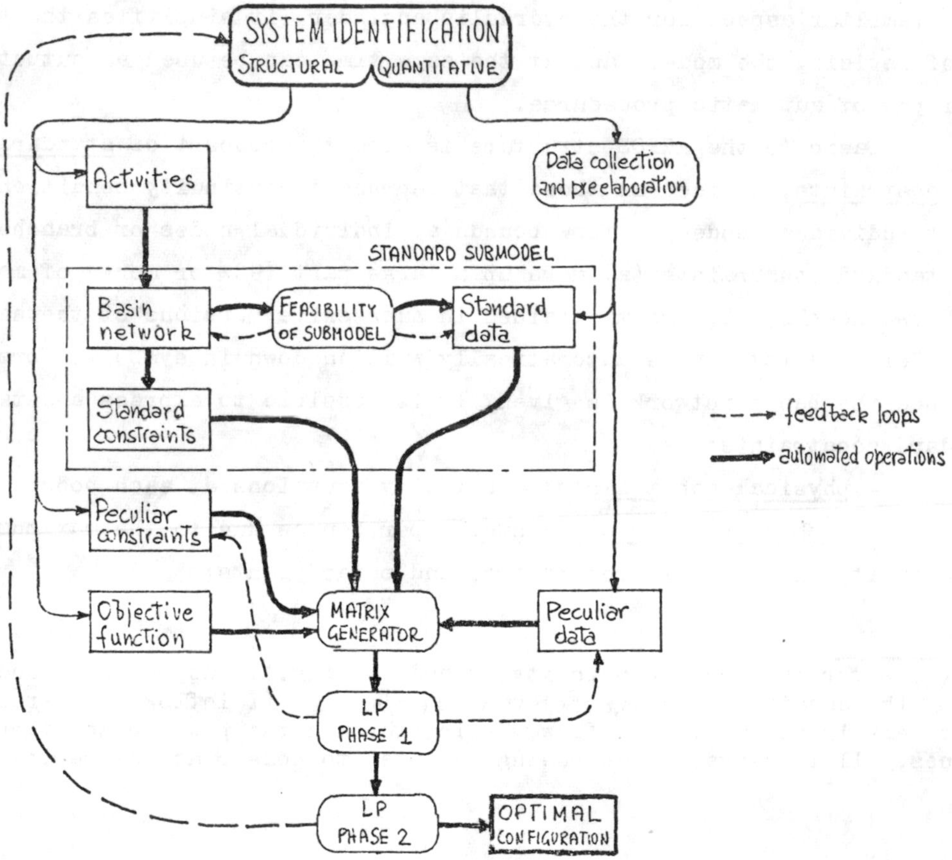

It should be pointed out that ARPOS performs the feasibility test of the standard submodel at an earlier stage than the usual procedure (Phase 1 of the Simplex Algorithm for the entire model). This allows the user to enter the first LP run with a higher degree of confidence on the model consistency.

3. Certain techniques employed in ARPOS.

3.1 Feasibility test. Checking feasibility of the standard subsistem may be viewed as a problem of existence of feasible flows in the network E (+), which is defined as follows, starting from a given basin network B (fig.1):

1) We produce p identical copies $B_1,..,B_p$ of B, where p is the number of periods, e.g., p = 12 for a monthly model. Thus to each node x of B corresponds a node x_t in B_t. For each reservoir node x, we draw an arc from x_t to x_{t+1} (t = 1,...,p-1). This device allows us to convert a network with memory into a memoryless network.

2) Moreover, in order to get an all-activities-in-arcs representation, we add to each B_t a special node u_t and draw an arc to u_t from all nodes which are sinks or have some consumptive activities; since our only concern is feasibility, we can always aggregate all the consumptive activities of each node into a single one.

3) It is also convenient, in order to have a simpler algorithm, to avoid sources and sinks. To this purpose, we add one more node v and draw arcs from v to all input nodes of $B_1,.....,B_p$ and from each u_t to v. It is convenient as well to eliminate multiple arcs by inserting, when necessary, a fictitious node in some arc. The network that is so obtained is, by definition, the extended network E (see fig.3).

(+) - Terminology is as in Ford and Fulkerson, 1962.

Extended network for FEAS

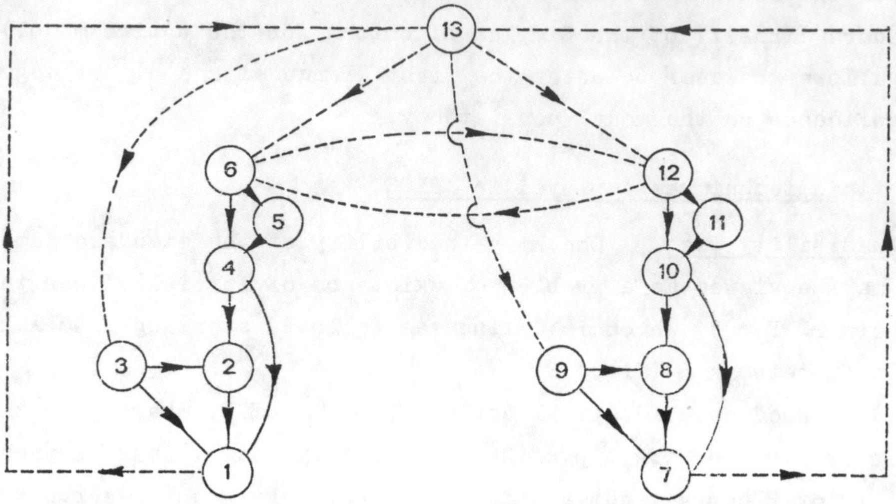

FIGURE 3

It is now straightforward to assign lower and upper capacities to the arcs of E. These are defined:

- for each reservoir arc (x_t, x_{t+1}) to be zero and, resp., the maximum reservoir capacity
- for each "consumptive" arc (y_t, u_t) as the minimum, and resp. the maximum requirement for the activity in node y in the period t
- for any input arc (v, z_t) both to be equal to the hydrological input in z in the period t, and so on.

In order to check the existence of feasible flows in E, we use a known algorithm of Hoffman and Fulkerson (Ford and Fulkerson, 1962 - pag.52). Without going into details, we only summarize the gross structure of the algorithm (called FEAS) in fig.4.

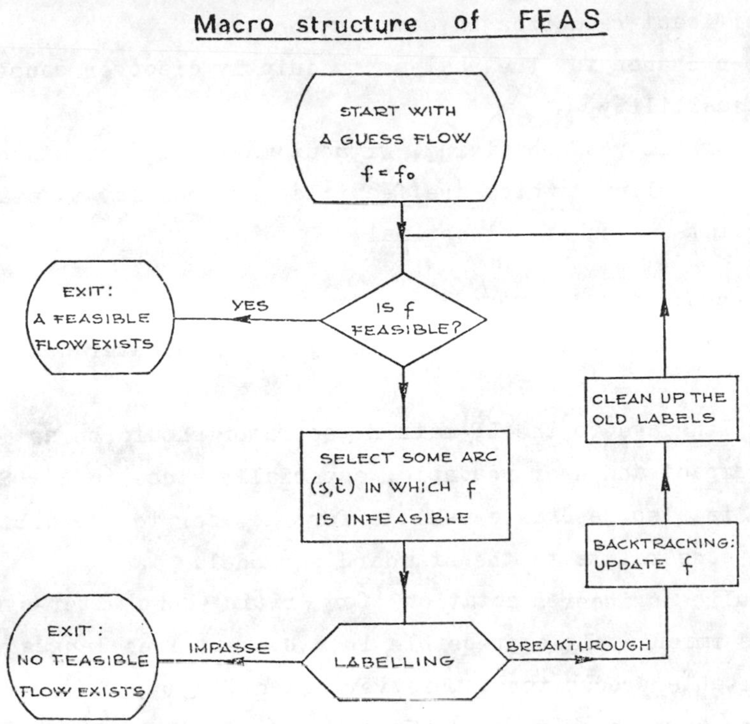

FIGURE 4

If the guess flow and the lower and upper capacities are all integral, it can be proved that, in a finite number of steps, the algorithm finds a feasible flow or detects the non-existence of such a flow. The assumption of integrality is not restrictive from a computational point of view.

It should be noted that the algorithm is flexible in three points: the initial guess flow, the selection of (s,t) and, inside the labelling procedure, the selection of a suitable labeled node whose neighborhood is to be explored in order to create new labeled nodes. We have implemented a version of the algorithm which makes use, in these three points, of heuristics which rely on the special structure of E. So, for instance, (s,t) is selected as downstream as possible in the hope that at each step many infeasibilities will be reduced.

The following advantages may be obtained with respect to the normal procedure (Phase 1 of the Simplex Algorithm):

a) a significant reduction in computation time,

b) a better chance for the analyst to quickly discover causes of infeasibility,

c) the possibility of obtaining, without making use of standard LP codes, policies which are feasible with respect to standard constraints, even if not optimal.

3.2 Symbolic-constraint language and matrix generator. This section of ARPOS is described elsewhere (Toffoli, 1973) in greater detail. In brief:

The input language to the LP-matrix generator should be as much as possible compact and user readable, especially since in ARPOS this language is also used as a vehicle for the user to add arbitrary additional constraints to the standard submodel.

The hydraulic engineer's notation for writing down water-system constraints is immediately expressible in a context-free language defined by suitable productions. Moreover, such language is everywhere LR1 (Look-Right-n, with n=1). In other words, in order to know what production generated a given syntactical element, it is sufficient to scan at most one symbol ahead in the input string. LR1-ness makes it easy to construct a fast single-pass parser for the input language.

The formal definition of the symbolic-constraint language can be given in such a way that there is a strict correspondence between syntactic (Backus Normal Form productions) and semantic (matrix-generation operations) features. In this way, scanning of the symbolic-constraint list and matrix generation proceed jointly in an interpretive manner.

4. A simple Operating system for ARPOS

In order to provide a working example of the ideas discussed above, a Simple Operating System (SOS) for ARPOS was built in its entirety. In spite of its small size (it works well on an IBM/1130 with 16K and one disk), SOS is fully able to support the construction of a model up to the point where standard LP codes take over.

SOS has been extensively used in developing LP models for the Tiber basin.

SOS consists of a set of procedures called by an interpreter. The stream of commands generated by user is translated on-line by the interpreter into a sequence of procedure calls. Thus, an SOS programs is dynamically constructed by joining together modules (procedures) selected from the following repertoire:

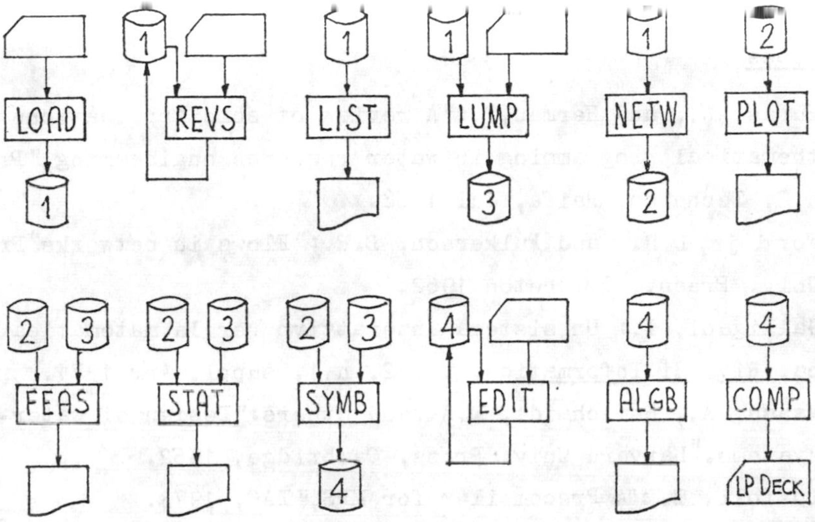

LOAD - Load input deck in input data file.

REVS - Edit or update input data file.

LIST - List input data file.

LUMP - Lump numeric data for n·m elementary time intervals into data for m seasons.

NETW - Generate river-system network.

PLOT - Map river-system network.

STAT - Give statistic tables on network and activities.

SYMB - Generate symbolic equations for standard constraints.

EDIT - Edit symbolic equations. Add additional constraints and objective function.

ALGB - List symbolic equations.

FEAS - Test feasibility of current version of the submodel.

COMP - Compile LP matrix for input to MPS.

A warning is given if a meaningless combination of modules is attempted.

In this way, commands can be added or repeated with different data, in an interactive manner, and at any moment, on the basis of the information currently available about the model's status, the user will be able to revise previous operations or proceed with the construction.

References

(1) Buras, N., and Herman,T.:"A review of some applications of mathematical programming in water resource engineering,"Prog.Rep. n.2, Technion, Haifa, Jul 1968.

(2) Ford jr, L.R. and Fulkerson, D.R.:"Flows in networks,"Princ. Univ. Press., Princeton 1962.

(3) Galligani, I.: Un sistema interattivo per la matematica numerica, Riv. di Informatica, vol.2, n.1, suppl. Apr 1971.

(4) Maass, A., Hufschmidt, M.A.,and others:"Design of water-resource systems,"Harvard Univ. Press, Cambridge, 1962.

(5) Toffoli, T.:"A Precompiler for MPS,"IAC, 1973.

Acnowledgements

The authors are indebted to Prof. I. Galligani for his suggestions and to Mr. A. Bonito for having written some of the programs.

ON THE OPTIMAL CONTROL ON AN INFINITE PLANNING HORIZON OF CONSUMPTION, POLLUTION, POPULATION AND NATURAL RESOURCE USE

Alain Haurie
Ecole des Hautes
Etudes Commerciales
and Ecole Polytechnique
Montreal, Quebec, Canada

Michael P. Polis
Ecole Polytechnique
Montreal, Quebec, Canada

Pierre Yansouni
Ecole Polytechnique
Montreal, Quebec, Canada

ABSTRACT

In this paper a five state variable economic planning model is presented. A Malthusian hypothesis is employed which gives rise to a zero-growth argument. The asymptotic behavior of the model is studied. Classical results involving Turnpike theory are used in conjunction with recently published results on the infinite time optimal control problem to show the convergence towards a Von Neumann point of the optimal trajectory on an infinite planning horizon.

1. INTRODUCTION

Recently J.W. Forrester in "World Dynamics" [1] proposed a simulation model to study the interactions between population growth, the use of natural resources, pollution and consumption. The simulation results showed the catastrophic effects which would result from a policy of "laissez-faire". Forrester's work has been widely commented upon and criticized [2], [3]. Two of the most repeated criticisms concern the high level of aggregation of the model ("five state variables to describe the world?"), and the Malthusian philosophy which gives rise to the fundamental hypotheses upon which the model is based. Although these are valid criticisms they do not negate the utility of the model even though the conclusions drawn from the simulations may be questioned.

Independently of Forrester, economists have developed a theory of optimal economic growth based largely on techniques of optimal control [4]-[7]. Effectively the complex phenomena of interaction simulated by Forrester can be studied under a form similar to that used by the economists. Further, optimization techniques may be used on the aggregated models to establish a direct relation between the conditions of the problem and the form of the solution which must be sought.

In this paper an economic planning model having the same level of aggregation as Forrester's model, and under the Malthusian hypothesis is presented. The techniques of optimal control on an infinite time planning horizon are used in conjunction with the model to provide information on the nature of the solution which must be found. Thus, a five state variable model of economic growth is proposed to study, at a macro-economic level, planning policies concerning consumption, the use of replenishable and non-replenishable natural resources, pollution emmission and population growth. Such a model can be useful in two ways:

(1) A price mechanism sustaining the optimal policy can be defined and eventually an optimal fiscal policy leading to optimal or suboptimal paths can be proposed (See Arrow and Kurz [8]).

(2) The asymptotic behavior of the optimally controlled economy can be studied.

In this work only the second point is considered. A Malthusian model gives rise to a zero-growth argument. In the context of an economy without marked technological progress the objective of zero-growth has already been proposed [9]. The planning model studied in the following paragraphs permits the rational choice between various equilibrium states which are all characteristic of zero growth. Further, these equilibrium states indicate how the capital structure of the economy can be transformed in order to attain a desired objective following an optimal trajectory consistent with the objective.

The principal theoretical result presented here concerns the convergence to a steady-state-zero-growth-economy (Von Neumann path) of an optimal trajectory on an infinite planning horizon. This result is established by using a definition of optimality recently proposed by Halkin [10] as well as classical results concerning the well known "Turnpike" theorem for a multisector economy [11], [12].

2. THE SYSTEM'S MODEL

Models used in the neo-classical theory of economic growth are highly aggregated. Most often only one state variable is considered: the capital stock; labor is an exogeneous variable assumed to grow exponentially with time; and production is generated by the combination of labor and capital stock.

In order to account for such mechanisms as pollution, population control, extraction and exhaustion of natural resources, a model is considered with the following five state variables:

$K \triangleq$ capital stock.

$X_1 \triangleq$ non-replenishable resource i.e. minerals, oil, etc...

$X_2 \triangleq$ replenishable resource i.e. forests, fish stock, etc...

$P \triangleq$ pollution stock.

$L \triangleq$ total labor force.

It is also assumed that direct manipulation of the following economic variables, is possible:

$Y \triangleq$ production level in the economy.

$C \triangleq$ total consumption.

$N \triangleq$ part of the revenue allocated to control the birth-rate.

$R_1 \triangleq$ extraction rate of the non-replenishable resource.

$R_2 \triangleq$ extraction rate of the replenishable resource.

$Q \triangleq$ level of economic activity dedicated to the regeneration of X_2.

$G \triangleq$ pollution emission rate.

These in fact are the control variables in the model. The relationship between state and control variables is imbedded in the state equations describing the dynamic behavior of the economy.

$$\dot{K} = Y - C - N - \sigma K \tag{2.1}$$

$$\dot{X}_1 = - R_1 \tag{2.2}$$

$$\dot{X}_2 = \phi (X_2, P, Q) - R_2 \tag{2.3}$$

$$\dot{P} = G - \nu P \tag{2.4}$$

$$\dot{L} = D(L, N) \tag{2.5}$$

- Equation (2.1) represents the capital accumulation in the economy; σ is the capital depreciation rate.
- Equation (2.2) represents the extraction of the non replenishable resource.
- Equation (2.3) represents the extraction of the replenishable resource; $\phi(\cdot)$ is the rate of regeneration of X_2.
- Equation (2.4) represents the accumulation of pollution. ν is a natural elimination rate of pollution by the environment.
- Equation (2.5) represents the population dynamics. $D(\cdot, \cdot)$ is the population "growth" (it can be negative) rate.

To a certain extent the mathematical description of the mechanisms involved in the economic sectors modeled by equation 2.1 - 2.5, is inspired by the work in Ref. [13] - [20].

A more detailed analysis of the mechanisms involved in the manipulation of the control variables is necessary to underline the physical constraints restraining the set of possible choices for the control variables. Stating again the neo-classical hypothesis that economic activity is generated by the combination of capital and labor allocated to each sector; it is assumed that a fraction of the capital K and of the labor L is allocated to each of the areas of economic activity defined by the state equations. Each of these fractions is identified by a subscript referring to the respective control variable.

They are:

$$K_Y, K_1, K_2, K_Q, K_G$$

$$L_Y, L_1, L_2, L_Q, L_G$$

Two physical constraints arise immediately, they are:

$$K_Y + K_1 + K_2 + K_Q + K_G \leqslant K \tag{2.6}$$

$$L_Y + L_1 + L_2 + L_Q + L_G \leqslant L \tag{2.7}$$

Denote the maximum† obtainable rate of "activity" in each sector, compatible with the allocated fraction of K and L, by the variables $\widetilde{Y}, \widetilde{R}_1, \widetilde{R}_2, \widetilde{Q}, \widetilde{G}$. A set of "technological" constraints arises from the fact that the preceding variables constitute upper‡ bounds to the control variables. They are:

$$0 \leqslant Y \leqslant \widetilde{Y}(K_Y, R_1, R_2, P, L_Y) \tag{2.8}$$

$$0 \leqslant R_1 \leqslant \widetilde{R}_1(K_1, L_1) \tag{2.9}$$

$$0 \leqslant R_2 \leqslant \widetilde{R}_2(K_2, L_2) \tag{2.10}$$

$$0 \leqslant Q \leqslant \widetilde{Q}(K_Q, L_Q) \tag{2.11}$$

$$0 \leqslant \widetilde{G}(K_Y, K_1, K_2, K_Q, K_G, L_Y, \ldots, L_G) \leqslant G \tag{2.12}$$

In all the following it is assumed that $\widetilde{Y}(\cdot), \widetilde{R}_1(\cdot), \widetilde{R}_2(\cdot), \widetilde{Q}(\cdot)$ are concave functions of their arguments, and that $\widetilde{G}(\cdot)$ is convexe.

<u>Remark 2.1</u>:

These restrictions are necessary to the rigorous mathematical developments in part 3. It is likely however, that the results obtained in part 3 will remain valid in most cases when topological restrictions are relaxed. That is to say, that when direct numerical solution of the optimal control problem is attempted, convergence to a Von Neumann path can be expected and can be simply verified by numerical experimentation; see Ref. [19].

<u>Notation</u>: Let $x \in E^n$ be the state vector, and $u \in E^m$ the control vector.

---

† For $\widetilde{G}$ read: "The minimum", since K_G an L_G are dedicated to the reduction of pollution rate.

‡ For G read: lower bound. Same remark as above.

Equations 2.1 - 2.5 can be written

$$\dot{x} = f(x, u) \tag{2.13}$$

and constraints 2.6 - 2.12:

$$\Psi_j(u, x) \geqslant 0 \qquad j = 1, \dots, m \tag{2.14}$$

$$u \geqslant 0$$

Obviously the physical definition of x implies: $x \geqslant 0$.

Remark 2.2:

The mathematical development in part 3 requires that the velocity vector $\dot{x}$ be enclosed in a convex set. More rigorously, the following result is required: If there exist:

$$\dot{x}^1 = f(x^1, u^1) \quad \text{with} \quad u^1 \geqslant 0 \ , \ \Psi_j(x^1, u^1) \geqslant 0$$

$$\dot{x}^2 = f(x^2, u^2) \quad \text{with} \quad u^2 \geqslant 0 \ , \ \Psi_j(x^2, u^2) \geqslant 0$$

then $\forall \alpha \in [0,1] \ , \ \exists \ u^\alpha \geqslant 0 \ , \ \Psi_j(\alpha x^1 + (1 - \alpha)\, x^2, u^\alpha) \geqslant 0$

such that:

$$\alpha \dot{x}^1 + (1 - \alpha)\, \dot{x}^2 = f(\alpha x^1 + (1 - \alpha)\, x^2, u^\alpha)$$

This result is easily demonstrated with the assumptions that $\emptyset(\cdot)$ and $D(\cdot)$ are concave, and further, that they are linear with respect to the state variables. Without the linearity assumption no demonstration has yet been found. For the more general functions $\emptyset(\cdot)$ and $D(\cdot)$ the result stated above has to be considered a basic assumption to which can be applied the comments contained in Remark 2.1.

The economic model is to be completed with a welfare criterion defined on an infinite planning horizon. It is assumed that the objective of the economic policy is to maximize a functional:

$$J = \int_0^\infty W(C, L, P, G, X_1, X_2)\, dt$$

where $W(\cdot)$ is restricted to be a concave function of its arguments. A particular choice of $W(\cdot)$ is interesting because of its simple implementation:

$$J = \int_0^\infty (\lambda_1 C + \lambda_2 X_1 - \lambda_3 P - \lambda_4 L)\, dt$$

with $\lambda_i \geqslant 0$ and $\sum_i \lambda_i \leqslant 1$.

Assuming convergence to the Von Neumann path, (see section 3) the equilibrium solution of the optimal control problem is parametrized as a function of the weights λ_i attributed to the respective variables in the welfare criterion. Solving the algebraic equations defined simultaneously by the necessary conditions for optimality (on the infinite horizon, Ref. [10]) and the equilibrium (steady state) condition, is a relatively simple task.† The solution yields in fact the equilibrium levels of consumption, population, pollution, and the remaining stock of non replenishable resource. This procedure gives the means of evaluating the long term consequences of a given choice of values for the weights λ_i. Furthermore a systematic exploration of the domain of equilibrium solutions as a function of the parameters λ_i will give a basis for assigning values to λ_i's compatible with the long range objective of the economy. i.e. expected per capita consumption $\frac{C}{L}$, population and pollution levels allowed, and amount of resource X_1 to be protected.

Remark 2.3:

The use of a weighted performance criterion may also correspond to the search for Pareto-optimal solutions arising from a vector valued criterion. Consideration of a non-scalar criterion is indeed most likely to occur in an actual planning situation.

3. OPTIMALITY ON AN INFINITE TIME HORIZON AND CONVERGENCE TOWARD A STEADY-STATE

Consider the dynamic system:

$$\dot{x} = f(x, u) \tag{3.1}$$

$$x(0) = x^o \text{ given}$$

$$x \in X \subset E^n, \; u \in U(x) \subset E^m$$

f is C^1 with respect to x and continuous with respect to u. The accumulated utility x_o satisfies:

$$\dot{x}_o = w(x, u) \tag{3.2}$$

where w is C^1 with respect to x and continuous with respect to u.

Definition 3.1:

The control $\tilde{u}^* : [0, \infty] \to E^m$ is optimal if the following conditions are satisfied:

(i) it generates a trajectory $(\tilde{x}^*, \tilde{x}_o^*) : [0, \infty) \to E^{n+1}$ such that

$$\forall t \; \tilde{u}^*(t) \in U[\tilde{x}^*(t)], \; \tilde{x}^*(t) \in X$$

† This is equivalent to solving a standard mathematical programming problem.

(ii) $\forall \tilde{u} : [0, \infty) \to E^m$ generating $(\tilde{x}, \tilde{x}_o) : [0, \infty) \to E^{n+1}$ such that

$$\forall t \ \tilde{u}(t) \in U[\tilde{x}(t)] \ , \ \tilde{x}(t) \in X$$

then $\forall \varepsilon > 0 \ \forall t \geqslant 0 \ \exists \ \tau > t$ such that $\tilde{x}_o(\tau) < \tilde{x}_o^*(t) + \varepsilon$ (3.3)

This definition of optimality on an infinite time horizon has been proposed by Halkin [10]. ■

The following assumptions are inspired by the Turnpike theory [11], [12] in mathematical economics.

<u>Assumption 3.1</u>:

The set $\Omega \triangleq \left\{(x, \dot{x}, \dot{x}_o) : \exists x \in X \ , \ \exists \ u \in U(x) \text{ such that: } \dot{x} = f(x, u) \ , \ \dot{x}_o = w(x, u)\right\}$ is closed and convex.

<u>Assumption 3.2</u>:

If $(x, \dot{x}, \dot{x}_o) \in \Omega$ and $x' > x$, $\dot{x}_o' \leqslant \dot{x}_o$, $\dot{x}' \leqslant \dot{x}$ then $(x', \dot{x}', \dot{x}_o') \in \Omega$ †. This assumption is equivalent to the free disposition of goods.

<u>Assumption 3.3</u>:

The following propositions hold

- $\exists x \in X \ , \exists u \in U(x)$ s.t. $f(x, u) > 0$ (3.4)
- $\forall \varepsilon > 0 \ , \exists \ \eta(\varepsilon)$ s.t. $\|x\| \leqslant \varepsilon \Rightarrow \forall u \in U(x), \ \| f(x, u) \ , \ w(x, u)\| < \eta(\varepsilon)$ (3.5)
- $\exists \zeta$ s.t. $\|x\| \geqslant \zeta \Rightarrow \forall u \in U(x), \frac{1}{2} \frac{d}{dt} \|x\|^2 = \langle x, f(x, u) \rangle < 0$ (3.6)

The first main result can thus be established. ■

<u>Lemma 3.1</u>:

There exists $w^* \in R$ and $p^* \in E^n$, $p^* \geqslant 0$ such that

(i) $w^* = \{\text{Max } \dot{x}_o : (x, \dot{x}, \dot{x}_o) \in \Omega \ , \ \dot{x} \geqslant 0\}$ (3.7)

(ii) $\forall (x, \dot{x}, \dot{x}_o) \in \Omega \ \dot{x}_o + \langle p^*, \dot{x} \rangle \leqslant w^*$ (3.8)

■

<u>Proof</u>: Define the sets:

$$V \triangleq \{(\dot{x}, \dot{x}_o) : \exists x \in X \text{ s.t. } (x, \dot{x}, \dot{x}_o) \in \Omega\} \tag{3.9}$$

$$V^+ \triangleq \{(\dot{x}, \dot{x}_o) \in V : \dot{x} \geqslant 0\} \tag{3.10}$$

---

† For this assumption to hold Equation (2.4) must be written $-\dot{P} = -G + vP$.

It is clear from Assumption 3.1 and Equation (3.4) that V is non-empty closed convex, and that from Assumption 3.3, Equation (3.6) V^+ is compact. Thus:

$$w^* = \mathrm{Max}\{\mathring{x}_0 : (x, \mathring{x}_0) \in V^+\} = \mathrm{Max}\{\mathring{x}_0 : (x, \mathring{x}, \mathring{x}_0) \in \Omega, \mathring{x} \geqslant 0\}$$

can be defined on V^+ which establishes (i).

If $(\mathring{x}^*, w^*) \in V^+$, then $(\mathring{x}^*, w^*) \in \partial V$, the boundary of the set V. If not, it would be possible to find $\mathring{x} \geqslant 0$ and $w > w^*$ s.t. $(\mathring{x}, w) \in V^+$ which is impossible. Since V is closed convex there exists $(\pi, p) \in E^{n+1}$ such that:

$$\forall (\mathring{x}, \mathring{x}_0) \in V \quad \pi w^* + \langle p, \mathring{x}^* \rangle \geqslant \pi \mathring{x}_0 + \langle p, \mathring{x} \rangle \tag{3.11}$$

From Assumption 3.2, $p \geqslant 0$ and it is always possible to choose $\mathring{x}^* = 0$. Thus π is positive and can be choosen equal to one; p^* is the corresponding value of p. Equation (3.8) is then a direct consequence of Equation (3.11). ∎

Definition 3.2:

Following the economist's terminology a "point of Von Neumann" is a vector $(x, \mathring{x}, \mathring{x}_0) \in \Omega$ such that

$$\mathring{x}_0 + \langle p^*, \mathring{x} \rangle = w^* \tag{3.12}$$

The system defined in (3.1) and (3.2) is said to be regular if the set of Von Neumann points has only one element:

$$(x^*, 0, w^*)$$

Remark 3.1:

w^* is also the maximal utility flow which can be maintained at steady-state, i.e. (x^*, w^*) is the solution to the following optimization problem:

$$\begin{cases} \mathrm{Max}\ w(x, u) \\ f(x, u) = 0 \quad u \in U(x) \ x \in X \end{cases}$$

Now consider the convergence towards a Von Neumann point of an optimal trajectory on an infinite time horizon. Once again a classic result from the Turnpike theory [11] is used.

Lemma 3.2:

Let F be the set of all Von Neumann points, $F \subset E^{2n+1}$, for $y \triangleq (x, \mathring{x}, \mathring{x}_0) \in \Omega$ define the distance:

$$d(y, F) \triangleq \underset{z \in F}{\mathrm{Min}} \|y - z\|$$

Thus:

$$\forall \varepsilon > 0, \exists \delta(\varepsilon) \text{ s.t. } d(y, F) > \varepsilon$$

$$\|x\| \leqslant \zeta \Rightarrow \mathring{x}_o + \langle p^*, \mathring{x} \rangle < w^* - \delta(\varepsilon) \tag{3.13}$$

Proof:

If Equation (3.13) does not hold there exists a series $\{y_n\}$ such that $y_n \in \Omega$ and $d(y_n, F) > \varepsilon$, at the same time $\|x_n\| < \zeta$ and $\lim_{n \to \infty} \mathring{x}_o + \langle p^*, \mathring{x}_n \rangle = w^*$. Because $\{y \in \Omega : \|x\| \leqslant \zeta\}$ is compact there exists $\bar{y}$ such that:

$$d(\bar{y}, F) > \varepsilon \quad \text{and} \quad \mathring{\bar{x}}_o + \langle p^*, \mathring{\bar{x}} \rangle = w^*$$

Thus $\bar{y}$ is a Von Neumann point at a non-zero distance from F, which is a contradiction. ■

It is now possible to prove the principal result:

Theorem 3.1:

For an initial state x^o such that $\|x^o\| < \zeta$ and such that there exists an admissible control $\tilde{u}^T$ generating an admissible trajectory $\tilde{x}^T$ such that for $t_1 > 0$:

$$\tilde{x}^T(t_1) = x^*$$

Then under Assumptions 3.1 - 3.3 if $\tilde{u}^* : [0, \infty] \to E^m$ is an optimal control generating an optimal trajectory $(\tilde{x}_o^*, \tilde{x}^*) : [0, \infty) \to E^{n+1}$ then this trajectory will converge towards the set F in the following sense:

$$\forall \varepsilon > 0 \lim_{T \to \infty} \mu[\{t \in [0,T] : d(\tilde{y}^*(t), F) > \varepsilon\}] < \infty \tag{3.14}$$

where:

$$\tilde{y}^*(t) \triangleq (\tilde{x}^*(t), f(\tilde{x}^*(t), \tilde{u}^*(t)), w(\tilde{x}^*(t), \tilde{u}^*(t)))$$

and $\mu[\cdot]$ denotes the Lebesgue measure.

Proof: Let $\tilde{u}^S : [0, \infty) \to E^m$ be the control defined by:

$$\forall t \in [0, t_1) \; \tilde{u}^S(t) = \tilde{u}^T(t)$$

$$\forall t \geqslant t_1 \; \tilde{u}^S(t) = u^* \text{ s.t. } f(x^*, u^*) = 0$$

Thus:

$$\forall t \geqslant t_1 \; \tilde{x}_o^S(t) = \tilde{x}_o^T(t_1) + w^*(t - t_1) \tag{3.15}$$

Now consider the optimal trajectory; $\hat{x}^* : [0, \infty] \to E^n$. Since $\|x^o\| < \zeta$ and from assumption 3.3, Equation (3.6):

$$\|\hat{x}^*(t)\| \leq \zeta \quad \text{for all} \quad t > 0$$

Thus, from Lemma 3.2 and Equation (3.8) the following must hold:

$$\hat{x}_o^*(t) = \int_0^t w[\hat{x}^*(t), \hat{u}^*(t)]\, dt$$

$$\leq tw^* - < p^*, (\hat{x}^*(t) - x^o) > - \delta\mu_t \tag{3.16}$$

where:

$$\mu_t \triangleq \mu[\{\tau \in [0,t] : d(\hat{y}^*(\tau), F) > \epsilon\}] \tag{3.17}$$

If 3.14 does not hold, then $\lim_{t \to \infty} \mu_t = \infty$, and thus (3.15) and (3.16), denoting the largest component of p^* as p^*_{max}, imply:

$$\exists \forall \epsilon > 0 \ \exists T \ \text{ s.t. } \forall t' > T$$

$$\hat{x}_o^*(t') + \epsilon \leq t'w^* + 2n\, p^*_{max}\, \zeta - \delta\mu_{t'} + \epsilon$$

$$< (t' - t_1)\, w^* + \hat{x}_o^S(t_1) \triangleq \hat{x}_o^S(t')$$

which contradicts the optimality of $\hat{u}^*$.

Corollary 3.1:

If the system defined in (3.1) and (3.2) is regular and if the Von Neumann steady-state x^* is reachable from the initial state x^o then:

$$\lim_{t \to \infty} \|\hat{x}^*(t) - x^*\| = 0 \tag{3.18}$$

Proof: Direct consequence of Assumption 3.3, Equation (3.5).

4. CONCLUSION

This work is essentially composed of two parts: in the first part a five state variable planning model taking into account consumption, pollution, population and natural resource use has been presented; in the second part a general theorem is proved concerning the optimal convergence on an infinite time horizon of a "regular", concave dynamic economic model towards a steady-state. The interest of such a result in relation to the debate on "limits to growth" and "zero growth arguments" is clear. Indeed it was important to show that under suitable conditions the optimal control of the economy leads to a stabilization around a steady-state; this has been shown to hold under general concavity-convexity assumptions and under the more restrictive assumption that the state equations are linear in the state variables. Work is currently in progress towards relaxing this linearity assumption. Furthermore, the economic choices involving simultaneously consumption, pollution and other external effects are in fact optimization problems with vector-valued performance criterion. The scalarization process which may be used to characterize Pareto-optimal solutions involves the choice of weightings which are of the utmost importance in expressing the fundamental trade-offs between the various factors affecting the quality of life. The convergence property of the optimal trajectory makes it possible to perform on the steady-states a sensitivity analysis leading to the correct choice of the weighing. This involves a simpler mathematical programming problem than if the entire trajectory was to be considered.

ACKNOWLEDGEMENT

This work was supported in part by: CNRC Grant n° A-8670 and n° A-8796; Canada Council Grant n° S-72-0513 and by a Ministère de l'Education du Québec, DGES grant.

REFERENCES

[1] FORRESTER, J.W., World Dynamics, Wright-Allen, Cambridge, Mass., 1971.

[2] WARFIELD, J.N., Book Review - World Dynamics, IEEE Transactions on Systems, Man and Cybernetics, Vol. Smc-2, No. 4, Sept. 1972, pp. 558-559.

[3] BOYD, R., "World Dynamics: A Note", Science, Vol 177, August 1972, pp. 516-519.

[4] RAMSEY, F., A Mathematical Theory of Saving, Economic Journal, Dec. 1928, pp. 543-559.

[5] KOOPMANS, T.C., "On the Concept of Optimal Economic Growth" in "The Econometric Approach to Development and Planning", North Holland; 1965.

[6] KOOPMANS, T.C., "Objectives, Constraints and Outcomes in Optimal Growth models", Econometrica 35, 1, Jan. 1967, pp. 1, 15.

[7] ARROW, K.J., "Applications of Control Theory to Economic Growth" in Mathematics of the Decision Sciences, G.B. Dantzig and A.F. Veinott edit., A.M.S. 1968.

[8] ARROW, K.J. and KURZ, M., Public Investment, the Rate of Return, and Optimal Fiscal Policy, The Johns Hopkins Press, 1970.

[9] MEADOWS, D.L., The Limits to Growth, Universe Books 381, Park Avenue, N.Y.C. 10016.

[10] HALKIN, H., "Necessary Conditions for Optimal Control Problems with Infinite Horizon", Core-Discussion Paper. 1971, Core, De Croylaan 54, Heverlee, Belgium.

[11] McKENZIE, "Accumulation Programms of Maximum Utility and the Von Neumann Facet" in Value, Capital and Growth (J.N. Wolfe edit.) Chicago Aldine, 1968, pp. 353-383.

[12] BURMEISTER, E. and DOBELL, A.R., "Mathematical Theories of Economic Growth", Macmillan, N.Y., 1970.

[13] LOTKA, A.J., Elements of Mathematical Biology, Dover 1956.

[14] SMITH, V.L., "The Economics of Production from Natural Resources", American Economic Review, Vol. LVIII (June 1968) pp. 409-431.

[15] PLOURDE, C.G., "A Simple Model of Replenishable Natural Resource Exploitation" Amer. Econ. Rev., June 1970, pp. 518-522.

[16] KOLM, S.C., "La Croissance et la Qualité de l'Environnement", Analyse et Prévision, VIII, 1969, pp. 445-452.

[17] KEELER, E., SPENCE, M., ZECHAUSER, R., "The Optimal Control of Pollution" Journal of Economic Theory 4, 1971, pp. 19-34.

[18] HAURIE, A., POLIS, M. and YANSOUNI, P., "On optimal Pollution and Consumption Control in a Macro-economic System", Proceeding 1972, IEEE Conference on Decision and Control.

[19] HAURIE, A., POLIS, M. and YANSOUNI, P., "On Optimal Convergence to a Steady-State-Zero-Growth Economy with Pollution and Consumption Control", to appear Proceeding 1973, IFAC/IFORS Conference on Dynamic Modelling and Control of National Economies.

[20] PITCHFORD, J., "Population and Optimal Growth", Econometrica, Vol 40, No 1 (January 1972).

LIMITED ROLE OF ENTROPY IN INFORMATION ECONOMICS

Jacob Marschak
University of California
Los Angeles+)

SUMMARY.

"Information transmitted" is defined as the amount by which added evidence (or "message received") diminishes "uncertainty". The latter is characterized by some properties intuitively suggested by this word and possessed by conditional entropy, a parameter of the posterior probability distribution. However, conditional entropy shares these properties with all other concave symmetric functions on the probability space.

Moreover, a given transmission channel (or, in the context of statistical inference, a given experiment) yields a higher maximum expected benefit than another, to any user, if and only if all concave functions of the posterior probability vector have higher values for the former channel (or experiment). Hence one information system (channel, experiment) may be preferable to another for a given user although its transmission rate, in entropy terms, is lower.

But only entropy has the economically relevant property of measuring, in the limit, the expected length of efficiently coded messages sent in long sequences. Thus, while irrelevant to the value (maximum expected benefit) of an information system and to the costs of observing, estimating, and deciding, entropy formulas are indeed relevant to the cost of communicating, i.e., of storing, coding and transmitting messages.

+) Acknowledgements of support are due to the Alexander von Humboldt-Foundation (Special Program for the Co-operation of Research Institutes of F.R.G. and U.S.A.) and to the U.S. Office of Naval Research.

SOME DEFINITIONS

Sets (assumed finite for simplicity)

$A = \{a\}$: benefit-relevant <u>actions</u>;

$Z = \{z\}$: benefit-relevant <u>events</u>.

$Y = \{y\}$, $Y' = \{y'\}$, ...: <u>messages</u> received (or experimental results.

$\beta = \left[\beta_{az}\right]$: <u>benefit matrix</u> (function).

<u>Probabilities</u>: joint, p_{zy};

$\pi_z = \sum_y p_{zy}$ ("prior");

$p_y = \sum_z p_{zy}$ (probability of message);

$\eta_{zy} = p(y|z) = p_{zy}/\pi_z$ ("likelihood");

$\varepsilon_{yz} = p(z|y) = p_{zy}/p_y$ ("posterior").

<u>Probability vectors</u>: $\pi = \left[\pi_z\right]$; $\varepsilon = \left[\varepsilon_{yz_1}, \varepsilon_{yz_2}, \ldots\right]$;

general: $q = (q_1, \ldots, q_n)$.

<u>Markov matrix</u> $\eta = \left[\eta_{zy}\right]$, interpreted as channel,or experiment, or "information system"; the latter is equivalent to a chain or network of Markov matrices, each stochastically transforming symbol-inputs into symbol-outputs, with z the initial input and y the terminal output.

"Uncertainty functions": $\{U\}$ = class of all real-valued concave symmetric functions U on the space of probability vectors. (When $q = (q_1, \ldots, q_n)$ we also write $U^n(q)$, following Aczél).

Examples: $U(q) = H(q) = \sum q_i \log (1/q_i)$, entropy;

$U(q) = D(q) = 1 - \max_i q_i$ (De Groot).

"Prior" uncertainty $U(\pi)$, also written $U(Z)$;

"Posterior" uncertainty $U(\varepsilon_y)$, also written $U(Z|y)$;

"Expected posterior" uncertainty $\Sigma p_y U(\varepsilon_y)$ also written $U(Z|Y)$;

"Uncertainty removed" $U(Z) - U(Z|Y)$.

Message y noiseless: $\eta_{zy} = 1$ or $0, \forall z$;

perfect : $\eta_{zy} = 1$ for unique z.

Information system noiseless (deterministic): all y noiseless; example:

$$\eta: \quad \begin{matrix} & y_1 & y_2 \\ z_1 & 1 & 0 \\ z_2 & 1 & 0 \\ z_3 & 0 & 1 \end{matrix}$$

Information system perfect: all y perfect, hence η is a permutation matrix or, without loss of generality, identity matrix I.

Null-information system: all rows of η identical. Canonical form (equivalent with respect to "value": see below):

$$\eta^{o} = \begin{pmatrix} 1 \\ \vdots \\ 1 \end{pmatrix}$$

Let $\eta = [\eta_{zy}]$, $\eta' = [\eta'_{zy'}]$. We say that η is quasi-garbled into η' iff $\exists$ a Markov matrix $g = [g_{yy'}]$ with

$$\eta' = \eta g$$

(If, in addition, $g_{yy'} = p(y|y'), \forall y,y'$, η is said to be garbled (cascaded) into η'). Since $I\eta = \eta$ and $\eta\eta^{o} = \eta^{o}, \forall \eta$, all information systems, given the number of elements of Z, form a lattice induced by the quasi-garbling (a partial ordering), with

$$\eta^{max} = I, \quad \eta^{min} = \eta^{o}.$$

We say that η' is coarser than η iff η is quasi-garbled into η' and, in addition, g is noiseless. Then each column of η' is a sum of some columns of η.

(Pure) strategy: a function $\alpha: Y \to A$. It is an element of $\{\alpha\}$.

Expected benefit: $\sum_{zy} p_{zy}\, \beta_{az} = \sum_{zy} \eta_{zy}\, \pi_z\, \beta_{az} = B_{\pi\beta}(\eta,\alpha)$,

where the functions π (his "beliefs") and β (his "tastes", "production conditions", "market conditions") are given to the chooser and user of the information system η. If he applies an optimal strategy he will obtain the maximum expected benefit, also called

Value of the information system,

$$\max_{\alpha\in\{\alpha\}} B_{\pi\beta}(\eta,\alpha) \triangleq V_{\pi\beta}(\eta),$$

where again the givens are written as subscripts. Where all pure strategies have equal cost, only the set $\{\alpha\}$ of all pure strategies need be considered when maximizing expected benefit. For an average (the expected benefit of a mixed strategy) cannot exceed all of its components (the expected benefits of pure strategies), to use the late L.T. Savage's pungent formulation, which applies to all costless one-person games. The assumption of equal (or nil) strategy costs is usual in the theory of statisticians, though not in their practice (linear unbiased estimates are cheapest to compute!).

We say that η is more informative than η' and write $\eta > \eta'$ iff

$$V_{\pi\beta}(\eta) \geq V_{\pi\beta}(\eta'), \quad \forall \ \pi \text{ on } Z, \forall \ \beta \text{ on } A \times Z.$$

SOME THEOREMS.

1. Blackwell Theorem. η is more informative than η' iff η is quasi-garbled into η'.
   Corollaries: 1) If η' is coarser than η, then η is more informative than η'.
   2) The lattice induced on the set of information systems by the relation "more information than" is identical with the lattice induced by the relation "quasi-garbled into".

2. Some "desirable" properties of concave symmetric functions on the probability space. $U^n(q_1,\ldots,q_n)$ has a unique maximum on the n-simplex, and by symmetry

2.1 $\quad U^n(\frac{1}{n},\ldots,\frac{1}{n}) \geq U^n(q_1,\ldots,q_n)$;

and since

$$U^{n+1}(\frac{1}{n},\ldots,\frac{1}{n},0) = U^{n}(\frac{1}{n},\ldots,\frac{1}{n}),$$

it follows that

2.2 $$U^{n+1}(\frac{1}{n+1},\ldots,\frac{1}{n+1}) > U^{n}(\frac{1}{n},\ldots,\frac{1}{n}) \text{ ; and}$$

$$U^{n}(\frac{1}{n},\ldots,\frac{1}{n}) > U^{m}(\frac{1}{m},\ldots,\frac{1}{m}) \text{ if } n>m.$$

(<u>Note</u>: 2.1 and 2.2 hold also for <u>quasi</u>-concave symmetric functions).

3. <u>Non-negativity of "uncertainty removed"</u>: for all $U \in \{U\}$,

$$U(Z) - U(Z|Y) \geq 0,$$

("added evidence cannot increase uncertainty").(<u>Note</u>: 3. holds for all concave functions. See De Groot.)

4. Comparative "<u>informativeness</u>" vs. comparative "<u>information transmitted</u>":

$$\eta > \eta' \text{ iff } U(Z) - U(Z|Y) \geq U(Z) - U(Z|Y') \quad \text{for all } U \in \{U\}.$$

(Note: 4. holds for all concave functions. See Marschak and Miyasawa .)

<u>Corollary</u>: Suppose $H(Z) - H(Z|Y) > H(Z) - H(Z|Y')$, but there exists $U \in \{U\}$ such that

$$U(Z) - U(Z|Y) < U(Z) - U(Z|Y').$$

Then there exist π, β such that

$$V_{\pi\beta}(\eta) < V_{\pi\beta}(\eta').$$

5. <u>Properties exclusive to entropy H.</u>

5.1 <u>Additivity</u>.

$$U(p_1q_1,\ldots,p_1q_n,p_2q_2,\ldots,p_mq_n) = U(p) + U(q) \quad \text{iff } U = H.$$

(This permits to "measure information". <u>What for</u>?)

5.2 Minimum expected length of decodable code word. Let q_i = probability (prior or posterior) of event z_i and l_i = length of the word, encoding it in alphabet of size r. For the sake of exposition, neglect the integer constraint on l_i, a constraint which makes the result an asymptotic one (Shannon; see also Feinstein, Wolfowitz). But retain the constraint that the code should be decodable. For this, the "Kraft inequality"

$$\sum_i r^{-l_i} \leqslant 1$$

is sufficient and necessary. Write l = vector $[l_i]$, $L = \{l: \sum r^{-l_i} \leqslant 1\}$. Then

$$\min_{l \in L} \sum q_i l_i = U(q) \text{ iff}$$

$$U = \sum q_i \log_r (1/q_i) \triangleq H$$

since, by simple calculus, the optimal code word length $l_i^* = \log_r(1/q_i)$.

The result agrees with the characterization of entropy as a measure of disorder if high disorder is understood to require a large number of parameters to describe a state. Quite different appears to me the mathematical derivation of entropy via the logarithm of the Stirling approximation of the probability of a given frequency distribution: see, e.g., Schroedinger.

COMMUNICATION COST, OTHER SYSTEM COSTS, AND SYSTEM VALUE.

The cost of communication -- of storing the messages to accumulate long sequences encodable economically, then to encode, transmit and decode them --, does depend (at least asymptotically) on the entropy parameters of the pertinent probability distributions. But these steps of the communication process form, in general, only a part of the information system. We have already described such a system as a chain or, more generally, a network of (stochastic) transformers of symbol-inputs into symbol-outputs. The communication steps are such transformers.

But they are preceded by, for example, the transformation of (generally unobservable, e.g., future) benefit-relevant events into observed data, related to the events by a likelihood matrix; and the transformation of data into statistical estimates or predictions, possibly to be communicated to the decision-maker _via_ storing, encoding, transmitting, decoding. Each transformer is associated with the cost of acquiring and operating it, and the operation cost is in general dependent on the transformer's inputs and is therefore random. There is no reason why the costs of sampling or of manipulating data , for example, should be related to any entropy-like expression involving the relevant likelihood matrices and the input-probabilities. The expected cost of the information system η will depend on the prior distribution π and will be denoted by $K_\pi(\eta)$. If we neglect as before the costs of decision-making, the problem of the chooser and user of the system η is to maximize the expected net benefit, i.e., the difference between the expectations of benefit and of cost:

$$\max_\eta \ \max_\alpha \left(B_{\pi\beta}(\alpha,\eta) - K_\pi(\eta)\right)$$

$$= \max_\eta \left(V_{\pi\beta}(\eta) - K_\pi(\eta)\right) \quad ,$$

where $V_{\pi\beta}(\eta)$, as defined before, is the value of the information system η. We have shown that this value, too, like the system costs other than those of communication, does not depend on any entropy parameters. If the assumption of equal costs of all pure strategies α is dropped, the problem becomes:

$$\max_{\alpha,\eta} \left(B_{\pi\beta}(\alpha,\eta) - K_\pi(\eta) - C_\pi(\alpha,\eta)\right) \quad ,$$

where $C_\pi(\alpha,\eta)$, the expected cost of strategy α, depends on the probability of its input y, and hence on the given π and the chosen η. Now it becomes impossible to separate the expected benefit $B_{\pi\beta}(\alpha,\eta)$ when maximizing with respect to α. The "value of the information system", $V_{\pi\beta}(\eta)$ loses its significance. At the same time, mixed strategies, expressed by,generally "noisy", Markov matrices $\alpha=\left[\alpha_{ya}\right]$, where $\alpha_{ya}=p(a|y)$, must now be considered. But again, there is no reason to relate strategy costs to any entropy formula.

I feel therefore that the assertions made in the SUMMARY are justified.

BIBLIOGRAPHY

Aczél, J. On Different Characterizations of Entropies. Probability and Information Theory, 1-11. Behara, M. et al., eds., Springer (1969)

Blackwell, D. Equivalent Comparisons of Experiments. Ann. Math. Stat. 24, 265-72 (1953)

________ and Girshick, A. Theory of Games and Statistical Decisions, Mc Graw Hill (197o)

De Groot, M.H. Uncertainty, Information and Sequential Experiments, Ann. Math. Stat. 33, 4o4-419 (1962)

________ Optimal Statistical Decisions. Mc Graw Hill (197o)

Feinstein, A. Foundations of Information Theory. Mc Graw Hill (1958)

Marschak, J. Economics of Information Systems. J. Amer. Stat. Ass, 66, 192-219 (1971)

________ Optimal Systems for Information and Decision, Techniques of Optimization. Academic Press (1972)

________ and Miyasawa, K. Economic Comparability of Information Systems. Intern. Econ. Rev., 9,137-74 (1968)

Savage, L.J. The Foundations of Statistics, Wiley (1954)

Schroedinger, E. Statistical Thermodynamics, Cambridge Univ. Press (1948)

Shannon, C. The Mathematical Theory of Communication. Bell Syst. Tech. J. (1948)

Wolfowitz, T. Coding Theorems of Information Theory. Springer (1961)

ON A DUAL CONTROL APPROACH TO THE PRICING POLICIES OF A TRADING SPECIALIST

Masanao Aoki

Department of System Science

University of California

Los Angeles

I. INTRODUCTION

We consider a market in which there is a specific economic agent or authority who sets the price, and trading takes place out of equilibrium. We call him a marketeer following Clower [8]. A trading specialist is one example of a marketeer whose pricing policies are the main concern of this paper. We assume further that he does not know the exact demand and supply condition that he faces in the market, i.e., he has only imperfect knowledge of the market response to a price he sets. He does have some subjective estimate of the market response as a function of the price he sets. See Arrow [2] for a related topic.

Denote by $f(p;\theta)$ his subjective estimate of the market response (for example, excess demand for the commodity), where θ is an element of a known set Θ . In other words, in the opinion of the marketeer $\{f(p;\theta), \theta \in \Theta\}$ represents a family of possible responses to his setting p . By specifying θ, a specific response is chosen as his estimate. Assume that the true response (unknown to him) corresponds to $f(p;\theta^*) + \xi$, $\theta^* \in \Theta$, where ξ is a random variable to be fully specified later. We take Θ to be time-invariant.

Therefore, the actual response may deviate from $f(p;\theta)$ by: (i) the agent's estimate of the parameter θ, being different from the true parameter θ^*; and by (ii) the random variable ξ which may be used to represent the effects being unknown to be systematic to the agent. Various anticipated or systematic trends[†] such as price expectation on the part of the buyers could be modeled. The probability distribution of ξ is assumed to be known to the agent. The type of consideration to be presented below can be easily extended to the case where the distribution is known imperfectly; for example, up to certain parameter values specifying the distributions uniquely. This added generality is not included in the paper, since it represents a straightforward extension of this paper.

[†] To consider these, p in $f(p;\theta)$ must be replaced by the history of past prices. We do not consider this case explicitly in this paper.

The model to be discussed in this paper could arise in several economically interesting contexts; for example, in considering optimal price setting policy of a monopolist *, (Barro [5]) or a trading specialist or inventory-holding middleman in a flow-stock exchange economy. For a non-random treatment of a pricing question under imperfect knowledge, see Hadar-Hillinger [10].

In this paper we present the case of a trading specialist by assuming that the agent holds sufficient inventory so that he can complete all transactions. The non-price rationing behavior will not be discussed.

In Section 2, we formulate the model. The process by which the marketeer's subjective estimate is updated with additional observation is discussed in Section 3, and the derivation of optimal and suboptimal (second-best) pricing policies are derived in Section 4. The last Section discusses some extensions and approximations.

2. MODEL

Consider a trading specialist dealing in an isolated market who trades in a single good. As pointed out in [7], monopolistic or oligopolistic price adjustment can be formulated in an entirely analogous manner. The only main technical difference is due to the fact that the price and output rate must be treated as decision variables, rather than just the price.

We consider a Marshallian "short period" market. At the beginning of each trading period (day, say), he posts the price at which the good is to be traded. There are several possibilities. His selling and buying price will be different by a fixed percentage, or he faces imprecisely known demand but known supply, and so on. In each case, it may be reduced to a situation with his subjective estimate of excess demand being given by $f(p;\theta)$ with appropriately chosen θ . Since he trades in disequilibrium, he is not certain what current price p will clear the market. When he changes the price of the good, he has only his subjective estimate of the effects of the price change. Faced with this uncertainty, however, the trading specialist sets the price which, in his estimate, will maximize some chosen intertemporal criterion function.

We assume, therefore, that the trading specialist's criterion function over the next T periods is a function of x_t, $t = 0, \ldots, T$ where x_t is the excess demand at time t

$$x_t = d_t - s_t$$

with d_t and s_t being the market demand and supply at the t-th marketing day. Note that x_t becomes known only at the end of the t-th market day.

---

*Barro's approach may be considered to be a special case of this paper where Θ is a singleton. We do not consider price adjustment cost, however, in the criterion function of the specialist. This is done for the sake of simplicity of presentation.

We examine an optimal price setting policy by formulating this problem as a parameter adaptive (or a parameter learning) optimal control problem, since he learns about the unknown parameters which specify the excess demand function. We then apply the dual control theory to derive the equation for the optimal price policy [3].

This paper therefore represents one possible mathematical formalization of price setting mechanisms discussed by Clower [6-8], who emphasized the subjective nature of economic agents' decision making processes.

Let S_t be the stock level of the commodity at time t . We have

$$S_{t+1} = S_t - x_t \quad . \tag{1}$$

Rather than assuming that the marketeer chooses p by setting $f(p;\hat{\theta}_t) = 0$, where $\hat{\theta}_t$ is his current estimate of θ, we consider optimal and suboptimal pricing policies by minimizing a certain criterion function explicitly.

The criterion function to be minimized is taken to be

$$J = k(S_{T+1}) + \sum_{t=0}^{T} \phi_t(x_t) \tag{2}$$

where ϕ_t is a function of x_t , and represents the cost of being out of equilibrium (non-clearing market days). The function k evaluates the cost of the stock S_{T+1} at the end of the current planning horizon deviating from a desirable stock level. One possibility is to set

$$k(S_{T+1}) = (S_{T+1} - S^*_{T+1})^2 , \quad \phi_t(x_t) = \lambda x_t^2 \tag{2'}$$

with S^*_{T+1} planned desired stock level; λ is the parameter giving relative weights of these two costs. The time discounting factor can easily be incorporated but set equal to zero here for simplicity.

Another possible criterion function at the s-th market day is

$$J_s = \sum_{t=s}^{T} \left[\psi_t(S_{t+1}) + \phi_t(x_t)\right] \quad , s = 0,1, \ldots, T$$

to express the fact that the specialist has a preferred level of stocks throughout the planned period, and to deviate from it in either direction is costly.

These criterion functions incorporate explicitly costs due to non-clearing markets and to deviation of stock levels from some desired levels, reflecting such considerations as perishable goods and storage costs.

A third possibility for J is to include cost due to price changes. The cost in this case is due partly to the price stability and predictability which is relevant in minimizing search cost, and not so much due to the set-up cost of price changes [1]. This additional cost is not incorporated here in order to focus our attention on the parameter learning aspect. The criterion function (2) will be assumed in this paper to be specific.

The criterion functions (2) express the possibility that the marketeer may wish to increase or decrease the terminal stock level by purposely setting price that makes his estimated excess demand flow at non-zero values.

The stock of the commodity held by the marketeer at time t, S_t, is therefore given by

$$S_{t+1} = S_o - \sum_{\tau=0}^{t} x_\tau(p_\tau) \quad ,$$

where the excess demand x_t is assumed to be a function of p_t only.

We assume the market excess demand function $x_t(p)$ is to be given parametrically as

$$x_t(p) = f(p;\theta^*) + \xi_t \quad , \theta^* \varepsilon \Theta \tag{4}$$

where θ^* is the parameter vector that specifies a particular excess demand curve out of a family of such curves, and where $\{\xi_t\}$ is taken to be a sequence of independent, identically distributed random variables[†] with

$$E\xi_t = 0 \ , \quad \text{all } t \geq 0$$

$$\text{var } \xi_t = \sigma^2 < \infty \ , \quad \text{all } t \geq 0 \ .$$

In this paper we assume σ to be known.

One example discussed later considers a linear excess demand function

$$x_t(p_t;\alpha,\beta) = -\alpha p_t + \beta + \xi_t \ , \quad \alpha > 0 \ , \quad \beta > 0 \ . \tag{4'}$$

The unknown parameters are α and β[††] in this case.

It is assumed that the trading specialist has a priori probability density function for θ, $p_o(\theta)$.

The problem may now be stated as:

Determine the sequence of prices $p_o, p_1, \ldots, p_T$ such that

$$E[J|H_t] \qquad t = 0,1, \ldots, T$$

are minimized subject to the dynamic equation

$$S_{t+1} = S_t - x_t(p_t) \qquad t = 0,1, \ldots, T$$

where

[†]Serially corrlated ξ's can be handled with no conceptual or technical difficulties. The independence assumption is chosen for the sake of simplicity.

[††]Theoretically, there is no difficulty to increase the number of unknown parameters more to include σ , say. See Sec.III.2 of [3]. The computation becomes more involved, of course.

$$H_t = \{x_{t-1}p_{t-1}, H_{t-1}\}$$

$$H_o = \{S_o \text{ and a priori knowledge of } \theta\} \quad .$$

As stated above, we do not impose the constraint $S_t \geq 0$ for all t explicitly, assuming that S_o is sufficiently large.

At time t, S_t is a known number. However, $S_{t+1}, \ldots, S_{T+1}$ are random variables.

3. DERIVATION OF CONDITIONAL PROBABILITY DENSITY FUNCTION OF θ

At the beginning of the t-th market day, the specialist knows the past and current stock levels, S_τ, $\tau = 0,1, \ldots, t$ and from these past excess demands $x_\tau(p_\tau)$, $\tau = 0,1, \ldots, t-1$; and he computes the posterior probability density function of θ, $p(\theta|H_t)$ by the Bayes rule recursively from $p_o(\theta)$:

$$p(\theta|H_{t+1}) = \frac{p(\theta|H_t)\, p(x_t|H_t,\theta,p_t)}{p(x_t|H_t,p_t)} = \frac{p(\theta|H_t)\, p(x_t|\theta,p_t)}{p(x_t|H_t,p_t)} \tag{6}$$

where

$$p(\theta|H_o) = \frac{p(x_o|\theta,p_o)\, p_o(\theta)}{\int_\Theta p(x_o|\theta,p_o)\, p_o(\theta)\, d\theta}$$

where we compute $p(x_t|\theta,p_t)$ from our knowledge of the probability density function for the noise ξ_t .

The trading specialist's uncertainty or his knowledge about the unknown market excess demand function at time t is summarized by $p(\theta|H_t)$.

As a function of H_t, he sets p_t for the market period t .

For example, if ξ_t is Gaussian, with mean 0 and standard deviation σ, then we have

$$p(x_t|\theta,p_t) = \frac{1}{\sqrt{2\pi}\,\sigma} \exp - \frac{1}{2\sigma^2}\left[x_t - f(\theta,p_t)\right]^2 \quad .$$

Unless $p_o(\theta)$ is such that it admits sufficient statistics, the size of H_t linearly grows with t .

The calculation involved in updating the posterior density function by (6) may be very large. Only a few density functions have the pleasant property of "reproducing" the functional form after the transformation (6). The normal density function and Beta density functions are two well-known ones [3]. To illustrate, suppose that $f(p;\theta)$ is linear as given by (4'), with $\theta = \begin{pmatrix} -\alpha \\ \beta \end{pmatrix}$.

Suppose further that $p_o(\theta)$ is given by the normal density function with mean vector $\hat{\theta}_o$ and the covariance matrix Λ_o . By the repeated application of (6), we obtain the normal posterior density function $p(\theta|H_t)$ with:

$$\hat{\theta}_t = E(\theta|H_t) = \frac{1}{\sigma^2}\Lambda_t C_t X_t + \Lambda_t\Lambda_o^{-1}\hat{\theta}_o \tag{7}$$

where

$$X_t = \begin{bmatrix} x_o(p_o) \\ \vdots \\ x_{t-1}(p_{t-1}) \end{bmatrix}, \quad C_t = \begin{bmatrix} p_o, & p_1, & \dots, & p_{t-1} \\ 1 & 1 & & 1 \end{bmatrix}$$

where

$$\sigma^2\Lambda_t^{-1} = \sigma^2\Lambda_{t-1}^{-1} + \begin{pmatrix} p_{t-1} \\ 1 \end{pmatrix}(p_{t-1},1)$$

$$= \begin{bmatrix} \lambda_1 + \sum_{\tau=0}^{t-1} p_\tau^2 & , & \lambda_2 + \sum_{\tau=0}^{t-1} p_\tau \\ \lambda_2 + \sum_{\tau=0}^{t-1} p_\tau & , & \lambda_3 + t \end{bmatrix}$$

and where $\lambda_1 \sim \lambda_3$ are defined by

$$\sigma^2\Lambda_o^{-1} = \begin{bmatrix} \lambda_1 & \lambda_2 \\ \lambda_2 & \lambda_3 \end{bmatrix}$$

It is instructive to write (7) as

$$\hat{\theta}_t = (I - K_t)\left[\hat{\theta}_{t-1} + \frac{1}{\sigma^2}\Lambda_{t-1}\begin{pmatrix} p_{t-1} \\ 1 \end{pmatrix} x_{t-1}(p_{t-1})\right] \tag{7'}$$

where

$$K_t = \frac{\dfrac{\Lambda_{t-1}}{\sigma^2}\begin{pmatrix} p_{t-1}^2 & p_{t-1} \\ p_{t-1} & 1 \end{pmatrix}}{1 + (p_{t-1},1)\dfrac{\Lambda_{t-1}}{\sigma^2}\begin{pmatrix} p_{t-1} \\ 1 \end{pmatrix}}$$

which shows explicitly how the latest piece of data $x_{t-1}(p_{t-1})$ influences the trading specialist's estimate of θ at time t, $\hat{\theta}_t$.

A similar, slightly more involved expression obtains if the Beta density function is used. This function may be suitable to be used for α in the linear excess demand function, since it is quite likely that $0 \leq \alpha \leq 1$.

4. DERIVATION OF THE OPTIMAL PRICE POLICY

In this section, we derive a sequence of optimal prices. We follow the developments in Sec. III of Ref.[3].

Last Stage

Suppose $p_o, p_1, \ldots, p_{T-1}$ has been determined, and only p_T remains to be chosen. p_T is chosen to maximize

$$E\left[k(S_{T+1}) + \phi_T(x_T) | H_T\right] .$$

The excess demand at time T, x_T, hence S_{T+1} is predicted by

$$\begin{aligned} E\left(k(S_{T+1}) | H_T\right) &= \int k(S_{T+1})\ p\left(S_{T+1} | S_T, \theta, p_T\right)\ p(\theta | H_T)\ dS_{T+1} d\theta \\ &= \int k\left(S_T - f(p_T;\theta) - \xi\right)\ p(\xi)\ p(\theta | H_T)\ d\theta d\xi \end{aligned} \tag{8}$$

where $p(\xi)$ is the known probability density function for ξ, and

$$E\left(x_T(p_T) | H_T\right) = \int f(p_T;\theta)\ p(\theta | H_T)\ d\theta \quad .$$

When $k(\cdot)$ and $f(\cdot)$ are specified, (8) can be evaluated explicitly, at least numerically.

The optimum p_T is determined by

$$\operatorname*{Max}_{p_T} E\left[k(S_{T+1}) + \phi_T(x_T) | H_T\right] .$$

Denote the maximum value by γ_N^*.

Next to Last Stage

The prices p_{T-1} and p_T are to be chosen to maximize

$$E\left[k(S_{T+1}) + \phi_{T-1}(p_{T-1}) + \phi_T(p_T) | H_{T-1}\right] .$$

The computation of $E[\phi_{T-1}(p_{T-1}) | H_{T-1}]$ can be carried out similarly to the last stage case.

The other two terms are computed as

$$E\left\{E\left[\left(k(S_{T+1}) + \phi_T(p_T)\right) | H_T\right] \Big| H_{T-1}\right\} .$$

At time T-1, Λ_{T-1}, C_{T-1}, and X_{T-1} are known. x_{T-1} is unknown, and p_{T-1} is to be determined.

Thus at time T-1, we need $p\left(x_{T-1} | p_{T-1}, H_{T-1}\right)$ and $\hat{\theta}_{T-1}$ is known.

Thus, the determination of optimum p_{T-1} can be expressed as

$$\max_{p_{T-1}} \gamma_{T-1}$$

where

$$\gamma_{T-1} = \lambda_{T-1} + E(\gamma_T^* | H_{T-1})$$

with

$$\lambda_{T-1} = E\left[\phi_{T-1}(p_{T-1}) | H_{T-1}\right] \ .$$

In evaluating $E(\gamma_T^* | H_{T-1})$, we use (7)' to express $\hat{\theta}_T$ in terms of $\hat{\theta}_{T-1}$ and $x_{T-1}(p_{T-1})$. Thus γ_{T-1} is a function of $\hat{\theta}_{T-1}$ and p_{T-1}, and the indicated maximization can be carried out either analytically or approximately numerically.

We sketch some approximation techniques for the case where an analytical solution is not available in Section 5.

General Case

Defining the maximum value as γ_{T-1}^*, etc., the determination of the optimum price at time k can be expressed as

$$\max_{p_k} \gamma_k$$

where

$$\gamma_k = E\left[\phi_k(p_k) | H_k\right] + E(\gamma_{k+1}^* | H_k) \ .$$

5. DISCUSSIONS

The process of determining the optimal sequence of prices described in the previous section is generally rather complex, unless the marketeer chooses a simple criterion function such as setting p_t by $E(x_t | H_t) = 0$. For example, in the next to last stage, $E(\gamma_T^* | H_{T-1})$ cannot be expressed in an analytically closed form, generally. The situation is exactly analogous to that of Example C of Sec. III.3 of Ref [3]. α, which multiples p_t, appears as an unknown gain in the equation for S_t when f is linear and J is given by (2').

One way to simplify the price determination is to separate the price determination from the estimation; for example, θ is replaced by $E(\theta | H_t)$ at time t (certainty equivalent pricing policy), or θ appearing in J is estimated using $p(\theta | H_k)$ at time k. Then p_k is determined using $\hat{\theta}_k$ as the parameter. In other words, at time k, the demand function is approximately taken to be

$$x_t(p_t) = f(p_t, \hat{\theta}_t) + \xi_t \ .$$

Another approximation is based on the idea of predictive control [11]. Assume for example that at time t the price p is maintained throughout the remainder of the planning period (static price expectation). This enables the specialist to have his estimated sequence of stochastic excess demands for the remainder of the periods $f(p,\hat{\theta}_\tau) + \xi_\tau$, $\tau = t, t+1, \ldots, T$. Expand $k(\cdot)$ and $\phi(\cdot)$ about this predicted excess demand value up to quadratic terms, say. This gives an approximate value to

$$E\left[k(S_{T+1}) + \sum_{s=t}^{T} \phi(x_s) | H_t\right] \tag{+}$$

as a function of p. (This is a purely stochastic problem.)

Repeat the above for two other p values, say $p + \Delta p$ and $p - \Delta p$ with some fixed Δp. Pass a quadratic curve through these three points to obtain a quadratic approximation to (+) as $Q_t(p)$. Then perform

$$\max_p Q_t(p) \quad .$$

The maximizing p is chosen to be p_t. Repeat the whole process at $t+1$.

In some dynamic systems, this approximation is known to give better results than open-loop feedback approximations. See for example Sec. VII of [3] and [9,11].

The details of various pricing policies such as certainty equivalent or open-loop feedback pricing policies are left to a separate paper, where an intuitively appealing price adjustment mechanism can be shown to be derivable from (sub) optimization of certain criterion functions. See [4].

We now briefly indicate modifications that are necessary when we drop the synchronized trading assumption.

Let R be the basic period that the trading specialist uses in revising his price, i.e., after each R period, he evaluates the past history of the excess demands and decides to revise his price or not.

Arrivals of buyers (and sellers) can be modeled by a stochastic point process, for example by a Poisson process with intensity ρ. The market demand then is modeled by a compound Poisson process.

During the period R, suppose that the trading specialist observes N buyers.

Let $d(p) + \xi$ be the demand schedule where ξ is a mean zero random variable with finite variance, independent of the point process.

The market excess demand, now written as z, is then

$$z = N_1 d(p) - N_2 s(p) + \sum_{i=1}^{N_1} \xi_i - \sum_{j=1}^{N_2} \eta_j$$

where ξ's and η's are assumed to be independent.

Since

$$E(N_1) = E(N_2) = \rho T$$

we have

$$E(z)/\rho T = x(p) \ , \quad Var\{z/\rho T\} = \{\sigma_1{}^2 + \sigma_2{}^2\}/\rho T$$

where $x(p) = d(p) - s(p)$ is the "representative" customer's excess demand for the commodity.

Therefore, if the specialist merely wants to devise a pricing policy which achieves zero excess demand in some probabilistic sense, he could use

$$p_{t+1} = p_t + \lambda z_t$$

where λ is related to k, small positive adjustment coefficient by $\lambda = k/\rho T$.

For sufficiently small k and when $d(p)$ and $s(p)$ are linear in p with incompletely specified coefficients, it can be shown that $E(p_t)$ converges to p^* such that $E\{x(p^*)\} = 0$ with finite $Var\{x(p^*)\}$ under suitable assumptions. This aspect will not be pursued further here.

Another possibility is to treat the change in price as a statistical hypothesis testing, zero hypothesis being that $x(p) = 0$ and the alternate hypothesis being $x(p) \neq 0$. This is especially easy to apply when ξ's and η's are assumed to be normally distributed. See for example Ferguson, Lehman and Wilkes.

When the specialist wishes to use the criterion function (2), then (N_1, N_2) must be added to H_t. The recursion formula for $p(\theta|H_t)$ requires only a minimal amount of modification, since the point process generating N_1 and N_2 and the noise processes ξ and η are independent by assumption. Thus the formula for $\hat{\theta}_t$ quite analogous to (7) is obtained for this more general case.

Under the constant future price expectation, one can express the probability distribution for the future excess demands which are used to obtain quadratic approximation to $E(J|H_t)$ as outlined in Section 4. Minimization of this approximate $E(J|H_t)$ generates an approximation to the optimal p_t.

The details must be left to a separate paper, again because of the lengthy development.

REFERENCES

1. Alchian, A.A., "Information Costs, Pricing and Resource Unemployment," in Phelps et al, Micro-economic Foundations of Employment and Inflation Theory, 27-52, W.W. Norton and Company, Inc. (1970).

2. Arrow, K., "Towards a Theory of Price-Adjustment," in The Allocation of Economic Resources, Abramovitz, M. et al (ed.), Stanford University Press (1959).

3. Aoki, M., Optimization of Stochastic Systems, Chapter III, Academic Press (1967).

4. Aoki, M., "On Some Price Adjustment Schemes of a Marketeer," presented at the 2nd Stochastic Control Conference, NBER Conference on the Computer in Economic and Social Research, Chicago (June 1973).

5. Barro, R.J., "A Theory of Monopolistic Price Adjustment," Rev. Econ. Stud. 34 (1), 17-26 (January 1972).

6. Clower, R.W., "Oligopoly Theory: A Dynamic Approach," Proc. 34th Ann. Conf. Western Economic Association (1969).

7. Clower, R.W., "Some Theory of an Ignorant Monopolist," Economic Journal, 705-716 (December 1959).

8. Clower, R.W., "Competition, Monopoly, and the Theory of Price," Pakistan Economic Journal, 219-226 (September 1955).

9. Dreyfus, S.E., "Some Types of Optimal Control of Stochastic Systems," SIAM J. Control 2, 120-134 (January 1964).

10. Hadar, J., and Hillinger, C., "Imperfect Competition with Unknown Demand," Rev. Econ. Stud. 36, 519-525 (1969).

11. Tse, E., Bar-Shalom, Y., and Meier, L., "Wide-Sense Adaptive Dual Control for Nonlinear Stochastic Systems," to appear in IEEE Trans. Aut. Control.

PROBLEMS OF OPTIMAL ECONOMIC INVESTMENTS WITH FINITE LIFETIME CAPITAL

Bernardo Nicoletti
Istituto Elettrotecnico
Università di Napoli
Napoli, Italy.

Luigi Mariani
Istituto di Elettrotecnica e Elettronica
Università di Padova
Padova, Italy.

1. INTRODUCTION

The theory of optimum economic dynamic investments has been extensively studied. However, apart from simple models, the finite lifetime of capital has not been considered. As a matter of fact, in many allocation problems, either in the context of macro- or micro-economics, the effects of some decisions may have only finite duration in time. Aim of this paper is to focus this problem and to present a method for dealing with the problem in the discrete-time case.

Mathematically, when one considers finite lifetime L, each decision affects not only the state of the systems at the following stage but also at a number of stages later. The usual prescription for handling problems of this type is to increase the original state space by L additional variables: as a consequence, even for problems with relatively short delays, the dimensionality of the problem is increased beyond the reach of present-day computing capabilities and the new system is not controllable.

D. G. Luenberger suggested an approach for exploiting the inherent structure of this problems in order to get a low-dimensional algorithm: the "cyclic dynamic programming".

In the present paper, a similar approach is used to reformulate the original problem. However, rather than aiming directly to a computer algorithm, a "quasi-periodic discrete maximum principle" is derived, which allows one to write general canonical optimum conditions. From these, it is also possible to derive computer algorithms.

In section 2 and 3, as an example of applications, two models for systems with finite lifetime investments, in the context of macro- and micro-economy respectively, are discussed. In Section 4, the general problem is stated. Using a mathematical programming approach, necessary conditions for the optimality are derived in

Section 5 for the general nonlinear case. For systems with linear dynamics, concave inequality constraints and convex performance index, the conditions derived are both necessary and sufficient and can be formulated as an extended discrete maximum principle. This is done in Section 6. Finally, in Section 7 the method is applied for deriving the optimality conditions for the models discussed in Section 2 and 3.

2. A MACROECONOMIC MODEL

To show the basic ideas of this paper, the simplest model for the accumulation of capital in one-sector closed economy will be considered. As it has been suggested, (J. Tinbergen and H. C. Bos), when one considers economic growth, it is desiderable to distinguish between the stock of equipment or capital goods and the stock of capital. The difference lies in the fact that an industrial machine remains a constant volume of equipment until it is scrapped, whereas its contribution to the capital stock decreases because of its depreciation. As a consequence, the structure of the model is different from the classical one.

Using the variable:

t =increasing ordinator, indicating the discrete-time instant considered;

k_t =capital stock, at time t;

y_t =national income (net product), at time t;

i_t =gross investment, at time t;

c_t =consumption, at time t;

b_t =capital goods or volume of equipment, at time t;

L =finite lifetime;

$r = \frac{1}{L}$ =constant proportionate rate of depreciation,

(where k_t, y_t, i_t, c_t, and b_t are expressed in the same unit), the following relationships hold

(1) $y_t = c_t + i_t$ income identity

(2) $k_{t+1} = k_t + i_t - r\, b_t$ gross investment identity

(3) $b_{t+1} = b_t + i_t - i_{t-L}$ capital goods identity

(4) $y_t = f(b_t)$ production function

(5) $c_t \geqslant 0$, $i_t \geqslant 0$ and b_o, i_{-1}, ... , i_{-L} given.

Dynamic economic models are considered here as a tool in planning development. Particularly, the viewpoint considered is that of a government which is in a position to control the economy completely by choosing the gross investments and to plan perfectly so as to maximize a given social welfare function over the time T

$$(6) \quad \sum_{t=0}^{T-1} w_t(c_t) \quad .$$

T is the planning horizon: in a logical sense, the relevant period of time is the entire future. However, development plans are often chosen to maximize a welfare function over a finite horizon T, subject to terminal stocks constraints which represent a weight attached to the welfare of the generation beyond the horizon

$$(7) \quad b_T \geqslant \bar{b}_T .$$

3. A MICROECONOMIC MODEL

In many dynamic allocation problems for a single firm, the effect of some decisions may have only finite duration in time: this is the case, for example, in the planning of production plants with a finite period of economical operation, the replacement of Faculty members in educational institutions, the schedule for cutting and planting trees in a given area, etc. Problems of this type may be described by a discrete-time model of the following type.

Using the variables:

u_t = capacity built, at time t;

$c(u)$ = cost of building capacity u;

a = interest rate;

x_t = total capacity operating at time t;

d_t = demand at time t;

T = planning horizon;

L = plant lifetime;

(where u_t, x_t, and d_t are expressed in the same unity), the following relationship hold

$$(8) \quad x_{t+1} = x_t + u_t \qquad t = 0, \ldots, L-1$$

(9) $x_{t+1} = x_t + u_t - u_{t-L}$ $\quad t = L, \ldots, T-1$

(10) $x_t \geqslant d_t$ $\quad t = 1, \ldots, T$

(11) $x_o = 0.$

It is desired to minimize the costs over the planning horizon T

(12) $$\sum_{t=0}^{T-1} \frac{1}{(1+a)^t} c(u_t)$$

by choosing the building policy.

4. STATEMENT OF THE PROBLEM

The models discussed in Section 2 and 3 can be reformulated in a general way as follows.

It is given a dynamic system governed by a set of nonlinear difference equations

(13) $x_{k+1} = f_k (x_k, u_k, k)$ $\quad k = 0, \ldots, L-1$

(14) $x_{k+1} = f_k (x_k, u_k, u_{k-L}, k)$ $\quad k = L, \ldots, T-1$

where k is an increasing ordinator (the stage number), x_k and u_k respectively the the state and control variables, and L (the finite lifetime) is less than T (the time horizon). L and T are positive integers. (At a slight increase in notational complexity, the whole discussion can be trivially extended to the case where both x_k and u_k are vectors). The problem is to find a sequence of $\{u_k\}$ $(k = 0, \ldots, T-1)$, subject to belong to the constraint set represented by

(15) $e_k (x_k, u_k, k) \geqslant 0$ $\quad k = 0, \ldots, T$

in such a way to minimize the performance index

(16) $$J = \sum_{k=0}^{T-1} g_k (x_k, u_k, k)$$

starting from a given initial condition x_o.

Following Luenberger's approach, first one define the integer n such that

(17) $(n-1) L \leqslant T \leqslant nL$

and the vectors (the prime' denotes transpose)

(18) $X'_k = (x_k, x_{k+L}, \ldots, x_{k+(n-1)L})$ $\quad k = 0, 1, \ldots, T-(n-1)L$

(19) $X'_k = (x_k, x_{k+L}, \ldots, x_{k+(n-2)L})$ $\quad k = T-(n-1)L+1, \ldots, L$

(20) $U'_k = (u_k, u_{k+L}, \ldots, u_{k+(n-1)L})$ $\quad k = 0, 1, \ldots, T-(n-1)L-1$

(21) $U'_k = (u_k, u_{k+L}, \ldots, u_{k+(n-2)L})$ $\quad k = T-(n-1)L, \ldots, L-1$

with (18) and (20) of dimension n and (19) and (21) of dimension n-1.

With these position, equations (13) and (14) can be expressed by the equivalent set

(22) $X_{k+1} = F_k(X_k, U_k)$ $\quad k = 0, 1, \ldots, L-1$

while the constraints of type (15) are replaced by

(23) $E_k(X_k, U_k) \geqslant 0$ $\quad k = 0, 1, \ldots, L-1$

and the performance index to be minimized

(24) $$G = \sum_{k=0}^{L-1} G_k(X_k, U_k)$$

where

(25) $$G_k(X_k, U_k) = \sum_{i=0}^{n-1} g_{k+iL}(x_{k+iL}, u_{k+iL}) \qquad k = 0, 1, \ldots, T-(n-1)L-1$$

(26) $$G_k(X_k, U_k) = \sum_{i=0}^{n-2} g_{k+iL}(x_{k+iL}, u_{k+iL}) \qquad k = T-(n-1)L, \ldots, L-1$$

In addition, there is the consistency condition, which is reflected in the boundary conditions

(27) $$\begin{bmatrix} x_o \\ X_L \end{bmatrix} = X_o \quad , x_o \text{ given}$$

which couples the initial and the final states. In the case T = nL the boundary condition is

(28) $$\begin{bmatrix} x_o \\ X_L \end{bmatrix} = \begin{bmatrix} X_o \\ x_T \end{bmatrix}, x_o \text{ given.}$$

The function F_k, G_k, and E_k are supposed to be continuous and differentiable.

In the case of infinite horizon the preceding vectors have an infinite number of components; what follows may be extended to this case by using functional analysis (Canon et al.).

5. MATHEMATICAL PROGRAMMING APPROACH

This problem can be reformulated as a mathematical programming problem, by defining the vectors

$$Z' = (X_o, X_1, \dots, X_L, U_o, U_1, \dots, U_{L-1})$$

$$F(Z)' = (F_o - X_1, F_1 - X_2, \dots, F_{L-1} - X_L)$$

$$E(Z)' = (E_o, E_1, \dots, E_{L-1}).$$

The optimization problem becomes that of minimizing the scalar function $G(Z)$ subject to $F(Z)=0$ and $E(Z)\geqslant 0$ and to (27).

By direct application of mathematical programming theory, it is possible to derive the following theorem.

Theorem 1

In order that $\{U_k\}$ furnishes an optimal solution to problem (22)÷(24), it is necessary that there exist a real valued number η, vectors λ_k and μ_k (of the same dimension of X_{k+1} and X_k respectively), such that the following equations hold

(29) $\lambda_{k-1} = \left(\frac{\partial F_k}{\partial X_k}\right)'\lambda_k + \eta\frac{\partial G_k}{\partial X_k} - \left(\frac{\partial E_k}{\partial X_k}\right)'\mu_k$ $\qquad k=0,..,L-1$

(30) $\eta\frac{\partial G_k}{\partial U_k} + \left(\frac{\partial F_k}{\partial U_k}\right)'\lambda_k - \left(\frac{\partial E_k}{\partial U_k}\right)'\mu_k = 0$ $\qquad k=0,..,L-1$

(31) $X_{k+1} = F_k(X_k, U_k)$ $\qquad k=0,..,L-1$

(32) $E_k(X_k, U_k) \geqslant 0$ $\qquad k=0,..,L-1$

(33) $\mu_k' E_k(X_k, U_k) = 0$ $\qquad k=0,..,L-1$

(34) $\mu_k \geqslant 0$ $\qquad k=0,..,L-1$

(35) $(0,I)\, X_o = X_L$

(36) $\lambda_{-1} = (0,I)'\,\lambda_{L-1}$

(37) $\eta \geqslant 0$

(38) $(\eta, \lambda'_{-1}, \lambda'_{o}, \ldots, \lambda'_{L-1}, \mu'_{o}, \mu'_{1}, \ldots, \mu'_{L-1}) \neq 0.$

Constraint (38) in the statement of this theorem may be omitted, and the variable η be set equal to 1, if the constraint qualification holds: for a general discussion of the conditions under which they hold, see Canon et al.

In order to obtain sufficient conditions for the problem under consideration, further conditions over F_k, E_k, and G_k must be imposed. Using mathematical programming theory, it is possible to show that the following theorem holds.

Theorem 2

Given F_k linear in X_k and U_k, given the initial condition x_o, given the performance index to be minimized (24), with G_k convex in X_k and U_k, and given the constraints (23) with E_k concave in X_k and U_k, in order that $\{U_k\}$ furnishes an optimal solution, it is necessary and sufficient that there exist a real valued number η and vectors λ_k and μ_k such that equations (29) to (38) hold.

6.A QUASI-PERIODIC DISCRETE MAXIMUM PRINCIPLE

The results of theorem 2 can be posed in the form of a discrete maximum principle. Define the Hamiltonian

(39) $H_k(X_k, U_k, \lambda_k, \eta) = \eta G_k(X_k, U_k) + \lambda'_k F_k(X_k, U_k)$. $k=0,..,L-1$

Assume that the conditions of theorem 2 are satisfied and that $(X^o_k, U^o_k, \lambda^o_k, \mu^o_k, \eta^o)$ is the optimum solution. Proceeding in a way similar to Pearson and Sridhar, it is possible to show that (30) may be substituted by

(40) $H_k(X^o_k, U^o_k, \lambda^o_k, \eta^o) = \min_{U_k \in \Omega} H_k(X^o_k, U_k, \lambda^o_k, \eta^o)$

with

(41) $\Omega = \{U_k : E_k(X^o_k, U_k) \geqslant 0\}$.

In terms of Hamiltonian, (29), (30), and (31) may be rewritten as follows

(42) $\lambda_{k-1} = \frac{\partial H_k}{\partial X_k} - (\frac{\partial E_k}{\partial X_k})' \mu_k$ $k=0,..,L-1$

(43) $\frac{\partial H_k}{\partial U_k} - (\frac{\partial E_k}{\partial X_k})' \mu_k = 0$ $k=0,..,L-1$

(44) $X_{k+1} = \frac{\partial H_k}{\partial \lambda_k}$. $k=0,..,L-1$

If H_k is linear in U_k, eqn. (43) must be substituted by (40).

Therefore, it is possible to conclude that conditions (42), (43) or (40),

(32) to (38), and (44) represent a discrete maximum principle. The particularity of this case lies in the consistency conditions (27) and (36); which account for the name of "quasi-periodic discrete maximum principle" for its similarity to the principle which holds for periodic systems (Bittanti, Fronza and Guardabassi).

7. EXAMPLES

7.1. An Economic Growth Problem.

As a first example of application, the macroeconomic model outlined in Section 2 is considered.

The production function $f(b_t)$ and the welfare function $w(c_t)$ are supposed to be concave monotone increasing. We also assume, for simplicity, that T=NL, so that the time horizon is integrally divided by the finite lifetime. Let

(50) $C'_k = (c_k, c_{k+L}, \ldots, c_{k+(N-1)L})$; $B'_k = (b_k, b_{k+L}, \ldots, b_{k+(N-1)L})$;

$$I'_k = (i_k, i_{k+L}, \ldots, i_{k+(N-1)L}) \qquad k=0,I,\ldots,L-1$$

(51) $$G = \sum_{k=0}^{L-1} W_k(C_k) \quad ; \quad W_k(C_k) = \sum_{i=0}^{N-1} w_{k+iL}(c_{k+iL})$$

(52) $$I^{o\prime}_k = (i_{k-L}, 0, \ldots, 0) \quad , i_{k-L} \text{ given} \quad . \qquad k=0,I,\ldots,L-1$$

The problem ca be re-stated as follows

(53) $$B_{k+1} = B_k + A\, I_k - I^o_k \qquad k=0,\ldots,L-1$$

(54) $$I_k \geqslant 0 \qquad k=0,\ldots,L-1$$

(55) $$F(B_k) - I_k \geqslant 0 \qquad k=0,\ldots,L-1$$

(56) $$(0,\ldots,0,1)\, B_L \geqslant \bar{b}_T$$

(57) $$\begin{bmatrix} b_o \\ B_L \end{bmatrix} = \begin{bmatrix} B_o \\ b_T \end{bmatrix} \qquad \text{or} \qquad (0,I)\, B_o = (I,0)\, B_L$$

(58) $$W_k(C_k) = W_k(F(B_k) - I_k)$$

with

(59) $$A = \begin{bmatrix} 1 & 0 & 0 & . & . & . & . & 0 \\ -1 & 1 & 0 & . & . & . & . & 0 \\ 0 & -1 & 1 & 0 & . & . & . & 0 \\ . & . & . & . & . & . & . & . \\ 0 & 0 & 0 & 0 & . & 0 & -1 & 1 \end{bmatrix}$$

Using theorem 2, with λ_k, μ_k, and ς_k vectors of dimension N, ν vector of dimension N-1 to take into account condition (57), η scalar and δ the scalar multiplier to take into account the condition (56), it is possible to derive the following necessary and sufficient conditions (canonical equations)

(60) $\lambda_{k-1} = \lambda_k + \eta\left(\frac{\partial F(B_k)}{\partial B_k}\right)'\left(\frac{\partial W_k(C_k)}{\partial C_k}\right) + \left(\frac{\partial F(B_k)}{\partial B_k}\right)'\mu_k$ $\qquad k=0,..,L-1$

(61) $B_{k+1} = B_k + A\ I_k - I_k^o$ $\qquad k=0,..,L-1$

(62) $\lambda_{L-1} - (0,0,..,0,1)'\delta = -(I,0)'\nu\ ;\lambda_{-1} = -(0,I)'\nu$

(63) $0 = -\eta\left(\frac{\partial W_k(C_k)}{\partial C_k}\right) + A'\lambda_k - \mu_k + \rho_k$ $\qquad k=0,..,L-1$

(64) $F(B_k) - I_k \geqslant 0$ $\qquad k=0,..,L-1$

(65) $\mu_k'(F(B_k) - I_k) = 0$ $\qquad k=0,..,L-1$

(66) $\mu_k \geqslant 0$ $\qquad k=0,..,L-1$

(67) $I_k \geqslant 0$ $\qquad k=0,..,L-1$

(68) $\rho_k'\ I_k = 0$ $\qquad k=0,..,L-1$

(69) $\rho_k \geqslant 0$ $\qquad k=0,..,L-1$

(70) $(0,0,..,0,1)\ B_L - \bar{b}_T \geqslant 0$

(71) $\delta(\ (0,0,..,0,1)\ B_L - \bar{b}_T) = 0$

(72) $\delta \geqslant 0$

(73) $(0,I)\ B_o = (I,0)\ B_L$.

For the presence of the boundary condition on the final value (70), the condition (36) becomes now (52).

For actually computing the optimal solution, it is possible to employ an algorithm of the following type.

Let us consider first of all the case that no one of the constraints (64) and (67) are binding. As a first trial

(74) $B_o^{(1)}$, $\lambda_o^{(1)}$, $0 \leqslant I_o^{(1)} \leqslant d\ B_o^{(1)}$, $\mu_o^{(1)} \geqslant 0$ (with $\mu_o^{(1)} = 0$ if $I_o^{(1)} < dB_o$),
$\rho_o^{(1)} \geqslant 0$ (with $\rho_o^{(1)} = 0$ if $I_o^{(1)} > 0$)

are fixed. From (60) one gets $\nu^{(1)}$. From (61) one gets $B_1^{(1)}$. Then, from (60) and (63), (65) and (66) one gets: $\lambda_1^{(1)}$, $I_1^{(1)}$, $\mu_1^{(1)}$, and $\rho_1^{(1)}$. Similarly, it is possible to compute the state and control values for all successive steps up to k=L-I. Once computed: $\lambda^{(1)}_{L-I}$, $\mu^{(1)}_{L-I}$, $I^{(1)}_{L-1}$, $B^{(1)}_{L-1}$, and $\rho^{(1)}_{L-1}$, with eqn. (61) one computes B_L and can verify if the (70) and (73) are satisfied and with eqn. (60) one can verify if (62) is satisfied. If they are not, it is necessary to modify $\lambda^{(1)}_o$, $B^{(1)}_o$, and $I^{(1)}_o$ in a suitable way and to repeat all preceding steps.

If some of the constraints (64) and (67) is binding, then I_k is immediately found since it is equal to the value of the constraint, $F(B_k)$ and I_k respectively. The adjoint variables λ_k "jumps" of the term $(\frac{\partial F(B_k)}{\partial B_k})'\mu_k$ or eqn. (63) is modified, respectively.The initial guess (74) must then be suitably modified.

7.2.A Plant Investment Problem

Consider the problem of optimum investment described in Section 3 and governed by eqn. (8) to (12), with $L<T<2L$. Following positions (18) to (25), let

(75) $X'_k = (x_k, x_{k+L})$ k=0,..,T-L ; $X_k = x_k$ k=T-L+1,..,L

(76) $U'_k = (u_k, u_{k+L})$ k=0,..,T-L-1; $U_k = u_k$ k=T-L,..,L-1

(77) $D'_k = (d_k, d_{k+L})$ k=0,..,T-L ; $D_k = d_k$ k=T-L+1,..,L-1

(78) $C_k(U_k) = c(u_k) + \frac{1}{(1+a)^L} c(u_{k+L})$ k=0,..,T-L-1 ; $C_k(U_k) = c(u_k)$ k=T-L,..,L-1

The problem becomes that of minimizing

(79) $$G = \sum_{k=0}^{L-1} \frac{1}{(1+a)^k} C_k(U_k)$$

subject to

(80) $X_{k+1} = I_k X_k + A_k U_k$ k=0,..,L-1

(81) $X_k - D_k \geqslant 0$ k=0,..,L-1

where

(82) $I_k = \begin{bmatrix} 1 & 0 \\ 0 & 1 \end{bmatrix}$ k=0,..,T-L-1 ; $I_k = (1,0)$ k=T-L ; $I_k = 1$ k=T-L+1,..,L-1

(83) $A_k = \begin{bmatrix} 1 & 0 \\ -1 & 1 \end{bmatrix}$ k=0,..,T-L-1 ; $A_k = 1$ K=T-L,..,L-1 .

The canonical equations for the optimum solution are

(84) $\lambda_{k-1} = I'_k \lambda_k - \mu_k$ k=0,..,L-1

$$(85)\quad \eta \frac{1}{(1+a)^k} \frac{\partial C_k(U_k)}{\partial U_k} + A_k' \lambda_k = 0 \qquad k=0,..,L-1$$

$$(86)\quad X_{k+1} = I_k X_k + A_k U_k \qquad k=0,..,L-1$$

$$(87)\quad X_k - D_k \geqslant 0 \qquad k=0,..,L-1$$

$$(88)\quad \mu_k' (X_k - D_k) = 0 \qquad k=0,..,L-1$$

$$(89)\quad \mu_k \geqslant 0 \qquad k=0,..,L-1$$

$$(90)\quad (0,1)\, X_o = x_L$$

$$(91)\quad \lambda_{-1} = (0,1)' \lambda_{L-1}$$

$$(92)\quad (\eta, \lambda_{-1}, \dots, \lambda_{L-1}, \mu_o, \dots, \mu_{L-1}) \neq 0$$

$$(93)\quad x_o \geqslant d_o \quad \text{given}$$

with λ_k of dimension 2 for $k=0,..,T-L-1$ and of dimension 1 for $k=T-L,..,L-1$; μ_k of dimension 2 for $k=0,..,T-L$ and of dimension 1 for $k=T-L+1,..,L-1$.

A computational algorithm, similar to the one discussed for the economic growth model, may be used for actually finding the optimal solution.

8. CONCLUSIONS

The optimum economic dynamic investments problem has been investigated in the case of finite lifetime investments. The problem is reformulated in such a way that, by using mathematical programming, it is possible to derive general canonical optimum necessary conditions. From these, a computer algorithm can be derived. For linear systems with convex performance index and concave constraints, the conditions found are both necessary and sufficient and may be posed in the form of a discrete maximum principle.

As examples of application, a macroeconomic growth model and a microeconomic plant investment model are considered.

Acknowledgment

This work was supported by the Consiglio Nazionale Ricerche, CNR, Roma, Italy.

REFERENCES

Bittanti, S., Fronza, G., and Guardabassi, G., Discrete Periodic Optimization, Internal

Report 72-13, Ist. Elettrotecnica ed Elettronica, Pol. Milano (1972).

Canon,M.D., Cullum,C.D., and Polak,E., Theory of Optimal Control and Mathematical Programming, Mc Graw Hill, New York (1970).

Luenberger,D.L., Cyclic Dynamic Programming: a Procedure for Problems with Fixed Delay, Operations Research, 19, 1101-1110 (1971).

Pearson,J.B., and Sridhar,R. A Discrete Optimal Control Problem, IEEE Trans. on Auto. Cont., AC-11, 171-174 (1966).

Tinbergen,J., and Bos,H.C. Mathematical Models of Economic Growth, Mc Graw Hill, New York (1962).

SOME ECONOMIC MODELS OF MARKETS

by M. J. H. Mogridge

Centre for Environmental Studies, London*

1. Introduction

In this paper I am drawing together a theme from a whole series of papers which I have written over the last few years. Several other authors have also contributed work along these lines, and I will be incorporating such work in this exposition.

Indeed, the models which I am using have a very long and distinguished history, both in mathematics, and more specifically in economic systems.

The gamma distribution, first extensively tabulated by Pearson in 1928 has been used in its collapsed form of the negative exponential distribution in an economic system at least as far back as 1892 by Bleicher, while the logarithmic normal distribution has the distinction of having a book written about its manifold uses in 1957 by Aitchison & Brown.

What justifies a new look at these models, in my view, is the wealth of data now becoming available about the prices in markets with the advent of computers in governmental systems. The three major economic systems that I am going to discuss have all undergone, or are undergoing, a remarkable change in data availability and accuracy. This has made possible much more sensitive evaluations of the use of these models.

In the first place we have the labour market. With the advent in 1964 of the graduated pension scheme in the United Kingdom, the Department of Health and Social Security now has a complete set of annual earnings records of everyone on one computer, and has been taking a ½% completely random sample of these records annually since then. Since 1970, the Department of Employment has also been taking a 1% sample of the same records and obtaining details from employers of all matters concerning the earnings of each such employee in a given week. This again is now done annually. This has given us entirely new precision in our data on the labour market.

In the second place, the car market, all licencing records both of cars and of individuals are currently being placed on one computer by the Department of the Environment. The data which will be available here will hopefully include

*Now with the Greater London Council, Department of Planning & Transportation.

not only the current ownership but also the previous car owned, but at present these details are only voluntary and we do not yet have any indication of response rates.

The third market that I wish to discuss is that of space, or more specifically the spatial location of households, worker residences and worker workplaces within urban areas. Here the key advance is that of geo-coding, or coordinate referencing, whereby each zone, ward or district in the urban area can be located with respect to all others. While such work has been done by hand before, the advent of the computer has made it possible to handle the enormous matrices that one gets when connecting worker residences with workplaces, eg. in London we have 1000 x 1000 matrices.

In London, coordinate referencing systems have been developed from the 1966 Census, both by the Department of the Environment and the Greater London Council for Census areas, and by Blumenfeld of University College London for the Traffic areas of the 1962 London Traffic Survey as applied to 1966 Census data.

Let us see then what statements we are able to make about the applicability of these models to markets.

2. The Labour Market

In a number of papers, I have shown that:

(a) The total labour market, covering the annual earnings of all employees earning full-year, can be approximated by a gamma distribution. I assume that earners earning part-year are earning at the same rate as those working full-year, so that they will not affect the distribution, although not included in the survey data analysed. (One could include them if one knew how many weeks they were actually earning, rather than being registered "employed".)

The upper end of the distribution is not gamma, but can be modelled with a Pareto-like tail (or negative power function). Many previous authors have noted this latter function, as their data was usually tax data of the upper incomes, which includes several markets (investment, property earnings, etc.) and indeed often several earners in a tax unit, eg. husband and wife. We can now say the tail is Pareto-like, because we can disaggregate it also to the distributions of earnings by age and sex.

(b) These male and female earnings distributions by age can be approximated with logarithmic normal distributions. For males the spread (or relative shape, or inequality) of the log-normal distributions is the same for each age, but the mean of the log-normal distributions follows a discounted Gompertz relationship, ie. the rate of change of mean income with age decreases constantly with age.

There is, however, for males, a modification to this pattern at the upper end of the distribution where the Pareto-like tail appears, whose size increases with age in a manner which can be expressed by a single equation. This states that the earnings of anyone is increased by a constant (if minute) proportion of the earnings of everyone earning less. This has the remarkable property of shifting the log-normal distribution tails in a smooth transition to a power function tail with an asymptote as near as I can estimate of -1.92. (It also shifts the gamma model too to this value.)

The female distributions are seemingly a complex mixture of two families of lognormal distributions, one equal to the male in spread for single, or childless, women, but of much lower mean value, and one of much higher spread equal to that of the total gamma distribution for all earnings, but for all ages of these women it has a constant mean value. I presume these are post-child-bearing women, who obviously have different characteristics as far as employers are concerned.

(c) The unexpected feature one notices if one tries to combine the two log-normal distributions which one obtains for all males and all females separately (ignoring the upper tails) is that there is a very definite range of parameters which will combine to give a gamma distribution for the total and all others will not. In particular one can vary the proportions of males to females quite a lot for given spread values, but only by changing mean and spread values together in a definite relationship connected with the mean and spread of the total distribution can one obtain satisfactory models to fit all at once in the stated forms.

(d) The log-normal distribution has been applied to occupational earnings many times, notably by Lydall (1968). With the present surveys one is now able to obtain over 100 members of a sample in an occupation for over 150 male occupations and nearly 100 female, none of whom deviate substantially from the distribution.

3. The household income distribution

The household income distribution has been modelled by using a gamma distribution, firstly to my knowledge by Wootton & Pick (1967), and since then in a number of UK transportation studies. This is satisfactory, provided one is not particularly interested in either the top or the bottom of the distribution.

This can, however, be taken considerably further using the income distributions of households by number of workers (workers normally or regularly working for however many hours per week). I use here a simple property of gamma distributions. If one takes individuals from a gamma distribution at random, and puts them into new distributions containing one or two or three etc. individuals each per unit, again at random, these new distributions will again be gamma distributions, with mean and shape values given by the number of individuals per unit multiplied by the original values. This one can also do even if one takes individuals in such a way that the multipliers are not integral but real numbers.

In concrete terms for our purpose, one takes workers from the gamma distribution of employment income (neglecting the upper tail) such that the average earnings of the first worker is about 1.3 times average earnings, and that of all subsequent subsidiary workers about 0.53 times average earnings. Households are formed with 1, 2, 3, etc. workers in the correct proportions. Households with workers then all have gamma distributions of income, and all households with workers together also have a gamma distribution with a mean income equal to the average earnings of workers multiplied by the average number of workers per household and its shape parameter similarly defined.

We then add in households without workers, obviously not modellable by this technique, and various corrections for the upper tail, and surprisingly we obtain very close to the gamma distribution overall yet again.

I have examined this in detail for regional incomes as well as for national, and over a period where the number of workers per household changed quite considerably.

4. The Car Market

In examining the labour market one can ignore, as I have done above, the "birth and death" processes in the market. To be specific, jobs are being continually created and destroyed and employees are continually moving between them. More importantly for the structure of the distribution, employees enter and leave the labour market, for men over fifty years or so, for women several times in this period.

There is thus a 2% pa. turnover in males, and rather more in females. The effect of this is quite small since male earnings on retirement are falling again from their peak in the early forties, and for women as we noted earnings are stable for married women irrespective of age. One would have to modify this conclusion if the population were to grow substantially - the current UK value is, however, only about ½% pa.

In the car market on the otherhand, we know that cars last for a much shorter time - this used to be twenty years but is rapidly falling towards about twelve. Moreover, the rate of growth of the car population has changed considerably from its former value of 15% pa. to values now below 5% pa. New car production still, however, fluctuates by up to 40% in a 12 month period.

Let us look then at the mathematical models of supply and demand. Firstly for supply, we generate new cars of fixed mean price, according to a gamma distribution of high spread value, since new cars are produced and bought over a price range of at least a factor of ten, although the majority of new cars lie within a price range of say three. Evidence on this exists but is poor, since we want to know the exact price paid to a seller by a buyer, not some notional list price.

We can then depreciate cars over time by a constant depreciation rate at a given time, ie. all existing cars are assumed to have the same depreciation rate - this is not true but is a useful first approximation.

One can simply grow the car population with time by using exponential or logistic growth curves, and from this obtain the average car price, and the price distribution of all cars, the stock value distribution. To obtain the distribution of the price of cars in the market one needs to know how long cars are kept before resale, especially so for new cars.

To obtain the demand side of the market, I use a technique I call the purchasing solid. This takes a given income distribution, here the household income distribution, and expresses for every income the rate of expenditure on the commodity in question, here on car purchase. This functional relationship is the Engel curve.

Each expenditure level I call a buying power, because power is a term used to denote the rate of expenditure of energy, and here we have a rate of expenditure of money. This ensures that we keep time in the forefront of our attention.

To each buying power level, I apply a gamma distribution of the price of the commodity being bought, here cars, per unit time. Each of these gamma distributions will have a mean price equal to the buying power multiplied by the time period, and a constant relative spread for all the distributions. We can then add up all the prices demanded within the income level x price solid, and obtain the price distribution within the entire market. Obviously it must match that on the supply side.

Several interesting points emerge for an Engel curve of decreasing elasticity as income rises. Most notably, the rate of growth of the number of cars in the market slows as income rises, but depreciation rates rise to offset this, causing cars to fall in price to the level at which maintenance costs are substantial enough for the car to be scrapped sooner. There is no need to create a bogeyman industrialist racketeer making shoddy cars - it seems to be the only way the market clearing operation can operate with such an Engel curve. But this whole area requires the much better data now in the offing with computerised car registration systems before this model is anything like verifiable or falsifiable.

5. The Market for Space - the Land Market

We do, however, already have a considerable body of data about the final market I want to discuss, that of space or land. I wish to limit my remarks here specifically to urban land, and by that I mean land essentially contiguously built upon or in urban use. The most intensive investigations in this area have been undertaken by Bussiere and his colleagues, in Paris, and Latham and his in Toronto, but I will quote my own experience on London data, which became available from the 1966 Census.

The subject is immensely complicated and to make any progress in modelling at all, I feel we have to start in the following manner. We limit ourselves to workers and their journey to work between home and work-place. We assume first one worker per household and modify this assumption later to a more realistic assumption. We assume that the price of the journey to work determines how far workers travel from work to home. In reality this is all interlocked with land prices and rents, but we just take this one side of the equilibrium first. We have data on both the expenditure on journey to work and on housing costs (the buying powers). We ignore that different methods or modes of transport have different costs per mile and simply use an average cost per mile basis.

The first model that we construct is the traditional economists' single pole of employment at the city centre. Both Mills (1972) and Evans (1973) have recently described some economists' positions on this problem. All employment is concentrated at one single point, and we construct the distribution of distances travelled to this point from home using the purchasing solid technique I described in the previous section, which I first used on this very problem in (1968). The resulting gamma distribution of distance converts easily for radial symmetry by dividing by the circumference at that distance into the worker and thus population density at given distance from the city centre. This is thus still a gamma distribution which we know from the census evidence is very close to a negative exponential density distribution with respect to distance. By examining the population this gives us at each distance via the purchasing solid, we can obtain the income distribution at each distance. Obviously, the richer live further from the city centre, but at all distances one still finds all incomes represented, but with changing proportions.

We can move from this simple model into the complexities of the real world step by step. First, we spread employment through the city again by a gamma distribution. Angel & Hyman (1972) have done this for Manchester, and taken the model one step further with a continuous negative exponential radially symmetric field for the speed of travel. Balckburn (1970) has incorporated rental (and lot size) as his next step. Another approach is to incorporate the relation between workers and households using the technique I have discussed above. Still another would be to incorporate different costs for different modes of travel, using different buying powers for each income group and each individual worker in the household.

At this stage, despite the simplicity of the initial model, it becomes increasingly difficult to obtain a feeling for the way the various parts of the model are interacting.

Nevertheless, and this to me is of the utmost importance, the use of continuous models of urban space allows one to avoid all the problems associated with the traditional transportation-land use models of defining a whole series of parameters for every transport link and every land parcel. The problems of retaining control over the input and retaining understanding of the output have in my view completely gone beyond our capability. This is especially true in forecasting, where many independent predictions are made of separate parameters with no good estimates of their interaction. It is my belief that only by the development of the continuous spatial model, as I have briefly outlined above and which has been rigorously examined in the papers quoted, can we begin to make progress at the strategic level in the understanding of the interactions between land-use and transportation, between rents and transport costs, between income distribution and spatial demands.

6. Conclusion

I expect the development of the simple but powerful techniques I have discussed to shed a great deal of light on the processes involved in markets over the next few years, so that we can consciously begin to control the way our markets develop.

References

Section 1

Aitchison, J and Brown, J A C, "The Lognormal distribution", CUP (1957).

Bleicher, H, "Statistische Beschreibung der Stadt Frankfurt-am-Main und Ihrer Bevölkerung", Frankfurt-am-Main (1892).

Blumenfeld, D E, "Effect of Road System Designs on Congestion and Journey Times in Cities", University College London, PH.D. Thesis (unpublished), (1972).

Hackman, G A, and Willatts, E C, "Linmap and Colmap; an automated thematic cartographic technique", Sixth International Cartographic Association Conference, Canada (1972).

Mogridge, M J H, "Tables of the cumulative distributions of annual earnings of employees by sex, age, region and £50 income interval 1964/5 et seq", Centre for Environmental Studies, (unpublished) (1971).

"New Earnings Survey", Department of Employment, HMSO, 1968 and 1970 et seq.

Pearson, K, "The incomplete gamma function", CUP for Biometrika reissue (1946).

Section 2

Lydall, H F, "The Structure of Earnings", CUP (1968).

Mogridge, M J H, "Theories of the generation of employment income", Centre for Environmental Studies, Working Paper 33 (CES-WP-33) (1972).

- "Planning, incomes and increasing prosperity", CES-WP-34 (1969).
- "A systems theory of personal income distribution", CES-WP-35 (1972).
- "Occupational earnings", CES-WP-49 (1970).
- "Employment income and age", CES-WP-50 (1971).
- "Regional incomes", CES-WP-51 (1972).
- "A theory of the growth of employment income", CES-WP-74 (1972).
- "Dynamic equilibrium characteristics of economic systems; the labour market as an example", Proceedings of the IFAC/IFORS Conference on "Dynamic Modelling and Control of National Economies" (1973).

Section 3

Mogridge, M J H, "Household income and employment income", CES-WP-48 (1970).

- "The forecasting of consumer income distribution", CES-WP-75 (1972).

Wootton, H J and Pick G W, "Trip generation by households", Journal of Transport Economics and Policy 1 (1967).

Section 4

Mogridge, M J H, "The stability of the car market over time", Urban Traffic Models Conference, Planning & Transport Research and Computation Co Ltd, London (1972).

- and Eldridge, D, "Car ownership in London", Greater London Council, Research Report 10 (1971).

Mogridge, M J H, "Forecasting car ownership in London", Greater London Council, Research Memorandum 386 (1973).

Section 5

Angel, S, and Hyman, G, "Urban spatial interaction", CES-WP-69 (1971).

Blackburn, A J, "Equilibrium in the Market for Land; obtaining spatial distributions by change of variable", Econometrica 39 (1970) pp 641-644.

Bussiere, R, "Static and dynamic characteristics of the negative exponential model of urban population distributions", Centre de Recherche d'Urbanisme, Paris (1970), in London Papers in Regional Science, 3, Pion (1972).

- "Modele Urbain de Localisation Residentielle", Centre de Recherche d'Urbanisme, Paris (1972).

Evans, A W, "Residential location", Centre for Environmental Studies, Macmillan (forthcoming).

Latham, R F, and Yeates, M H, "Population density growth in metropolitan Toronto", Geographical Analysis 2 (1970), pp 177-185.

Mills, E S, "Studies in the Structure of the Urban Economy", Johns Hopkins Press, Baltimore (1972).

Mogridge, M J H, "A discussion of some factors influencing the income distribution of households within a city region", Centre for Environmental Studies, Working Paper 7, (1968), in London Papers in Regional Science, 1, Pion, 1970.

- "Some thoughts on the economics of intra-urban spatial location of homes, worker residences and workplaces", Urban Economics Conference Proceedings, Centre for Environmental Studies (1973).

UTILIZATION OF HEURISTICS IN MANUFACTURING PLANNING AND OPTIMIZATION

J. Christoffersen Electric Power Engineering Department
P. Falster The Technical University of Denmark, Lyngby, Denmark
E. Suonsivu Outokumpu Oy, Pori, Finland
B. Svärdson Institute of Production Engineering, The Royal Institute of Technology, Stockholm, Sweden

INTRODUCTION

Planning and optimization of the manufacturing of components for the machine tool industry is an extremely complex problem because of the many technically feasible but not optimal solutions which exists for a problem and because no unique or well defined algorithm exists. The solution method is mainly based on manual routines utilizing empirical data and experience.

A manufacturing system for automatic planning and optimization may include all the decision-parameters and data describing the problem and is characterized by having little input describing a component and comprehensive output consisting of detailed working instructions and time tables.

The purpose with this paper is to demonstrate how engineering design concept in connection with explicit and consistent formulation tools are utilized in designing the above mentioned system.

First, the principles for developing general heuristic problem-solving programs are described. Secondly, the manufacturing problem is described and related to the function of the firm in order to establish the boundaries between a manufacturing system and the other functions in the firm. Thirdly, the manufacturing process is theoretical considered as a dynamic model of the micro-economic model of production to determine the different elements of which it consists and to convert it into a practical system. Fourthly, with two examples it is illustrated how heuristics are formulated in decision tables and matrices. Finally, it is described how APL is used as a formulation tool in the design phase and as an experimental tool in order to determine the best way to operate the heuristics on a certain problem.

REPRESENTATION OF HEURISTICS

There exist many definitions of the term heuristic, but generally heuristic can be defined as any principle or device that contributes to reduction in the search for a solution of a problem and which may not give the solution in a finite number of steps. The solution may be optimal but often it is just feasible, that means it meets all the specifications of the problem statement. In accordance with that every search principles from the simplest one, e.g. rules of thumb utilized by a production planner, to the very complex, well-organized methods are considered as heuristics.

Depending on the types of the problem there are well-structured and ill-structured problems. By well-structured problems we understand problems which can be described in terms of numerical variables and for which the goal to be attained can be specified in terms of a well-defined objective function. All other problems are considered as ill-structured and are characterized by poor definition of the functional relationship among the variables as well as uncertainty in the behavior of any individual variable.

In order to apply heuristic principles to a problem it is necessary to program them for a computer because of the large number of searches and evaluations that these methods often require. Unfortunately, most heuristic programs to date have the heuristic built in as an integral part of the program. Therefore, it is difficult to decide which heuristics are used and what their effects are. It is difficult to change or modify them to the actual problem. It is not possible to experiment with different heuristics in order to find the best for the actual problem. The heuristics can not be applied to other problems without making an entire new program. All of this indicates that it is necessary to divide a program into the following three parts

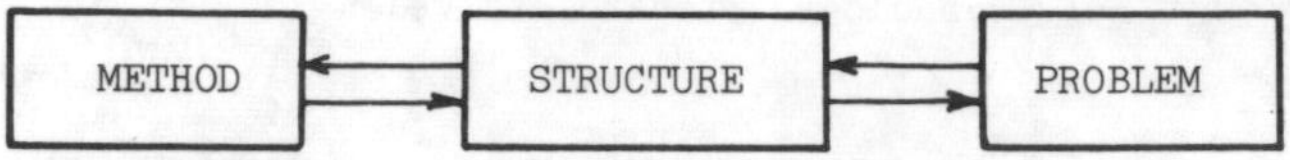

Fig. 1

where

METHOD, is a description of the different heuristics formulated in data files. That means, an exchange of heuristics just requires an exchange of datafiles without influencing the programs.

STRUCTURE, is the general program to perform the different heuristics and contains the common and invariant features of these.

PROBLEM, is a complete description of the actual problem. It can be exchanged without influencing the optimization program.

In order to describe the different heuristics in an explicit, unique, and logically way it is convenient to represent them in decisiontables and matrices.

In the remaining part of the paper it is shown how these principles are utilized on a problem, namely planning and optimization of manufacturing processes, which is an example of a complex, wellstructured problem.

MODEL OF A MANUFACTURING SYSTEM

The manufacturing problem consists in translation of design data and engineering specifications into manufacturing instructions, which are utilized in controlling the machines in the shop. That means, from a physical description of a product or a part and from knowledge about the shop, planning decisions are made concerning 1)machine selection, 2)tool selection, 3)operational sequence, 4)machining data, 5)time data, 6)operator instructions, in order to determine the best way to make the part. The results are given as operation planning instructions.

The best way to manufacture the part should be determined in respect to the optimal goal to the complete firm, e.g. as a result of an optimization of an economic model describing all the functions in the firm. However, because of the many data and difficult decision processes involved it is impossible in practice to design a total model which is detailed enough to describe the conditions in a shop in relation to the rest of the firm and to give the manufacturing instructions necessary to produce a part. This means, that it is necessary in some way to reduce the problem without loosing the overall view.

In order to solve the problem it is found useful to utilize the system concept (Franksen, 1965) which states that a system consists of an assembly of distinct parts, the so-called elements, which can be described independently of each other. The properties of the system are determined partly by the particular characteristics of each element, and partly by the manner in which the elements are interconnected.

When the subsystems are independent they can be optimized in such a way that the optimal goal for the subsystem will give the optimal goal in respect to the complete system.

However, no matter how a system is decomposed into subsystems it is not possible in practice to obtain these ideal conditions completely. In order to optimize a problem it is therefore important to decompose the system into subsystems so the influence of the interactions are minimized and then incorporate the interactions in the model of the subsystem, e.g. by transforming them into constraints or by modifying the objective function.

A firm can be defined by a simple model which describes the firm as a transformation process of input and output.

Following the system concept this model is decomposed in order to determine the subsystems, each of which describes a specific function, and their interconnections. The decomposition is divided into a vertical and a horizontal decomposition. By the vertical decomposition the system is considered as a feed-back system and is generally divided into two functions, namely a control function and a processing function, fig. 2.

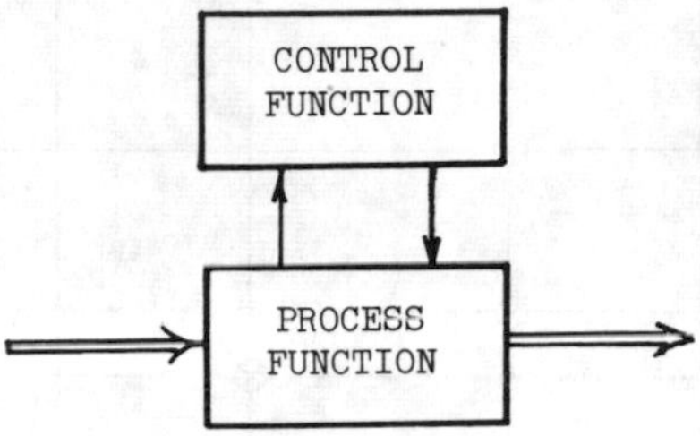

Fig. 2

Following this principle the system is divided into different levels where each level is considered as a controlling function for a system on a lower level and as a processing function for a system on a higher level. The vertical decomposition is performed in accordance with different criteria as types of decision variables, time-horizons, etc. By the horizontal decomposition the subsystems on every single level

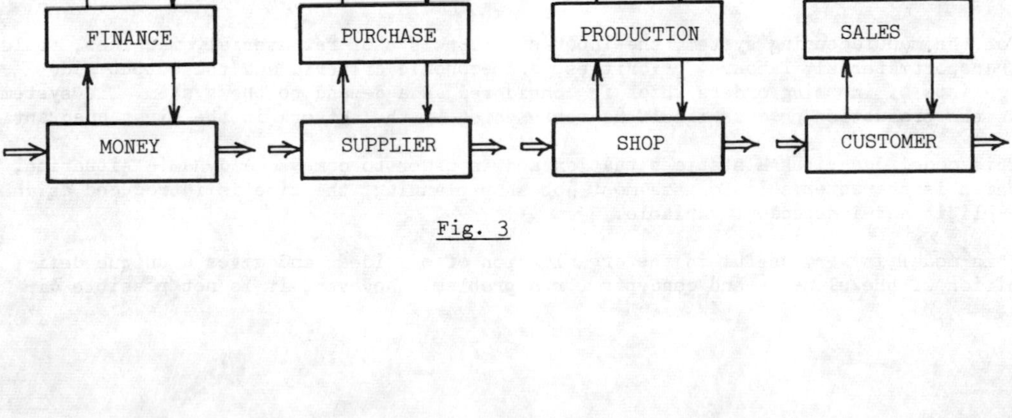

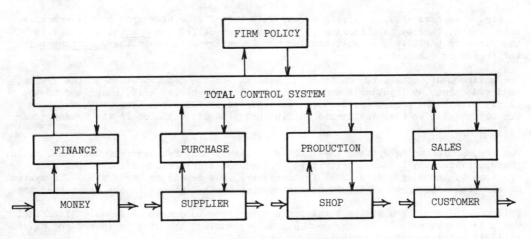

Fig. 3

are divided into functions which in the principles shall be considered as independent of each other. That means, the model of the firm is decomposed in subsystems which are interconnected in a hierarchical structure as shown in fig. 3. This figure contains the most important subsystems for a firm with small lots or series production.

The specific model which we have found most useful to represent the manufacturing system is the micro-economic process of production (Franksen, 1969) which depicts the transformation process of production factors into finished products. The structure of this model can be conceived as an economic network. The illustrative example in fig. 4 will be used to describe the model in more general terms.

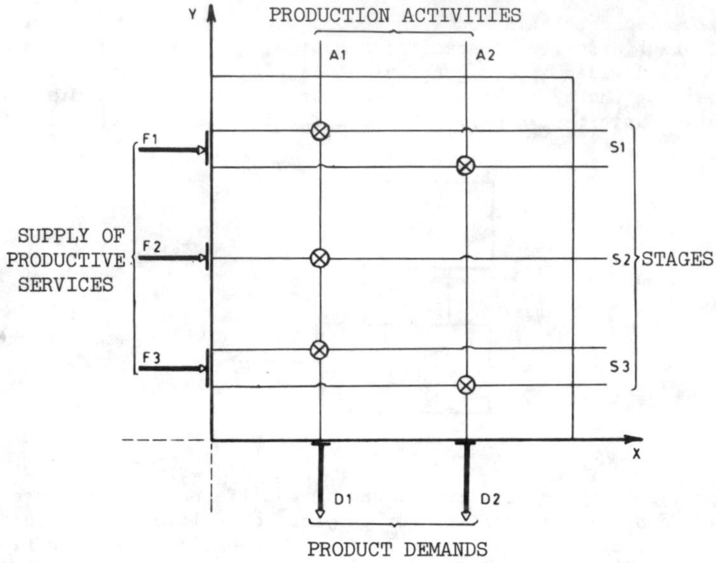

Fig. 4

In fig. 4 we have shown a mapping of our universe of discourse upon a (x,y)-coordinate system, thereby partitioning it into a subset called the system, situated in the first or positive quadrant, and its complementary set, named the environment in the remaining quadrants.

The environment, considered as a set of external influences impressed on the system, is divided into an input side and an output side. The input side, comprised of the supply of all the production factors, is depicted in the second quadrant, while on the output side we have the demand on finished products depicted in the fourth quadrant.

For the manufacturing system, the input side consists of rawmaterial, machines, tools, transportmaterial, labours, priorities, and economic criteria and the output side consists of incoming orders which is considered as a demand to the system. The system as the production process itself is represented by the network in the first quadrant.

This model describes a static situation and in order to come to a dynamic situation, which is characteristic for the most job shop planning, the time is introduced as an explicit and independent variable.

This model is very useful in the organization of our ideas and gives a unique definition of the elements and concepts for a problem. However, it is not possible direct to utilize the model to obtain a practical solution, so in order to convert the model into a practical system the model is heuristically decomposed into a sequence of subproblems which can be considered as independent of each other. Accordingly, the model can be divided into two main parts. First, a preparation part in which we determine the optimal network for producing one component consisting of the operations, their order, and the use of production factors relative to a time scale. Secondly, a scheduling part in which all the operations are coordinated and projected on an absolute time scale. A production schedule is in fact a timetable that tells us which production factor should be doing what, and when (Christoffersen & Falster, 1972).

The preparation part is further divided into four subsystems.

Decomposition: Determination of the operations to manufacture a component which is described as design data.

Sequencing: Determination of the order to perform the operations so technical precedence-requirements are met.

Optimization: Optimization of different alternatives for performing a single operation.

Optimal sequence: Selecting among the different alternatives for performing the operations in order to find the best way to manufacture the component.

The scheduling part is divided into two subsystems.

Operation selection: The set of operation for which all their predecessors have been scheduled is selected from the network of operations.

Operation allocation: Each operation is one by one allocated time and resource using a heuristic allocation method.

EXAMPLES OF HEURISTIC METHODS

This chapter describes how the general principles for heuristic programs outlined previously are utilized in the two parts of the manufacturing system, namely optimization of the machining data and scheduling of the operations.

The determination of optimal machining data for one operation is an example of a problem with non-linear objective function and non-linear constraints (Svärdson, 1972). In order to perform this optimization a general optimization-system utilizing direct search methods is developed. Direct search is a class of heuristic methods based on one heuristic principle, namely examination of the effects of small changes about the best point already known. The general formulation of the methods can be described by the vector expression

$$\overline{X}_B = \overline{X}_A + \overline{\Delta X}$$

where decision on acceptance of point B as the new searchpoint or a new search from point A are dependent on the difference between the values of the objective functions f_A in A and f_B in B. Dependent on the way $\overline{\Delta X}$ is chosen it is possible to get different search methods for which the effectivity is dependent on the type of problem.

To illustrate how a search method is completely described in decisiontables and matrices a simple search method is chosen, namely the univariate method called "one-at-a-time", fig. 5. The principles are the following:

Alter one variable, here X1, until no improvement is obtained. When the best point along the line has been located, then alter the other variable, here X2, until no improvement is obtained. A pattern is now established and a new pattern will be established using the same principles. This continues until the optimum is reached.

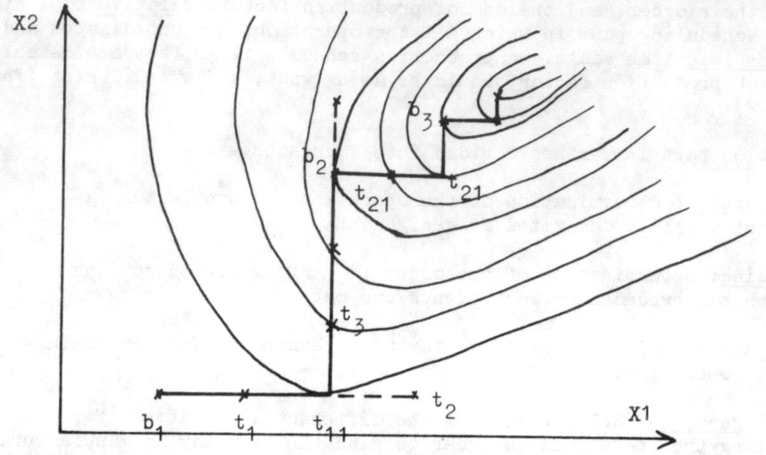

Fig.5

A search direction is described by a vector

$$\overline{\Delta X} = \sum_{i=1}^{m} a_i \cdot b_i \cdot \hat{x}_i$$

where

a_i - the wanted stepsize for the i'th variable dependent on the problem

b_i - contains information whether you want to search along the x_i-axis. Dependent on the method b_i can have the values 1, 0, or -1.

The reason to express $\overline{\Delta X}$ in this way is that we want to keep information belonging to the method and the problem separated.

The possible values of {b} for this example are described in the Directionfile, fig. 6. The possible values of {a} are described in the Deltafile, fig. 7, which contains more than one row because we want to use different stepsizes, depending on where we are on the surface. The value of {a} is dependent on the problem, so the Deltafile is changed every time a new problem is optimized.

	X1	X2
1	1	0
2	-1	0
3	0	1
4	0	-1

Fig. 6: Direction-file

	X1	X2
1	0.7	25
2	0.5	18
3	0.15	7

Fig. 7: Delta-file

The strategy is divided into two parts, namely one used for locating the searchpoint in the solution space and one containing the principles for choosing the next searchpoint.

To locate the searchpoint it is necessary to operate with three different points, namely

Base - The starting point for establishing a pattern. E.g. the points b_1 and b_2 in fig. 5.

Temporare Base - The latest accepted searchpoint. E.g. point t_{11}.

Test Point - The new searchpoint.

Besides this it is necessary to know which search-direction is used and whether it is an optimum condition.

When these elements are known it is possible to make the conditions which locate a searchpoint, and they are described in the condition part of the decision table, fig. 8.

RULES	1	2	3	4	5	6	7	8	9··20	21	22	23	24	25	26	27	28	29	30	31	32
TESTPOINT < BASE	0	0	0	0	0	0	0	0	0 1	1	1	1	1	1	1	1	1	1	1	1	1
TESTPOINT < TEMPORARE BASE	0	0	0	0	0	0	0	0	1 0	0	0	0	1	1	1	1	1	1	1	1	1
TEMPORARE BASE < BASE	0	0	0	0	1	1	1	1	0 0	1	1	1	1	0	0	0	0	1	1	1	1
LAST ROW IN DIRECTION FILE	0	0	1	1	0	0	1	1	0 1	0	0	1	1	0	0	1	1	0	0	1	1
LAST ROW IN DELTA FILE	0	1	0	1	0	1	0	1	0 1	0	1	0	1	0	1	0	1	0	1	0	1
REPLACE TEMP.BASE WITH TESTPT:	0	0	0	0	0	0	0	0		0	0	0	0	1	1	1	1	1	1	1	1
REPLACE BASE WITH TEMP.BASE	0	0	0	0	0	0	1	1		0	0	1	1	0	0	0	0	0	0	0	0
TRY NEW TESTPT. FROM TEMP.BASE	1	1	1	0	1	1	1	1		1	1	1	1	1	1	1	1	1	1	1	1
TAKE NEXT ROW IN DIRECT.FILE	1	1	0	0	1	1	0	0		1	1	0	0	0	0	0	0	0	0	0	0
TAKE 1. ROW IN DIRECT.FILE	0	0	1	0	0	0	1	1		0	0	1	1	0	0	0	0	0	0	0	0
TAKE SAME ROW IN DIRECT.FILE	0	0	0	0	0	0	0	0		0	0	0	0	1	1	1	1	1	1	1	1
TAKE NEXT ROW IN DELTA FILE	0	0	1	0	0	0	1	0		0	0	1	0	0	0	0	0	0	0	0	0
TAKE 1. ROW IN DELTA FILE	1	0	0	0	0	0	0	0		0	0	0	0	0	0	0	0	0	0	0	0
TAKE SAME ROW IN DELTA FILE	0	1	0	0	1	1	0	1		1	1	0	1	1	1	1	1	1	1	1	1
PATTERN MOVE	0	0	0	0	0	0	0	0		0	0	0	0	0	0	0	0	0	0	0	0
OPTIMUM	0	0	0	1	0	0	0	0		0	0	0	0	0	0	0	0	0	0	0	0

Fig. 8

The principles for choosing the next searchpoint are related to the same concepts as the location of a point, and the different possible actions for determination of the next searchpoint are listed in the action part of the decision table, fig. 8.

The decision tables used are limited entry tables, and it is seen, that the condition tests for the location of a point give a binary condition pattern, which identifies a rule and in that way a binary action pattern, which gives the actions necessary for decision of the next searchpoint. The complete search is in that way controlled by the decision table.

The table shown in fig. 8 contains for simplicity only the actions for univariate search, but by adding a few more actions it will also describe gradient methods.

The scheduling problem seems quite different from the above described problem because the proposed theoretical and practical methods to its solution are based on a variety of different heuristic principles. However, it is found that they have a general structure which is similar to the structure of direct search but further contains specific allocation features which are common and invariant for a variety of scheduling problems.

It is very easy to visualize the requirements of the scheduling problem as a sort of graphical puzzle in which one rearranges blocks representing the given set of operations on the machines in a shop, such that the respective due data are met.

Basically, scheduling can be conceived as an allocation problem in order to obtain the best allocation of limited means towards desired ends. From a planner's point of view two dual allocation problems can be distinguished, namely a time-allocation- and a resource-allocation-problem, where both problems originate from the abstract structure of the micro-economic model of production. That is, on the one hand the time allocation model depicts the causal relationships of the operations on the time scale and, on the other hand, the resource allocation model depicts the resource consumption for each operation on the production factors. According to this division of the problem the model analysis and thereby the optimization of the cost of the time- and resource-allocation can be considered as a geometrical search consisting of a local search where we search for the resource consumption of an activity and a global search where we search for its time delay. The formulation of an allocation strategy utilizes the concepts described in the first example and based on the following assumptions:

1. The operations are assigned different priorities
2. The objective function consists of two additive parts
3. The problem is considered as a multistage decision process

The concept of operation <u>priority</u> is inherent in many allocation methods. A priority is simply a numerical attribute of an operation on which selection is based whenever there is a potential conflict on a machine. Examples of priorities are operation slack, first in first served, random, length of next operation, etc. The list of waiting operations, i.e. the operations for which all the predecessors have been scheduled, are organized in the direction matrix ordered according to a priority function.

The formulation is based on a representation of the objective function as two parts, the one part relating to the time allocation model and the other part relating to the resource allocation model meaning that we can represent the function in the form:

$$OF = a \times O_1 + b \times O_2$$

In a specific planning situation we can evaluate the objective function on the bases of different values of a and b which are represented in the space matrix. The actual combinations of a and b can be:

$$(a,b) = \begin{cases} (1,1) \\ (1,0) \\ (0,1) \end{cases}$$

That means, a certain strategy in combination with a certain spacematrix control a search based on either comparing marginal changes in each part of the objective function or comparing the sum of the two parts of the objective function with a value evaluated earlier in the search process.

Because of the dimension of the problem the allocation procedure is presented as a multistage decision process both in the way it solves the precedence constraints exemplified in operation-selection and the way it period by period performs an optimization.

IMPLEMENTATION OF HEURISTICS

The major difficulties in applying heuristics are on the one hand to develop heuristics for a certain problem and by experimentation to prove that it "works" and on the other hand to operate a heuristic search to a given problem. Experience shows that the effectiveness of a heuristic search is mainly dependent on (1) finding the most suitable strategy for a problem, (2) the chosen startpoint for a search, (3) the criteria for termination of the search, and (4) the representation of the constraints.

Therefore, it is proposed that the use of prototype modelling and computer experimentation - an equivalent to the use of laboratory experiment in the physical sciences - can do much to benefit the designer in his research.

With references to the representation of heuristics presented in the preceding chapter it corresponds to exchange of data fields as decision tables, step-, direction- and space-matrices, problems, and parameters controlling a search.

The specific programming language we have found most useful as a tool and laboratory for representing such prototypes is *APL*\360 (Iverson, 1972; Falster, 1973). The APL language is used because it is an explicit and concise way of representing complex relationships between data and because it exists in the form of an interactive system which permits testing and experimentation with the prototypes. The data organization in APL is based on vectors and matrices which make it possible to describe the structure of the problem in APL in the same way it is described in graphs and tables.

The mentioned prototypes may range from simple and solely problem oriented ones that permit explorations of the basic elements and how they are interconnected, to complete copies of the large scale implemented system that permits simulation of alternative systemconfigurations.

Furthermore, it provides a vehicle for both documentation and explicit communication of ideas between system analysts and programmers as well as for education of users of the intended system.

CONCLUSION

In this paper it is shown how general heuristic programs are developed by separating the heuristic principles from the problems and representing the heuristics in the decision tables. This has been illustrated for two subproblems within the manufacturing system, i.e. optimization of machining data and scheduling. But it should be emphasized that the same principles are utilized in other parts of the system where complex decision- and optimization-processes exist, e.g. by the sequencing of the operations, determination of the optimal sequences etc.

The general heuristic program includes a number of features which is not necessary for the solution of a specific problem and therefore it may not be as efficient as a program developed specially to solve this problem. However, because of the way in which the heuristics are formulated the general program offers a number of advantages. On the one hand it is possible to apply the general program to different problems and on the other hand it is possible to utilize different heuristics to the same problem by exchanging datafiles. This allows experimentation in order to determine the best heuristics to a specific problem and the best way to operate them.

The development of the manufacturing system was organized as a cooperative work between different research disciplines in Scandinavia. The disciplines involved are manufacturing, system and economic sciences from Sweden, Denmark, and Finland respectively. The foundation behind the project has been The Scandinavian Council for Applied Research. Two companies are also involved as a laboratory for basic data, testing, and implementation. These are Outokumpu Oy in Finland, and Alfa Laval A/S in Sweden.

REFERENCES

Christoffersen, J. and Falster, P.: Decision tables used as a formulation tool in designing a manufacturing system. IV-CIRP Seminar on Manufacturing Systems, Ljubljana, 1972.

Falster, P.: A new approach in system design by utilizing APL. APL-Congress 73, Copenhagen, 1973.

Franksen, O.I.: Mathematical Programming in Economics by Physical Analogies. Part II: The Economic Network Concept. Simulation, No. 1, July 1969, pp. 25-42.

Franksen, O.I.: Closed and Open Design Projects in the Education of Engineers. IEEE Trans., Vol. PAS-84, No. 2, March 1965, pp. 228-231.

Iverson, K.E. and Falkoff, A.D.: *APL*\360 User's Manual GH 20-0906-0, 1972.

Svärdson, B.: The Adaptive Control Concept in Job Shop Scheduling. IV-CIRP Seminar on Manufacturing Systems, Ljubljana, 1972.

Wilde, D.J.: Optimum Seeking Methods. Prentice Hall, N.Y., 1964.

ECONOMIC SIMULATION OF A SMALL CHEMICAL PLANT

by

G. BURGESS, B.Sc. and G.L. WELLS, Ph.D., M.I.Ch.E.

Department of Chemical Engineering, University of Sheffield

The use of a generalised process simulator for economic simulation of chemical plant is discussed and illustrated by its application on a small Tar Plant. Essentially the system is used to produce a sound model of the plant which is open-ended and capable of further refinement according to the desired accuracy of the economic objective function.

1. DEVELOPMENT OF A SIMULATOR FOR CORPORATE PLANNING

1.1 Process Simulators

Most chemical process simulators are based on the modular approach. This consists of an executive or master program which supplies a flow of information to unit subroutines or modules, which modify such information in predetermined manner. Briefly the function of the executive is to read data representing the topology and parameters of the system; store the data input for ready access; store stream information for transfer between modules; execute the modules in the correct order and control data printout including debugging options.

The approach is not just replacing the 'unit operations' of a flow chart by modules. Subroutines essentially produce changes in information flow. They are class -ified as Unit procedures - which perform calculations for a specific unit operation such as simulating a heat exchanger, reactor or simply a junction of streams, i.e. mixer. Mathematical manipulation modules - which test for and promote convergence in recycle problems. Such modules may carry out simultaneous solution of equations. Decision making modules - which act upon the information generated earlier in the simulation study and adjust parameters or feed rates based on this information. Report-writing modules - which augment the normal printout of results by gathering together information generated in the study and write them in a form required by management.

A library of general purpose modules is prepared and the appropriate one's selected or custom built for use when required. Several executive systems exist mainly based on the same idea.[1] Some only simulate mass balance calculations while others also perform heat balance and rate equipment.

Primer

A version of the GEMACS process simulator[3] termed PRIMER has been used in this simulation. A full listing is given in the User's manual.[4]

1.2 Corporate Business Planning

Johnson[2] has applied the modular approach to planning aspects relevent to the operation of chemical plants. The business model can study the effect of many differemt factors on the profitability of a corporation. These include enhanced sales activity, raw material availability, efficiency of plant operation, changing price policy. The effect of introduction of new products or by-products, or closure of part of plant, can easily be integrated into the model such that long range plans can be studied for feasibility.

In process simulation, the flow of information is usually in the same direction as the material flow. The order of calculation follows the main flows of materials from feedstreams to final products. In corporate business simulation, information flow is the doniment consideration. Thus it may be preferable to have the products as the main inputs to the model. This demand orientated model outputs the intermediate product amounts and raw material requirements. Alternatively, when material availability is the planning criteria, the process sequence tends to be followed. In this case, intermediate raw materials and final product quantities are the outputs.

Considering only information flow frequently means any necessity for recycle can be avoided. Although the simulator will cope, recycle does increase computation costs. Thus on one complex we have simulated, mass flow of plant products is from sulphuric acid to nitrobenzene to aniline, with recycle between plants of spent acid and nitrobenzene. As shown in Fig.1 the use of information flow can generate an acyclic diagram.

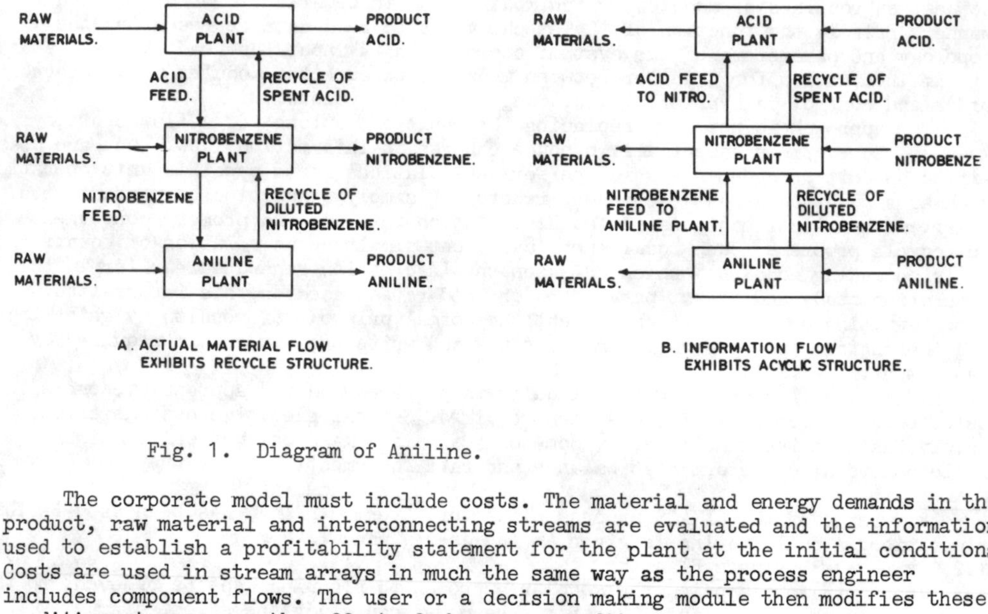

Fig. 1. Diagram of Aniline.

The corporate model must include costs. The material and energy demands in the product, raw material and interconnecting streams are evaluated and the information used to establish a profitability statement for the plant at the initial conditions. Costs are used in stream arrays in much the same way as the process engineer includes component flows. The user or a decision making module then modifies these conditions to measure the effect of change on profit.

1.3 Programming Considerations of Corporate Planning System

The executive requirements of the corporate planning model, although similar to that of the process simulator, differ as follows.

The stream information: In the process model, the stream information corresponds exactly to process stream information (i.e. stream number, total and component flows, temperature, pressure and enthalpy) and is usually directional as per material flow. In the planning model information such as temperature, pressure and enthalpy are frequently unnecessary and normally the total mass flow is all that is required. The stream information is thus

(a) Stream number:- identifies the stream in the model.
(b) Total mass flowrate:- the units used as appropriate to the system and time scale.
(c) Cost or Price per Unit:- permits evaluation of costs in report writing modules.

(d) Identification number:- when used in conjunction with a block data module permits writing the name of the stream.

The number of streams entering and leaving a module:
The corporate business model tends to have a large number of streams entering as input information such as fixed costs or product specification. An efficient data input method is essential. In many process simulators, a format 5F 10.4 or similar has been applied. But using this method for business planning would result in a large number of zero data spaces. This leads to the use of the NAMELIST unformatted input method to increase the efficiency of the data input. Also data which previously was read in as real and converted to integer for use, can now be read directly as integer. Additional stream output is of variable costs from process modules.

Having a large number of streams introduces considerations of storage and transfer of stream information. The method most prominently used on process simulators is a three or more array system. One array stores the stream information while others are used in the modules to store entering and leaving stream information. This method allows only the important streams to be permanently stored thus reducing the storage space required. In the planning model, it is necessary to store more stream information for efficient report writing. Therefore, it is better to store all the information in one array and use only this within the modules.

Flexibility of the System:
The executive has to be able to store and change input information so that more than one case can be considered per run. In the basic simulation executives, this has been achieved by the use of an ALTER routine capable of making changes in the initial data, stream information and certain module parameters. It is also possible to run several sets of data. In the planning model, a more flexible system is required so that the model topology can be changed as well as the stream and module information. This leads to a consideration of the storage of module parameters. The system best suited for ready access is one in which the module parameters are stored in one dimensional arrays as they are read in, before transfer to the main working array in named common blocks.

1.4 <u>Modules for Business Corporate Model</u>

When the initial models of the system are being developed, it is desirable to use simple modules. This initial simulation should highlight the need for more elaborate models and the areas in which these are required. More comprehensive models can then be integrated into the original so that the required aims of the study can be achieved. Although we have not found it necessary, a complete process simulation can be incorporated evaluating energy consumptions from chemical engineering unit operations. It is normally possible to use linear relationships between outputs and inputs, but in certain cases correlations and analytical models are required.

For the simple model only generalised routines are normally necessary. These are classified into the following groups.

<u>Mass Balance Group</u>

The mass balance group includes the modules which evaluate material flows through the system. This can often be achieved by the use of three or four modules, used singularly or in combination. These are:

Junction Module: This collects information from one or more input streams in the form of a total flow. It then relates this information to one or more output streams according to ratios which are stored either as module parameters or evaluated in the module itself. The junction can be used to simulate a manufacturing unit so that new demands and availabilities are obtained. The initial demand is given as an input stream, while the new demands and availabilities are calculated as output streams and the ratios given as module parameters. The module can also be used to split a demand or availability in two or more areas as required. It can be used to calculate the total demand and availability of a material from several sources.

Difference Module: This module finds the difference of two input streams. The difference is given as an output stream. The module can be used to examine the net availability or demand for a material. If these demands or availabilities arrive from more than one source, this module is used in combination with a junction module.

Constraint module: This module has a specific purpose as it compares a material flow with a constraint on that flow. If the flow is greater than the constraint (a conditional event), the original flow is split and the excess flow is given as a separate output stream from the module which may be processed elsewhere in the system. This module can be used to check the capacity of a manufacturing unit or the sales demand.

Control module: This module checks whether values of a stream give the same value in successive calculations through the system. It may include a convergence accelerator.

Costing Modules Group

The costing module group consists of routines which perform the cash balance of a model based on mass flow information. These are normally written in conjunction with a report writing routine so that the information obtained may easily be understood by management. Input streams from mass balancing modules containing mass flow information and price/cost information are evaluated and summed. Fixed charges are input directly. Costs sometimes are input as negative values by a negative sign in the stream price/unit. They are written as positive in the report by taking absolute values.

Decision Making Modules

These are modules required to perform changes in information flow which are not possible using the basic modules. This would include changes which depended on mass flowrate rather than fixed ratios. Where a material is being split between different centres, the split factors would depend on costs being optimised. There are no general modules that can achieve this type of decision, therefore special routines have to be written. The linear programming routine used in the case study is such an example.

2. DEVELOPMENT OF THE MODEL

2.1 Approach to Corporate Planning.

The first step in producing a model is to establish the purpose of the model. This helps to make decisions about the information required, the variables to be considered and the objectives the model must achieve. It leads to consideration of how accurate the model must be and how the data requirements depend on this. It will probably require the construction and consideration of a comprehensive process block flow diagram. This will highlight the areas which will require extra attention for modelling, which demands are set and where constraints exist. From this, an information flow diagram can be produced and initial modelling can be started. The next requirement is for data for the model. Discussion with management is necessary informing them of the type of data the model requires and the form of the results which the model will produce. The model may be derived from experience, from analysis of historical data or from the results of experiments. It may reproduce existing manual methods. In many ways, the ideal data for the model stems from experimentation, so that the manipulation of certain variables can be carried out while others, as far as possible are kept constant. In this way, cause and effect can easily be linked. It is essential to determine the relationships of variables in the model and by how much they change. The constants in the problem are normally determined by statistical analysis.

The first model will be a mass balance model. It is extremely important that this model is flexible. Once the material balance has been set up, solutions obtained from the model have to be compared with observed results. In most cases, accurate models for material balance are easily constructed. When this has been achieved, costs are introduced so that an overall picture of the system can be obtained including profitability. Both fixed and variable costs have to be considered. More uncertainty is introduced at this stage as price and cost variations can exist. Also data is sometimes difficulty to obtain where variable and fixed costs are not readily distinguished. When this model has been found to agree within limits with reality, work can begin on varying parameters so as to find the effect of certain changes in operating conditions. The model is normally set up for monthly or yearly average figures and does not involve any time effect. It is essentially a static model.

As modelling of the system is improved, it is possible to study the effects of seasonal, daily or monthly trends in operation. Also supply/demand price variations and time effects of storage availability can be included. The system then approaches closer the concept of a dynamic model.

2.2 Profit and Loss Account

The report writing routine can be structured such that all items in the report duplicate the profit and loss account, P & L, of the plant. Thus in the first place the simulator can produce a budget P & L for the process. This of course for business reasons forms an important basis for comparison of Company performance. If actual flows are incorporated by using a separate model, (separate because operating costs are generated automatically in the planning phase), the actual Company P & L account can be produced. A third, very useful version of this account is produced by inputing the actual raw material streams to the planning model. This permits producing a forecast P & L which the manager can use to directly compare with his actual results, thereby identifying actual plant problems and/or leading to ways of improving the computer simulation by highlighting the need for improved modelling of a section of plant.

Another feature of corporate planning and generation of P & L accounts is that it nullifies the effect of faulty transfer prices. When all cost centres are grouped together, simply varying one parameter such as demand for a product enables the entire effect, including intermediate activities such as reducing demand for steam etc., to be shown. This is of particular importance when considering plant closures. A common experience is the effect of shutting down an unprofitable plant has surprising repercussions on existing plant. The use of the model helps not only because side effects should be shown up, but also because the very effort of producing a model may result in the use of more accurate values at cost centres.

3. CASE STUDY. A NATIONAL COAL BOARD (U.K.) TAR PLANT

3.1 Purpose of the Model

The National Coal Board have permitted the simulation methods described above to be tested in an industrial environment. They also assisted in formulating those areas in which the model might be used to augment existing actions. In particular it was decided that the most important area should be the determination of the technical budget, which governs the years production and is also subject to change as a result of variation in sales demand, raw material availability and equipment failure. Simulation automates this procedure and can be used to determine the optimum route to achieve the sales demands. An extension of the model predicts the profit/loss account.

The plant permits a study of most aspects of economic simulation. Due largely to the stability of raw material quantities a number of optimisation techniques available to adjust parameters were not found necessary apart from in the batch blending plant. Instead incrementing values or direct input of alternative data was used.

3.2 Process Description

The Tar Plant consists of six small units as shown in Fig.2. Crude tar is feed to a primary fractionation unit and split into seven fractions. Light oil is not processed further. Tar acids are extracted from carbolic and naphthalene oils, the carbolic oil being further processed to give tar bases and naphthas. Naphthalene and anthracene are produced, drained oils being sent to the blending plant. Creosote, base oil and pitch are used for blending and direct sale. The blending plant produces road tars, fuels, etc. Seventeen blends are produced by batch mixing and the plant acts as a sink for disposal of unsold oils.

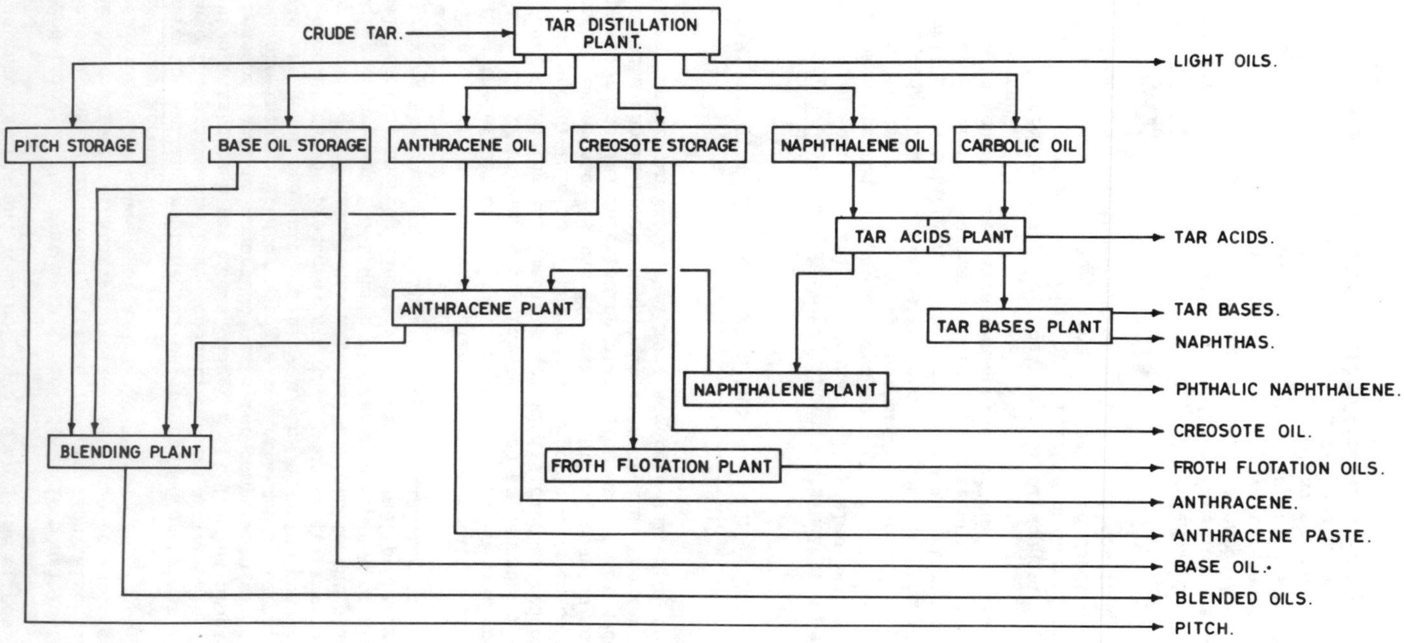

Fig.2. Block diagram of Tar Plant.

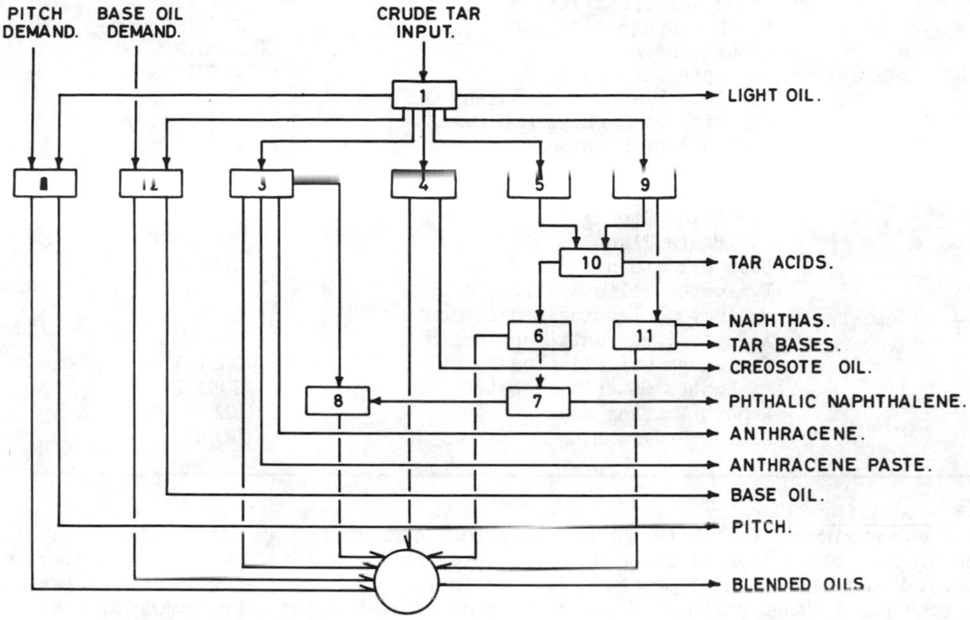

Fig.3. Initial Information Flow Diagram.

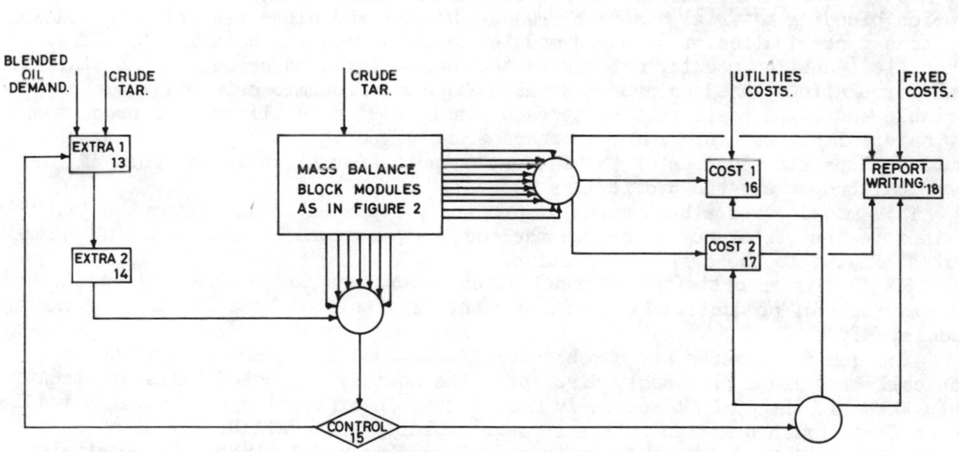

Fig.4. Final Information Flow Diagram.

TABLE 1 MODULE NAMES AND TYPE

NUMBER	NAME	TYPE
1	Crude Tar Distillation	JUNCTION
2	Pitch Storage	JUNCTION
3	Anthracene Plant	JUNCTION
4	Creosote Oil Storage	JUNCTION
5	Tar Acid Plant, Naphthalene Oil	JUNCTION
6	Capacity of Naphthalene Plant	CONSTRAINT
7	Naphthalene Plant	JUNCTION
8	Drained Naphthalene Oil Storage	DIFFERENCE
9	Tar Acid Plant: Carbolic Oil	JUNCTION
10	Tar Acids Storage	JUNCTION
11	Tar Bases Plant	JUNCTION
12	Base Oil Storage	DIFFERENCE
13	Evaluates Sales of Blended Oils	EXTRA 1
14	Evaluates blending requirements	EXTRA 2
15	Adjusts blend until converged	CONTROL
16	Evaluates utilities costs	COST 1
17	Evaluates materials costs	COST 2
18	Report Writing Mode	REPORT 1
19	Linear Programming	LINEAR

3.3 Modelling

For maximum flexibility the model is both demand and raw material orientated. The process block diagram is transformed into the initial information flow diagram, Fig.3. Module names and type are indicated in Table 1. Main flows were calculated in each plant based on known factors. Only standard modules were required, i.e. JUNCTION, DIFFERENCE and CONSTRAINT. The model is completed as shown in Fig.4, with the blending plant being calculated starting from the input of crude tar. The output from both parts are compared to give amounts of other sales material and excess of non-saleable material. Any differences were eliminated by using the module CONTROL to modify the mix at the blending plant (tar input being stable). The model requires two additional sub-routines. EXTRA 2 evaluates the net demand of each blending material based on product demands and blend ratios: This module replaces a combination of standard modules thereby reducing data input. EXTRA 1, a modified junction module, calculates the demand for each product using statistical correlations based on previous sales figures. Economic criteria includes variable and fixed costs related to each plant, cost of utilities and product and raw material prices. This information is transferred into a report writing subroutine. Separate sections of the latter, output information in the form of the technical budget and the profit/loss account.

This model gives a best value of profit by repeated trial of input data. To improve the efficiency of search the model was refined by incorporating various routes, Fig.5, to the desired objective.

ROUTE 2 is an optimisation model which determines for the year the most economic mix of products. The blending plant is simulated by a linear programming module, EXTRA 3.

The generated sales are checked for feasibility using ROUTE 1, which is basically the model previously discussed. The monthly projected sales are input into EXTRA 1. The CONTROL module is modified by incorporating a constraint module which contains information on the storage capacity and availability of oils in the plant. Oils may be deposited or withdrawn from this module. Should a constraint be violated, i.e. capacity exceeded or oil not available, the mix of oils at the blending plant is modified. The results of each month are added to give the revised yearly sales and technical budget.

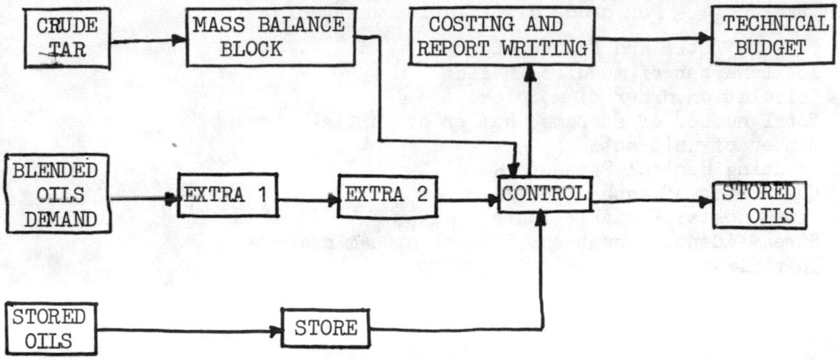

ROUTE 1 VERIFICATION OF TECHNICAL BUDGET

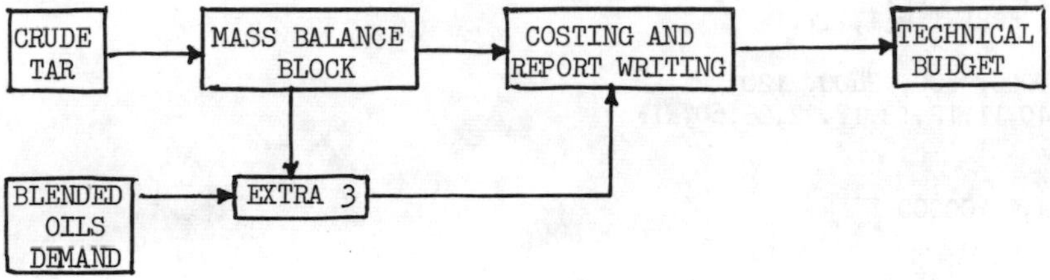

ROUTE 2 PRODUCTION OF TECHNICAL BUDGET

Fig. 5. ROUTES FOR OBTAINING TECHNICAL BUDGET

ROUTE 3 is indicated in Fig. 6. and generates a version of the profit and loss account of the plant based on what costs should have been incurred in producing a known range of products from crude tar, compared with the actual costs and consumptions. These are fed directly into the costing and report-writing section

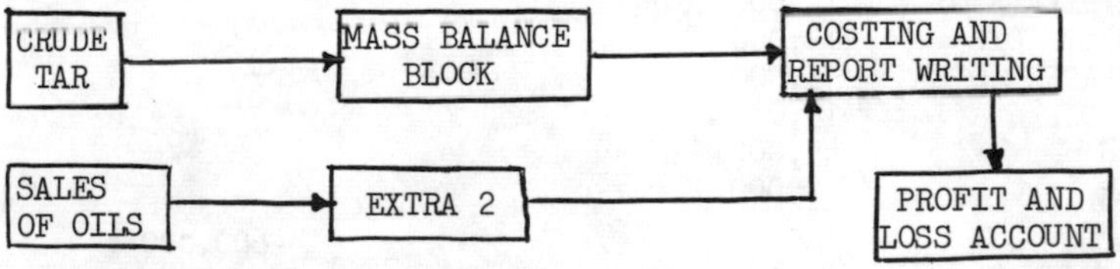

FIG.6. Production of Profit and Loss Account.

Data Input and Output

Naturally commercial considerations prevent giving detailed input and output data for this plant. Tables 2 and 3 indicate the type of information and output obtained using ROUTE 1, for the crude distillation section, module 1. For further information on procedure for data input see the PRIMER, User's Manual,[4] or Wells and Robson.[5]

TABLE 2. INPUT DATA - For crude distillation unit only

1. Program Title and Number of runs
2. Total number of modules in list
3. Calculation order of modules
4. Total number of streams; Number of initial streams
5. Number of unit sets
6. Printing Control Parameters
7. Cost/Price of each material
8. Fixed Costs; Variable Costs
9. Stream identity numbers; initial stream numbers
10. Initial streams
11. Module sets
12. Blend ratios

```
&ALIST
 TITLE = DISTILLATION, NRUNS = 1
&END
&BLIST
 NCALC = 1, LIST = 1, NFLOW = 9, IFLOW = 1
 NUNIT = 1, KPRINT = 1,1,1,1
 COST = -6
 VCOST = 20000, 600, 2400, 120
 IDFL = 1,10,11,12,13,17,22,25,60,NIF = 1
&END
&FLOWLIST
 FLOW 1 = 1,0,100000
&END
&UNITLIST
 UNIT 1 = 1,4,1,7,1,2,3,4,5,6,7,8,8*0,.01,.02,.15,.15,.05,.15,.45,.02
&END
```

TABLE 3: OUTPUT DATA

PLANT - CRUDE DISTILLATION SECTION ONLY
TECHNICAL BUDGET INFORMATION

STREAM NUMBER	COMPONENT NAME	MASS FLOW	PRICE/ UNIT	REVENUE £
1	CRUDE TAR	100,000	-6	-600,000
10	LIGHT OIL	1,000		
11	CARBOLIC OIL	2,000		
12	NAPHTHALENE OIL	15,000		
13	CREOSOTE OIL	15,000		
17	ANTHRACENE OIL	5,000		
22	BASE OIL	15,000		
25	PITCH	45,000		
60	WATER	2,000		
TOTAL REVENUE				-600,000

STREAM NUMBER	UTILITY	FLOW/ UNIT	COST UNIT	COST £
9	ELECTRICITY	0.02	20,000	.400
10	STEAM	1	600	600
11	FUEL OIL	9	2,400	21,600
12	WATER	3.5	120	420
TOTAL UTILITIES				23,020

4. DISCUSSION

The generalised simulator enabled the production of satisfactory models of the Tar Plant thus reducing future time in generating technical budgets and permitting comparison with predicted plant performance. Clearly as the equations in many modules are simple an one-off program could easily have been written but would not be readily upgraded as improved information became available. In building up the simulation as indicated it was easy for management to follow the steps taken and check for the possibility of error at any state. This is important as it has a vital bearing on the acceptability of a model and the proving of its claims for general use by the Company. Furthermore the model can now be refined to incorporate options such as replacing the JUNCTION module on crude tar distillation with an improved SEPARATION module based on the analysis of crude tar.

We regard this method of economic simulation essentially as building up an accurate model in stages appropriate to its purpose. If the model is flexible and open ended this is more readily possible. When the model is established the user then has the choice of simplifying the computation stages, or say integrating the algebraic expressions in a linear programming matrix, or using the model as produced.

ACKNOWLEDGEMENT

We would like to thank the N.C.B. Coal Products Division for permission to publish the case study and in particular Messrs. J.E.L. Smith, A. Beighton and S. Haywood for their assistance in this work.

REFERENCES

1. Flower, J. and Whitehead, B.D., The Chemical Engineer, May 1973.

2. Johnson, A.I., British Chemical Engineering, 17,119,1972.

3. Johnson, A.I., GEMACS Users Manual, University of Western Ontario, 1972.

4. Burgess, G.G., Robson, P.M., and Wells, G.L., PRIMER Users Manual, University of Sheffield, 1973.

5. Wells, G.L. and Robson, P.M., Computation for Process Engineers, Leonard Hill, London, October, 1973.

An Optimal Growth Model for the Hungarian National Economy

by I. LIGETI
Institute of Economic Planning
137o Budapest Postafiók 61o
Hungary

o./ Introduction

In 1968 a reform of economic management was introduced in Hungary. Since the national economic plan has contained condiderably fewer details than before. It contains,however, the major social, economic and technological objectives of development and the progressive tendencies of economic growth. Moreover the plan must establish in conformity with its objectives, the major economic instruments /regulators/, and the principles of their application.

On the base of these changes new vistas have been opened to the application of mathemathical models, too. In this field there had been a considerable amount of experiences in Hungary, but the models in accordance with the plan were mostly concentrated to preparing the plan-directives. In the field of economic models the first remarkable achievement was the large-scale natural, value and financial programming model of the fourth Five-year plan /1971-75/ developed in the Institute of Economic Planning and the Hungarian National Planning Office in order to improve the circumstances of decision-making for a medium-term period. Its results were published /7/, and applied in the actual planning work.

In the light of experiences gained by the research workers of the Institute of Economic Planning a new concept is needed for the application of economic models in national economic planning. According to this concept some smaller model should be developed to investigate and solve the various problems of planning. Evidently these models may be of different types and sizes and may cover different time horizon, too. At the second stage of the research work these models forming the elements of a system of models for medium-term national planning can be linked together. In the framework of such a system, in the prognosis phase of drawing up the plan a model being able to answer the most general and basic questions of the next few years can prove to be a useful tool of planning. Up till now the aggregated growth models have not been applied in the

Hungarian planning, despite of the fact they seem to be the most suitable for investigation of economy-wide development problems in a long-range period. The model following here is one of elements of system of models and designed to investigate some of these problems.

1./ The main task of the model

The allocation of investments among the various branches of national economy and the current state of the investment market have recently been in the lime light in Hungary. It is because we have to deal with the investments even in such an aggregated model as ours from various points of view.

This model is also focused on the analyses of investment policy and on its consequences. In the assumptions of the model three major aspects of investments are included:

/i/ relation between production of intermediary products and investments. /Materials and intermediary products may create a constaint for the realization of investment./

/ii/ relation between consumption and investments. It is well-known, that a number of growth models investigate the so-called golden path of growth. This model also comprises this relation facing the present with the future.

/iii/ the labour force as a constraint for the production, since this factor may also be a constraint of economic growth /e.g. in Hungary for the time being/. Of course the labour force is more important from the point of view of productivity than from that of number of population.

Summing up the task of the model the possible path of economic growth has to be investigated in the frame of the main aspects mentioned above.

2./Description of the model[x]

National economy is divided into three sectors in the model as follow:

/i/ sector producing intermediary products
/ii/ sector producing investment goods
/iii/ sector producing consumer goods.

As to the denomination of the sectors it should be noted that it does not exactly cover the real activity of the sector /branch or industry used in economic analyses/ because each of them produces both intermediary products, investment goods and consumer goods, too. The activity in the name is only the most characteristic feature of the sector.

2.1./ Investments in the model

Investment and the relations belonging to this sphere play a central role in the model. The investment can properly be characterised by

- the fixed assets coefficient /the ratio of total output to the fixed assets./
- the ratio of depreciation of fixed assets
- the increase of fixed assets
- the time lag of investments.

Relations based on these categories constitute the equations of the model following here.

Total output of sectors is linked to the fixed assets through the fixed assets coefficient.

$$x_i/t/ = k_i/t/\ g_i/t/ \qquad /1/$$

where: $x_i/t/$ the total output of sector i in year t,
$k_i/t/$ ratio of total output to the fixed assets in sector i for year t,
$g_i/t/$ the stock of fixed assets.

---

[x] A great number of models of this type have been developed. We mention only some of them containing the most familiar characteristics: /3/, /8/, /1o/, /11/.

Since each sector produces investment goods a certain ratio of total output of the sectors is used for the increase of investment. The output of investment goods in year $t-\tau$ is equal to the gross investment of the year t:

$$/2/ \quad \sum_{i=1}^{n} \beta_i x_i/t-\tau/ = I^G/t/ + K_2$$

where: β_i ratio of the goods produced for investment purposes to the total output,

$I^G/t/$ gross investment in year t,

K_2 foreign trade balance of investment goods,

τ the investment lag.

Total investment at the level of national economy can be formed as the sum of investment by sectors. Further on, supposing investments consist of two parts of different purpose, on the one hand investments substituting fixed assets depreciated in the last year; on the other hand the investments increasing the stock of fixed assets, that is the net investment, the total investment can be expressed in the form:

$$/3/ \quad I^G/t/ = \sum_{i=1}^{n} /I_i^W/t/ + I_i^N/t//$$

where: $I_i^W/t/$ investments substituting the depreciation of fixed assets

$I_i^N/t/$ net investment.

Going into the specification of the above categories investments substituting depreciated fixed assets can be linked to the fixed assets of the last year:

$$/4/ \quad I_i^W/t/ = \mu_i \, g_i/t-1/$$

where: μ_i the ratio of depreciation in sector i

$g_i/t-1/$ fixed assets in year /t-1/

A further assumption is the following: the net investment in every year is equal to the increase of fixed assets from year /t-1/ to year t:

$$/5/ \quad I_i^N/t/ = \Delta g_i/t/ = g_i/t/ - g_i/t-1/$$

2.2./ Introduction of control variables

The growth rate of the sectors depends on the share from the investments. If this share changes, so does the growth path of the sector concerned. In order to observe this connection let us consider the share of sector i in the net investment as a variable. The introduction of these variables makes possible to control the growth path of the sector and indirectly that of the economy.

The control variables can be defined by /6/ and because of the definition they are subject to /7/

$$/6/ \quad u_i/t/ = \frac{\Delta g_i/t/}{\sum_{i=1}^{n} \Delta g_i/t/}$$

$$/7/ \quad U = \left\{ u_i/t/ \;\middle|\; \sum_{i=1}^{n} u_i/t/ = 1;\ u_i/t/ \geqq 0 \quad \begin{matrix} i=1,2,..n; \\ t=1,2,..T \end{matrix} \right\}$$

2.3./ Balance of intermediary products

As the production of investment and consumer goods is limited by the intermediary products, we suppose, the input and output of intermediary products is equal up to a constant balance of foreign trade:

$$/8/ \quad \sum_{i=1}^{n} \alpha_i\, k_i/t/\, g_i/t/ = \sum_{i=1}^{n} m_i k_i/t/\, g_i/t/ + K_1$$

where: α_i the ratio of outputs in intermediary products to the total output of sector i

m_i the ratio of inputs in intermediary products to the total output of sector i.

K_1 foreign trade balance of intermediary products.

2.4. Labour force in the model

In case of Hungary the labour force is of special importance, because it sometimes appeares as one of the factors limiting the economic growth. In the model three major characteristics of labour force is built in:

- the productivity of labour force
- the total number of population at working age

- the possibility to convert labour force from one sector to the other:

$$/9/ \quad \ell_i/t/g_i/t/k_i/t/ - w_i/t/ \leq L_i/t/$$

$$/10/ \quad \sum_{i=1}^{n} w_i/t/ \leq K/t/$$

$$/11/ \quad \sum_{i=1}^{n} L_i/t/ + K/t/ = L/t/$$

where: $\ell_i/t/$ the labour force coefficient /labour force in sector i per total output/

$w_i/t/$ labour force convertible from a sector to the other

$L_i/t/$ labour force employed in sector i,

$K/t/$ the upper bound of convertible labour force.

2.5. Objective function

It is well-known, there is a multiplicity of goals at economy wide level, which are to be fulfilled. We have choosen the simplest and most evident objective: the maximization of cumulated consumption /private and public/ of the period considered:

$$/12/ \quad C = \sum_{t=1}^{T} \left(\sum_{i=1}^{n} \gamma_i k_i/t/ \, g_i/t/ + K_3 \right) \rightarrow \max$$

where: γ_i the ratio of consumer goods to the total output of sector i.

K_3 foreign trade balance of consumer goods

T the time horizon in year.

3./ Solution and analysis

To meet the requirements of planning two transformations of the model were applied.

The first transformation was designed to answer the problems directly connecting with the investment policy. Here the questions are: what is the feasible set of investment allocation; is it really the same as it was defined in /7/, or is it smaller; can we determine the feasible set from year to year, and so on.

At the first transformation two further assumptions were made:

1° $k_i/t/ = k_i$ x/

2° $\tau = 1$

In the next step using the above assumptions and having introduced some new notations,we can get the following form from the original //1/-/12// set-up of the model.

Eleminating $g_3/t/$ from /8/ the model can be reduced to a two-sector model.

$$\text{i.}\quad \begin{cases} g_1/t+1/ - g_1/t/ = u_1/t+1/\{\varrho_1 g_1/t/ + \varrho_2 g_2/t/ + \Psi\} \\ g_2/t+1/ - g_2/t/ = u_2/t+1/\{\varrho_1 g_1/t/ + \varrho_2 g_2/t/ + \Psi\} \\ g_1/o/ = g_{1o}\,; \\ g_2/o/ = g_{2o}\,; \end{cases}$$

$$\text{ii.}U= \begin{cases} u_1/t/\,,\ u_2/t/ > 0 \\ u_1/t/ + u_2/t/ < 1 \\ \dfrac{u_1/t/}{a_1} + \dfrac{u_2/t/}{a_2} = 1 \end{cases}$$

$$\text{iii.}\quad b_1/t/\, g_1/t/ + b_2/t/\, g_2/t/ \leq K^*/t/$$

$$\text{iv.}\quad C = \sum_{t=1}^{T} /f_1 g_1/t/ + f_2 g_2/t/ + K_2^*/ \longrightarrow \max$$

where: ϱ_1, ϱ_2, Ψ, a_1, a_2, $b_1/t/$, $b_2/t/$, f_1, f_2 are constans given by the transformation.

The set of control variables u can be drawn up, and it can be observed that its measure is well below that of the original set defined in /7/. If we draw up in addition the inequality iii, the feasible set of U' max be even smaller. If U' is not empty and the

x/ As in the reality the fixed assets coefficient is slowly increasing an adventageous case is considered. The form of the model gained by these assumptions allows the proportional increase of k_i, $i = 1,2,\ldots n$.

cutting by iii is in U' for every t , then exist investment policies, which can be maintained during the whole period.

In the <u>second transformation</u> of the model the exact solution was determined. This solution enables us to realize the opportunities remaining unutilized because of a constant investment policy.

After some transformations of /1/-/11/ we get the following form:

$$g_i/t+1/ - g_i/t/ = u_i/t/\left\{\sum_{j=1}^{n} \beta_j k_j/t/ - \mu_j \; g_j/t/ - K_2\right\}$$

$$\sum_{i=1}^{n} u_i/t/ = 1$$

$$\sum_{i=1}^{n} \ell_i/t/ \; k_i/t/ \; g_i/t/ \leq \sum_{i=1}^{n} L_i/t/ + K/t/$$

$$\sum_{i=1}^{n} /\alpha_i - m_i/ \; k_i/t/ \; g_i/t/ = K_1$$

$$\sum_{t=1}^{T} \sum_{i=1}^{n} \gamma_i k_i/t/ \; g_i/t/ \rightarrow \max$$

Let us introduce the notations:

$$v_i^* = /\alpha_i - m_i/ \; k_i$$
$$w_i^* = \beta_i k_i - \mu_i$$
$$\ell_i^*/t/ = \ell_i/t/ \; k_i$$
$$\sum_{i=1}^{n} L_i/t/ + K/t/ = L/t/$$
$$c_i^* = \gamma_i k_i$$

further on:

$$\underline{A} \; g/t/ = \begin{bmatrix} K_1 \\ G/t/ \\ \Delta/t/ + K_2 \end{bmatrix}$$

where:
$$A = \begin{bmatrix} \underline{v}^* \\ \underline{l}^* \\ \underline{w}^* \end{bmatrix}$$

and $\underline{l}^* \underline{g}/t/ = G/t/; \; - K_2 + \underline{w}^* \underline{g}/t/ = \Delta/t/$ the newly defined variables.

Using the new notations and the new variables the original set up can be reduced to a linear programming problem of the following type:

$$G\,/t+1/ - G/t/ = \Delta/t/$$

$$A^{-1} \begin{bmatrix} 0 \\ \Delta\,/t/ \\ \Delta/t+1/ \end{bmatrix} \geq A^{-1} \begin{bmatrix} 0 \\ 0 \\ \Delta/t/ \end{bmatrix}$$

$$\ell^*/t/\; A^{-1} \begin{bmatrix} K_1 \\ G/t/ \\ \Delta/t/+K_2 \end{bmatrix} \leq L/t/$$

$$\sum_{t=1}^{T} \underline{c}^* A^{-1} \begin{bmatrix} K_1 \\ G/t/ \\ \Delta/t/+K_2 \end{bmatrix} \rightarrow \max$$

From the optimal solution the value of original variables can be get back.

4./ The model and the system of models

From the above transformations of the original model the volume of fixed assets, the ratio of investment allocation and the labour force needed to production can be gained as the direct outcome of the model. Starting from these categories we are able to compute the total output, the net investment, the rate of accumulation, the national income,.... for drawing up the fifth Five-year plan of Hungary /1976-8o/. These latter make obtainable the values of aggregated macro figures of economy to the other elements /models/ of the system of models /2/. In this way the outcome of a model may form the input of other models in the framework of a system of models. At present the models operating in this system are the following:

- a less aggregated growth model of recursive type /4/
- the econometric model,
- the foreign trade simulation model,
- the energy programming model /9/,
- the the aggregated programming model /12/,
- the large-scale natural, value and financial programming model /7/
- the price model
- the micro-model of the population's consumption /5/,
- the income /component/ model /6/.

The optimal growth model's interconnected use in the system of models for medium-term national planning is now being tested in the Institute of Economic Planning in Hungary.

References:

/1/ Ámon,Zs: Népgazdasági mérleg modell a középtávu terv koncepciójának kidolgozásában. Statisztikai Szemle 1973.7-8.

/2/ Báger,G: Középtávu tervezési modellrendszer. Gazdaság 1973.1.

/3/ Biersach,B: Dynamisches Modell zur Prognoztizierung materiellen Grundproportionen ÖFI, 1971.2. Berlin.

/4/ Dániel,Zs-Jónás,A-Kornai,J-Martos,B: Plan Sounding. Economics of Planning 1973.

/5/ Enyedi,J: A lakossági fogyasztás modelljei. Tervgazdasági Intézet kiadványa. 1971.

/6/ Frigyes,E: The component model of personal income distribution Institute of Economic Planning of the Hungarian Planning Office, 1973.

/7/ Ganczer,S: ed. Népgazdasági tervezés és programozás K.J.K. Budapest, 1973.

/8/ Adam Lasciak, Teofil Gajarski: Principle of Maximum Application for the National Planning on the Basis of the Stoleru model European Meeting of the Econometric Society, Budapest, 1972.

/9/ Erdősi,P-Patyi,K: Energy Sector model for Analysing Economic Development Alternatives.
Symposion on Mathemathical Models of Sectors of the Energy Economy Alma-Ata, USSR, 1973.

/1o/ Szepesi,Gy-Székely,B: A gazdasági növekedés optimális pályái egy szabályozott gazdasági rendszerben.
Szigma, 1971.1.

/11/ Stoleru,L.G: An Optimal Policy for Economic Growth Econometrica 1965/4. Volume 33.

/12/ Ujlakiné,L: Összevont programozási modell alkalmazása a negyedik ötéves tervkoncepció kialakitásához.
Közgazdasági Szemle, 1969./9.

ON OPTIMIZATION OF HEALTH CARE SYSTEMS

John H. Milsum
Professor, Department of Health Care & Epidemiology
and
Director, Division of Health Systems
University of British Columbia, Vancouver

I. GENERAL

Living systems related to man can be arranged in a hierarchy, with the cell as the basic viable unit of human life, 10^0, ranging all the way to man at about the order of 10^{14} cells and world humanity in the order of 10^{24} cells.

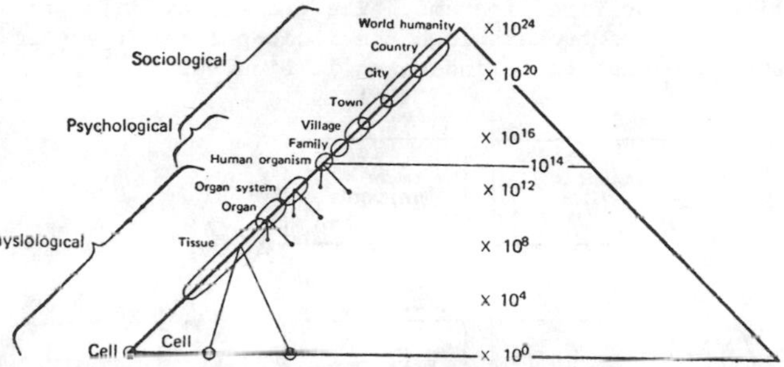

Figure 1. The human hierarchy. The multiple subsystems at any level are suggested by the several radiating lines. The variation in size at any level is suggested by the length of the link. From Milsum (1972)

At the "physiological" level a very plausible case can be made that evolution has operated so as to select those systems, which are optimal in some sense (Milsum (1968), Rosen (1967), Rashevsky (1973), Milsum & Roberge (1973)). The sense in which optimality occurs seems to be related to energy most typically, since this is a very expensive commodity for living systems. As we move to the "psychological" level of the individual organism and interaction with a few other individuals, the large variability inherent in man renders the search for optimality much more difficult. Nevertheless the concept seems generally acceptable, and the decision making of all individuals may well be rational or optimum-seeking, in terms of the personal internal performance criterion which applies, but may not be made known at the conscious level or to other individuals. As we move to the social level, including health systems as a particular sample, it may well be that the tendency to average out between large numbers of people may again make the idea of optimality more immediately practical. However, a basic problem in all these systems, and one especially difficult in the social and psychological systems, is how to correctly infer the particular criterion by which the optimality is achieved. Specifically, there is a uniqueness problem, in that while it may be possible to demonstrate that the optimization of a particular criterion is consistent with the behavior of the system, nevertheless this does not prove that the system in practice has achieved its optimality on this basis.

A health care system seems to be a particularly good example of a social system, and therefore seeking how a health care system can be optimized involves working in an area which may provide useful data for many other social systems. In this respect it may be worth noting that the integrated health of a society may, in fact, represent a reasonably good proxy for the quality of life of that society, in so far as increasing illness seems to be related to the stresses of living in a given society.

II. HEALTH, SICKNESS AND SOCIETY

The World Health Organization defines health as "a state of complete physical, mental and social well being and not just the absence of disease or infirmity". Even though the health so defined may be impractical of achievement, it appropriately emphasizes that a health care system must comprise much more than the practice of medicine. In this regard a number of pertinent points can be made:

1/ Health, Education and Welfare. The obtaining and maintaining of good health in a community involves an almost inextricable planning of appropriate health, education and welfare programs. The process by which these intercouple and ultimately feedback through the socio-political system to produce further adaptive changes is indicated in Figure 2.

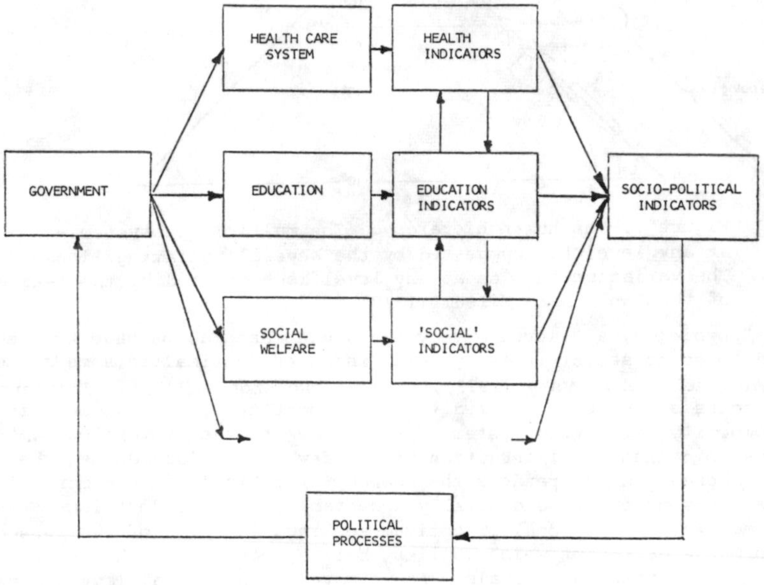

Figure 2. The Health Care System imbedded in the Government System.

2/ Sanitation. The direct and efficient application of good public "sanitation" measures has reduced or eliminated many diseases which were previously great scourges of mankind.

3/ Stresses. It is increasingly agreed that illness is highly correlated with the stresses of living (Selye (1950), Guirdham (1957), Holmes & Rahe (1967), Milsum (1973)). These stresses produce ubiquitous changes in the physiological systems through inappropriate neuro-endocrine changes. The stressfulness of particular events does not necessarily relate to whether they are perceived as being "good" or "bad", but rather only that they are dependent on the importance of the event. Table I has been abstracted from the work of Holmes and his colleagues showing the perceived stressfulness of various common-life events as averaged over a wide grouping of individuals.

TABLE 1. Social Readjustment Rating Scale
Adapted from Holmes & Rahe (1967)

Rank	Life Event	Mean value Life Change Units (LCU)
1	Death of spouse	100
2	Divorce	73
4	Jail term	63
5	Death of close family member	63
6	Personal injury or illness	53
7	Marriage	50
8	Fired at work	47
10	Retirement	45
12	Pregnancy	40
13	Sex difficulties	39
14	Gain of new family member	39
18	Change to different line of work	36
19	Change in number of arguments with spouse	35
20	Mortgage over $10,000	31
26	Wife begin or stop work	26
31-36	Change in work, residence, outside activities etc.	20-18
38-40	Change in home habits	16-15
41	Vacation	13
42	Christmas	12
43	Minor violations of the law	11

More generally, and as developed elsewhere (Milsum, 1973) we may usefully advance four major factors as affecting the onset of illness:

a/ Genetic Endowment, which predisposes us to certain types of illness, either congenitally (eg. sickle cell anemia) or later in life (eg. heart attack).

b/ Physical Stressors, which typically can affect our physiological systems directly and essentially involuntarily. Examples are disease vectors and various pollutions.

c/ Psychological Stressors, such as the events noted already and summarized in Table I.

d/ Endogenous Responses, by which the perceived stressfulness of a given situation is translated into stress in the body, amplified or attenuated according to the individual's responding characteristics (see De-stressing Techniques below). Guirdham emphasizes that there are strong socio-religious factors at play in determining our expectations and, generally therefore the higher the individual person's projection of his own individuality upon the world, then the higher his tendency to suffer morbidity in the central nervous system especially. This concept is consistent with the other idea that the degree of stressfulness is related to the inevitable loss of the individual's control over his personal environment in a complex technological civilization, and therefore over event outcomes (for example, being late because of unexpected traffic delays). Expectations and life style therefore play important roles, and may help explain differences of perceived stressfulness in various cultures.

4/ Risk-taking. Risk-taking seems to be an essential human activity (____, (1972)). It can, however, be either healthy or neurotic. Starr (1972) shows that there is a quantitative relationship between the degree of risk accepted by individuals and the benefit expected to be obtained thereby. Furthermore, there is a much higher level of voluntary risk which is accepted, (eg. in sports) than the risk at the involuntary level (e.g. in work and business travel). Indeed, this voluntary risk level is some three orders of magnitude greater than that accepted involuntarily (Figure 3). The background level of involuntary risk due to disease averages out to about one death per million hours exposure, and interestingly enough both commercial aviation and motor vehicles are operated at about this level of risk.

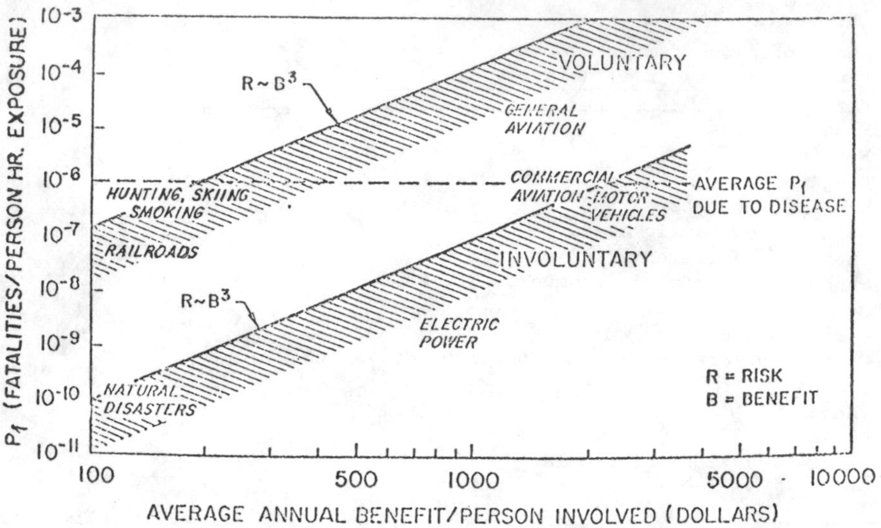

Figure 3. Risk vs. Benefit, Voluntary & Involuntary (From Starr, 1972).

Many accidents which produce trauma and death are undoubtedly due to neurotic risk-taking; for example, some fatal automobile accidents may be due to a sub-conscious wish to commit suicide. In consequence much more research is necessary on risk-taking, before large and expensive technological "solutions" are sought to prevent such accidents. In general, experience in this regard has shown that such technological "solutions" are at most only partially successful.

5/ Nutrition. Much of the world's population is thought to suffer from malnutrition. This may be either under- or over-nutrition. Informed guessing suggests that the total social costs attributable to malnutrition, in a country like Canada, may equal numerically about half the total costs incurred in the health care system (although these are not all the same costs).

Certainly developing countries are familiar with the vicious cycle that when workers are under-nourished they are sickly, and their productivity cannot rise enough to provide the good nutrition prerequisite for better productivity.

Of course there are many social correlates in nutrition so that neither the best nor the cheapest good diet will necessarily be acceptable. Perhaps this provides all the more reason why any attempt to optimize a health care system should certainly tackle the problem of nutrition as being one with large potential pay-off in the benefit:cost sense.

6/ Self-Abuse With Drugs (Alcoholic, Psychotropic and Tobacco). The above remarks on risk-taking, perhaps suggest the difficulty of "rational" health care optimization when the individual must "weigh immediate pharmacological and psychological benefits against long-term statistical risks" (Tamerin & Resnick(1972)) of the physiological kind. The following comment upon cigarette smoking presents a somewhat different viewpoint, which may also be pertinent for this whole class of self-abuse. "If there ever comes a time when cigarette smoking can be detoxified, I believe there will be very few people who will not smoke. One can readily suppose that were it not for the identification of smoking with dirt, disease and depravity, it would be a universal indulgence... Smoking can be used in the service of the id, the ego, or the superego, or any of these together...In brief, smoking is one of the lasting pleasures of life" (Marcovitz (1969)).

7/ Recreation. This subject leads easily into the whole gestalt of living including life style, physical environment and educational aspects. To restrict ourselves here to one comment, the best forms of recreation, from a health point of view, would seem to be those which tend to encourage psychological development, as for example the Being-needs described by Maslow (1962). Their satisfaction however presupposes adequate fulfillment of the Deficiency-needs, including the physiological need for "tuning-up" through exercise. Hence, sports of any kind should be beneficial to health provided they are tackled for love of the playing, rather than to satisfy obsessional needs to compete, and thus to beat opponents, whether humans or nature. In terms of the next section, healthful recreation tends to reduce a person's stress level.

8/ De-Stressing Techniques. Recreation, leisure, etc. can all play a major role in reducing the stress level in individuals. Further, an ideal life style would be that which did not induce stressful conditions. Unfortunately, most of us seem incapable of achieving this ideal. Thus, there is an important role for specific destressing or relaxation techniques. From the broad spectrum of possibilities we only mention here, yoga, transcendental meditation, autonomic learning, and biofeedback. Some experts in transcendental meditation have been tested for physiological correlates of this psychological state, and some interesting correlations have been found (Wallace (1970)). One particularily interesting correlation is that during transcendental meditation there is a statistically significant reduction in the level of arterial lactate, which is a substance known to rise in concentration with increasing anxiety level. Biofeedback tends to introduce technological monitors "to close the loop" back to the subject, so that he can monitor his own performance, in the sense of certain measurable physiological parameters. Thus the dominant frequency of the EEG rhythum may be measured as one example, (Karlins and Andrews (1972)). In general, much research on this subject is necessary before we can have any reasonable expectations as to the cost:benefit for society of such techniques, but it does seem possible that they will provide a valuable weapon in the armanentarium of the health scientist.

9/ Philosophical and Societal Aspects. As soon as a society accepts the principle that each living individual is entitled to equitable and accesible health care of good quality, then these rights inevitably require certain corresponding rights for society to determine enough basic criteria for ensuring that these promised services can be provided in practice, within the constraint of a society that would not be disastrously burdened economically. Specifically, this means that such previously taboo subjects as the desirable level of population in a given society and the desirable form of human life cycle, must at least be considered in principle. The latter requirement arises especially because the advances of biomedical science and technology make it possible on a statistical basis to extend life expectancy significantly. However, in general the cost of this extension is steeply rising in a non-linear way as a function of age. This

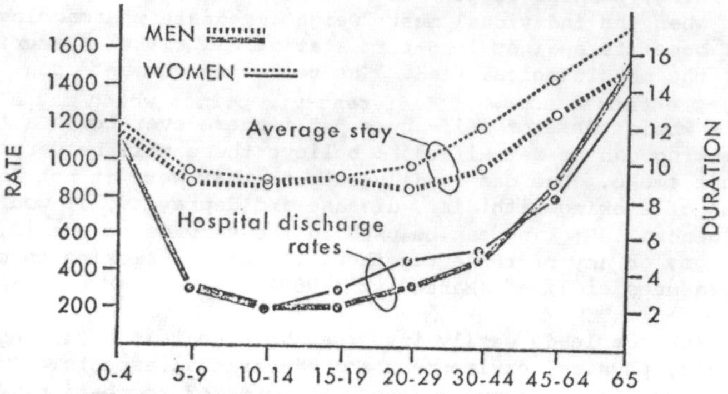

Figure 4. Hospital discharge rates per 10,000 population and average stays by age and by sex (excluding surgical operations, deliveries and live births) in Quebec hospitals, in 1967 (----, 1970)

is suggested in Figure 4 which shows the discharge rates and average stays in Quebec hospitals, particularized for the two sexes. Since the health care costs will vary approximately as the product of these two factors, it can be seen that it does indeed rise steeply in old age. Further, through the growth ethic of our technologically oriented society, there is a temptation to believe that we can and should continue to try to conquer "killer" diseases, even though the quality of the lives so saved may not be perceived by the surviver to be very satisfactory. However, we must bear in mind that in any balanced society the actuarial basis of the "insurance" system requires that on the average each individual will pay during his productive years for the amount that he will recover for health care especially during his old age. Since the productive years are usually also years of heavy family responsibilities, it may well be that faced with such a clear option, many would opt for somewhat less emphasis on life-extending health care in old age than is the common assumption.

On the issue of a desirable population level, there is usually considerable controversy. In practice the only politically acceptable way of establishing a desired level may be through the implicit decision-making whereby the general outlines of a birthrate and life-cycle policy are established. In particular, for example, it may become necessary to "license" births up to an agreed allowable level each year. Various schemes are possible whereby these licenses could be "marketed" to, or distributed among the population. Even though such a method may seem repugnant to some, we must bear in mind that if a society is to establish certain objectives, such as an optimization regarding quality of life, then this may imply certain apparently hard decisions regarding the actual number who can share in such a public good without destroying it (Hardin (1968)).

III HEALTH CARE, HEALTH CARE SYSTEM & CONTROL

The WHO definition of health, involving physical, mental and social components, makes it clear that a genuine health care system is inextricably involved with most of the other major "systems" of government and society. Unfortunately, this definition does not, in itself, indicate how to allocate relative priorities among the three sectors, although such allocation is a major requirement of any optimization program. For our discussion, it is convenient to divide health care into three major functional roles:

1. Care of Sickness
2. Prevention of Illness
3. Promotion (protection) of Health

Classically the area on which most medical attention has been placed is that of the care of sickness, but while this is still very important, societies generally are concluding that much greater emphasis must be placed upon the other roles.

1. Care of Sickness. National bodies continually recommend various reorganizations in hospitals and medical practice, especially, indicating a general perception that sick care is not yet optimal. As one major example, hospitals are often wastefully used under "third party" payment schemes because neither patient nor physician typically is penalized for longer stays than necessary. Indeed, the frequently-divided authority in hospitals and their "non-profit" nature, combine to make it difficult for hospitals to maintain high economic efficiency. From an optimization viewpoint, however, hospitals represent an important potential source of greater efficiency, since they consume about half the health care system budget in conventional sick-care-oriented systems.

Innovative ideas are now being tried in different countries which will hopefully overcome major drawbacks in the conventional systems. For example:

a. De-centralized, but coherently organized structure. The Community Health Centre concept embodies the idea of grouping the major health professionals necessary for primary care in conveniently-accessible locations for small communities. More broadly, a community social centre (CSC) embodies the "point of delivery" concept that all necessary governmental services should be made available to the consumer at a single location, again acknowledging the complex interactions of our modern society. Experience indicates that probably about 90% of initial health care contacts need proceed no further than primary care, as provided in community health centres. An integrated system then typically has two further hierarchical levels, namely, the community hospital and the specialist hospitals (often university teaching hospitals) operating to take care of successively more acute or complicated problems (----(1970)).

b. Division of responsibilities on health care team. Catalyzed in part by such structural reorganizations as mentioned above, innovations are also being tested in delegation of sick-care duties among the various health professionals. In particular, the trend towards a "health care team" concept involves a vastly increased communication between the various workers and considerable delegation of duties previously performed only by physicians.

c. Role of patient. We also note the slow evolution of the idea that the patient himself is a crucial member of the health care team, as such. More particularly, in specific operational procedures involving his own body the patient may have over-riding privilege and responsibility in decision-making.

d. Problem-oriented medical record. The problem-oriented medical record system developed by Weed (1969) provides an important example of innovation by which the basic document relating to the health and sickness of each patient can be based on a logical system. Many possibilities develop from it, notably; inter-professional communication, medical audit, teaching, generation of data bases, and record linkage.

e. Computerization. Health care in general, and sick care in particular, are informationally-intensive operations. Specifically, it is estimated that information handling typically costs between one-quarter and one-third of the hospital budget, thus representing around one-seventh of the total health care costs. Hence considerable savings could result from small increases in efficiency. The rationalizations implicit in the factors just discussed provide a sensible basis for computerization, and computer technology has now advanced sufficiently for computerization to be attractive economically in the long-term. However, as always happens when an industry is computerized, the vastly increased potential offered by the computerization of health information systems will result in many worthwhile new tasks being tackled, so that costs will not fall, but rather the quality of care will rise.

f. Telemedicine. The technical integration of computers, telephone and electronic displays permits in principle a vastly greater de-centralization of physical resources, without a corresponding reduction in the quality of health care offered. The good social return of such de-centralization, in terms of the WHO definition of health, may result in an overall increased benefit:cost situation even though the specific costs are increased.

2. Prevention of Illness. "Public Health" measures have in relatively recent history effected perhaps more dramatic improvements in morbidity patterns than almost any other aspect. These functions properly continue and improve, but we now concentrate more upon the ideas of mass screening for particular diseases, such as tuberculosis and cervical cancer, and upon multi-phasic screening for many possible diseases or malfunction in any one patient. The underlying concept is that of early detection of illness, which is very appealing socially. However, before applying the concept all the implications should be thought through systematically. For example, one current concern is whether effective actions can be taken at all the pertinent levels, medical, psychological and sociological, so that overall benefit does follow from early detection. On the other hand from the viewpoint of acquiring statistically-valid knowledge from data bases on morbidity patterns, on effectiveness of medical intervention and on health indicators, for example, there is unquestionably great value in obtaining such information. Undoubtedly the useful role for preventive measures will grow as biomedical science and technology increase, but careful review is continually needed in terms of overall systems optimization.

3. Promotion of Health. This subject implies a much more difuse, but positive approach to the promoting of health, rather than the negative aspects of preventing and curing illness. Thus it is legitimately concerned with such areas as education, nutrition, living environment, work habits, recreation and use of leisure time, and legal procedures, some of which were discussed in earlier sections. The difficulty in quantifying the benefits and costs in such admirable endeavors emphasizes the likely impossibility of a true rational optimization of a health care system, bearing in mind the inevitable constraints due to the processes occurring within any political system, with all that that implies regarding both competition of various group interests and a limited time horizon.

IV HEALTH CARE SYSTEM AND HEALTH INDICATORS

Basic to the idea of a health care system is that health care is under the direction of some more or less effective management, as implemented through a feedback control structure (Figure 5). Further, the idea of optimization presupposes

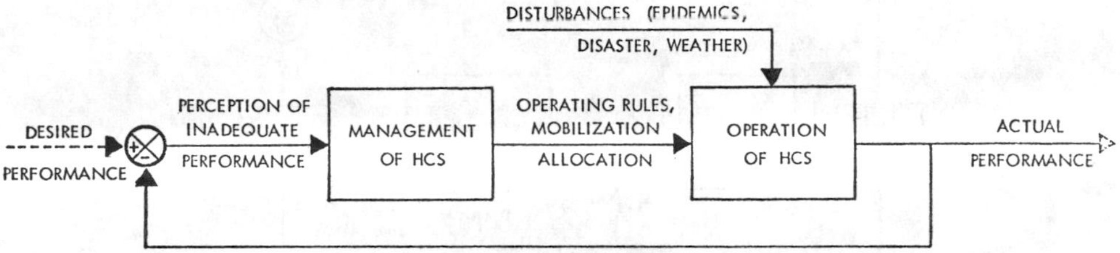

Figure 5. Basic Management of Health Care Systems (HCS)

some concept of desired performance and measurement of actual performance in order to evaluate present conditions and make plans for a more nearly optimal operation in future. An appropriate term for actual performance is "Health Indicators" and this will be discussed below. It should be noted that the management function may need to seek changes in many of the operations within the health care system. In order to explore this satisfactorily we need a mathematical model of the health care system, as suggested in Figure 6 where four main sub-models are indicated for the health care

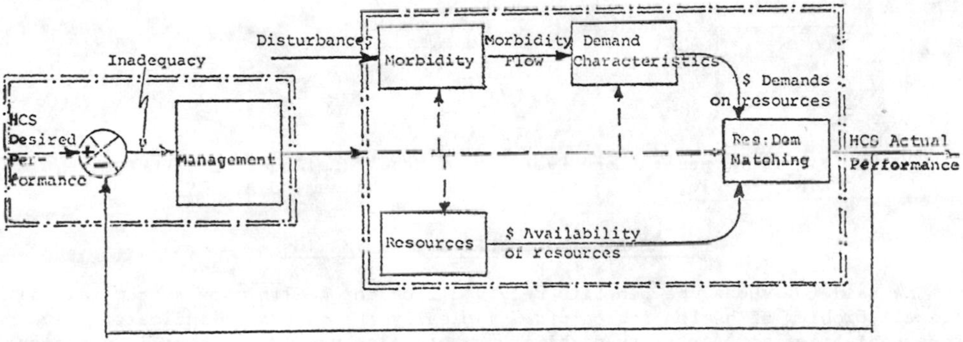

Figure 6. Management Influences on Health Care System Operation

system, under the general format of a resource:demand matching model. In greater detail such a macro-model of the health care system involves defining many more subprocesses, as indicated in Figure 7. It is not our purpose here to discuss this model in detail, but it is worth noting, in passing, that the morbidity flow needs to be aggregated into a number of groupings by age, sex, and illness, while the resources of personnel and equipment are simply aggregated by certain resource-type groupings. The potential usefulness of the model and its projections from computer simulations depends upon defining appropriate costs and benefits in order to evaluate the performance measures.

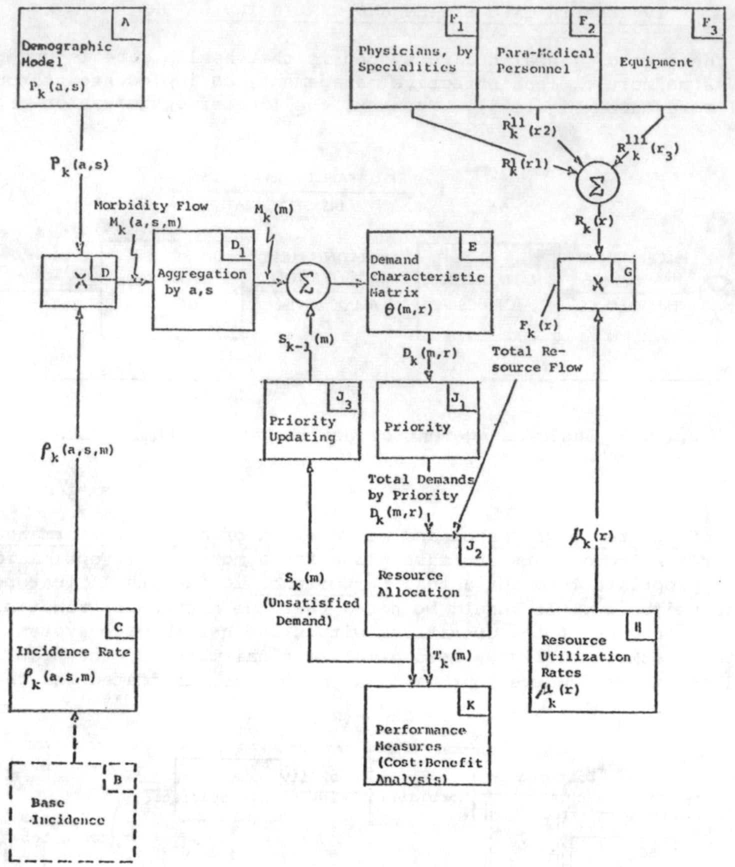

Figure 7. Health Care System Flow Model (Belanger, P., Hurtubise, A.B., Laszlo, C.A., Levine, M.D., and Milsum, J.H. 1972)

V HEALTH INDICATORS & OPTIMIZATION

The need to assess the benefits received from the health care system raises the difficult problem of health indicators. Typically the classical indicators have been life-expectancies and mortality/morbidity rates, although it has been long recognized that these are not necessarily very good proxies for the actual health quality of the population. More recently, more complex indicators have been developed, for example, the "Q-index" which takes into account with appropriate weightings, such aspects as periods in hospital, visits to physicians and days of work lost. Table 2 for the United States in the early 60's is notable in showing on this basis that accidents are second only to diseases of the circulatory system in total impact.

In another direction, the "activity levels" in the health care system represent services rendered to the population. These statistics are attractive candidates for health indicators, since they are relatively easily obtained, but unfortunately they tend to represent only the costs of the system, rather than the benefits received. To illustrate this with a specific example, it may well be, when programs for the promotion of health and prevention of illness are effectively applied, that the activity levels in the sick system can be reduced. The analysis here is further complicated by the fact that there will be different time lags between the applica-

TABLE 2. The Health Problem Index of the United States for Selected Diseases, 1963-64. Adapted from Michael, J.M., Spatafore, G., and Williams, E.R. (1968)

Major class of disease*	Q value
Diseases of the circulatory system	4,425
Accidents	2,781
Neoplasms	2,367
Diseases of the respiratory system	2,196
Diseases of the nervous system	1,344
Diseases of the digestive system	843
Infectious and parasitic diseases	575
Diseases of the genitourinary system	342
Complications of pregnancy	147
Diseases of bone and organs of movement	85

*From the International Classification of Diseases.

tion of funds and their subsequent benefits in various types of programs. Clearly it is much easier to appreciate the connection between cost and benefit in the short-term acute type of care, when the patient is desperately dependent psychologically upon the health care profession, than it is when there is a long advance-time as in educational and preventative programs. Obviously appropriate time-discounting procedures must be applied in the detailed economic analysis.

There is much current work aimed at quantifying the changes in health status of individuals (Goldsmith (1972)). The general concept behind this is that if the change in health status as a result of intervention by health professionals can be quantified then we may begin to have a measure of the effectiveness or benefit provided by the system. The costs are identified relatively easily, so that hopefully a cost:effectiveness or cost:benefit type analysis may be tackled.

Ideally for any analytical optimization procedure it is necessary to construct a scalar-type performance index in which a vector of component indices is appropriately weighted. Unfortunately, in view of the many factors described in this paper it seems unlikely that such weightings can yet be satisfactorily achieved, even if we can start to measure performance in commensurable units. (These units would most likely be money.) It may, therefore, be both helpful and necessary to develop decision-making aids in which the various multiple components are presented separately to the decision-makers who must combine them within the structure of their own perceptions about societal needs (Collins & Jain(1973)). The study of how this procedure is carried out by individuals over a sufficiently long sample of time and individuals should provide some valuable clues about value judgements in this complicated area of health care systems. Not incidentally it will also provide a major challenge to computer scientists in evolving new graphical techniques for data presentation. However, we should not leave such important judgemental matters to professionals alone, and much benefit would result from having computer simulation models, (as in Figure 7) available for the interested wider public to use. With such models, gaming procedures allow a wide range of inputs and parameter changes to be explored, as for example in disturbances due to epidemics, changes in the system "production function" due to innovations in the ways of delivering health care, and changes in perceptions of the values obtained from particular types of health care system performance. The resulting widespread public and professional interest should enable a good social attempt to be made at optimizing our increasingly expensive and powerful health care systems.

References

----, "Perspectives on Benefit:Risk Decision Making", Nat. Acad. Eng., Washington, D.C. (1972)

----, "Report on the Commission of Inquiry on Health and Social Welfare", Vol. IV, Government of Quebec, Quebec, Canada (1970)

Belanger, P., Hurtubise, A.B., Laszlo, C.A., Levine, M.D., and Milsum, J.H., MEDICS - An Effort to Model the Health Care Delivery System of Quebec, 4th Can. Med. Biol. Eng. Conf., Winnipeg (1972)

Collins, D.C. & Jain, A., Decision Aiding Tools in the Design of Health Care Delivery Systems, Proc. 6th Hawaii Int. Conf. on Sys. Sci., Western Periodicals Co. (1973)

Goldsmith, S.B., The Status of Health Status Indicators, Health Serv. Reports, 87: 212-220 (1972)

Guirdham, A., "A Theory of Disease" Neville Spearman, London (1957)

Hardin, G., The Tragedy of the Commons, Science, 162: 1243-1248 (1968)

Holmes, T.H., & Rahe, R.H., The Social Reajustment Rating Scale, J. Psychosomatic Res., 11: 213-218 (1967)

Jydstrup, R.A. & Gross, M.J., Cost of Information Handling in Hospitals, Health Serv. Res. I: 235-271 (1966)

Karlins, M. & Andrews, L.M., "Biofeedback", J.B. Lippincott Co. (1972)

Marcovitz, E., On the Nature of Addiction to Cigarettes, J. Am. Psychoanal. Assn., 17: 1074-1096 (1969)

Maslow, A., "Toward a Psychology of Being", D. Van Nostrand Co. (1962)

Michael, J.M., Spatafore, G. & Williams, E.R., A Basic Information System For Health Planning, Public Health Reports, 83: 1 (1968)

Milsum, J.H., On Optimization in Biological Control Systems, in "Advances in Biomedical Engineering & Medical Physics, I: 243-278 (Ed. S.N. Levine) J. Wiley & Sons (1968)

Milsum, J.H., The Hierarchical Basis For General Living Systems, in "Trends in General System Theory" 145-187, (Ed. G.J. Klir) J. Wiley & Sons (1972)

Milsum, J.H. & Roberge, F.A., Physiological Regulation & Control in "Foundations of Mathematical Biology", III: 1-95, (Ed. R. Rosen) Academic Press, New York & London (1973)

Milsum, J.H., What is Health & a Health Care System, Proc. Engineering in Health Care Symposium, Can. Med. Bio. Eng. Soc., Ottawa (1973)

Rashevsky, N., The Principle of Adequate Design, in "Foundations of Mathematical Biology", III: 143-175 (Ed. R. Rosen) Academic Press, New York & London (1973)

Rosen, R., "Optimality Principles in Biology", Butterworths, London & Washington, (1967)

Selye, H., "The Stress of Life", McGraw Hill Book Co., New York (1950)

Starr, C., Benefit-Cost Studies in Sociotechnical Systems, in "Perspectives on Benefit-Risk Decision Making", 17-42, Nat. Acad. Eng., Washington, D.C. (1972)

Tamerin, J.S. & Resnick, H.L.P., Risk Taking by Individual Option - Case Study: Cigarette Smoking, in "Perspectives on Benefit:Risk Decision Making", 73-84, Nat. Acad. Eng., Washington, D.C. (1972)

Wallace, R.K., The Physiological Effects of Transcendental Meditation, Ph.D. Thesis, Dept. of Physiology, UCLA. (Also pub. by Students International Meditation Soc., 1015 Gayley Ave., Los Angeles, Calif. 90024) (1970)

Weed, L.L., "Medical Records, Medical Education, and Patient Care: The Problem-Oriented Record as a Basic Tool", Press of Case-Western Reserve University, Cleveland (1969)

THEORETICAL AND OPERATIONAL PROBLEMS IN DRIVING A PHYSICAL MODEL OF THE CIRCULATORY SYSTEM

Abbiati, B., Fumero, R., Montevecchi, F.M.
Parrella, C. (*)

Foreword

The purpose of this communication is to deal with some of the problems met during the development of a physical simulator of the cardiovascular system.

The task of designing building and testing a reliable and significant moke up of blood circulation was assigned to the Bioengineering group of the CNPM by the CNR in 1971 and while the present paper is chiefly devoted to some adspects of the regulation involved a brief comment on the aims and scopes of the overall research program can possibly produce a better understanding of the difficulties and requisites encountered during the development of the control system, see also Abbiati (1972) et al. and Fumero (1972) et al.

The assigned research in the cardiovascular field had to perform the following specific tasks:

- give a faithful replica of vascular bed, blood and heart ventricles with particular regard to the fluidodynamic phenomena involved in blood circulation (large vessels)
- allow for system identification in circulation phenomena and carry out model optimizations
- explore the performance of the heart at different operating condi-

(*) Istituto di Macchine, Politecnico di Milano, Centro di Studio per Ricerche sulla Propulsione e sull'Energetica CNR; Laboratorio di Tecnologie Biomediche, CNR.

tions in wiew of comparative "in vivo" experiments and as a way to develop more sophisticated control modes

- and offer a reliable test bed for artificial devices both temporary and permanent to be inserted or applied to the cardiovascular system in order to modify flow conditions, heart work, oxigen consumption, blood oxigenation, etc.

This capability is intended to be applied in the following research areas:

- measurement tecniques and instrumentation
- artificial heart and artificial assist devices
- cardiovascular system control and regulation
- cardiovascular system modeling

The complexity of these aims determined from the very beginning of the work attentive examination and organization of the research system in order to match all the abovementioned requisites.

A processing computer with multiplexer digital and analog outputs was devised as the basic tool for driving the simulator and for carring out all the calculations dealing with portions of the cardiovascular system not to be simulated in a physical way; among these the control section of the heart and peripheral resistance was one of the most critical points to be satisfied.

Cardiac control and regulation

In the last years substantial improvements have been produced in basic physiological research on cardiac output and its regulation.

As a consequence some reliable tools for the analysis and prediction of the heart operation and the vascular system response have been commonly accepted and can be considered as a basis for further research.

In developing an heart control system we adopted and sometimes slightly modified the Sommenblick (1965) cardiac muscle model and utilized chiefly, Zatona, Martin (1970), Pikering (1969), Potil, Ghista (1972) relations and models.

In a somewhat simplified way we can consider the operations of the heart ventricle as determined in its pressure versus volume relation by the utilizing system-the vascular bed-; this dynamic matching is described by the position of the cardiac cycle, Fig. 1.

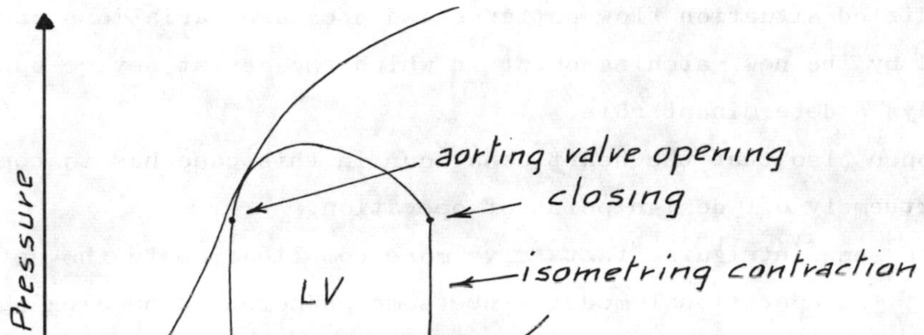

Fig. 1

The laws governing the cycle position are the stress/strain/time law of motion of the cardiac muscle, some parameters identifing the individual law within a large family of possible ones and the pressure/volume/time relations for the vascular bed.

A correct prediction of cardiac behaviour is a mandatory requisite for a simulating device of the cardiovascular system and the actuating system of the simulated ventricle has to follow very closely the natural pattern.

As an example we may consider the situation in which the hemodynamic variation has to be examined following an assist device application. In the modified situation flow patterns and pressure variations are determined by the new matching point in which the assist device operation plays a determinant rôle.

It may happen also that the heart behaviour in this case has to cope with an extremely off design point of operation.

Furthermore some intriguing facts give more complication to the determination of the operational model since some phenomena depending upon wave propagation and reflections and valve operation cannot be taken in proper account in the pressure/ volume relation for the cardiac cycle.

These facts determine the necessity of driving the ventricle by means of a transfer function applied on the output of the operational control model.

Procedure adopted in starting the simulation experiment

On the beggining of the experiment since mechanical accomodation may undergo changes the actuating system is tested in order to determine is actual transfer function.

The configuration of the device is sketched in Fig.2 and the procedure is shown in Fig. 3.

An imput signal is fed into the system with a continuos power spectrum as shown in Fig. 4 and the outcoming pressure signal power spectrum is that one of Fig. 5 in which low frequencies are filtered in order to obtain a zero medium value in the outcoming signal.

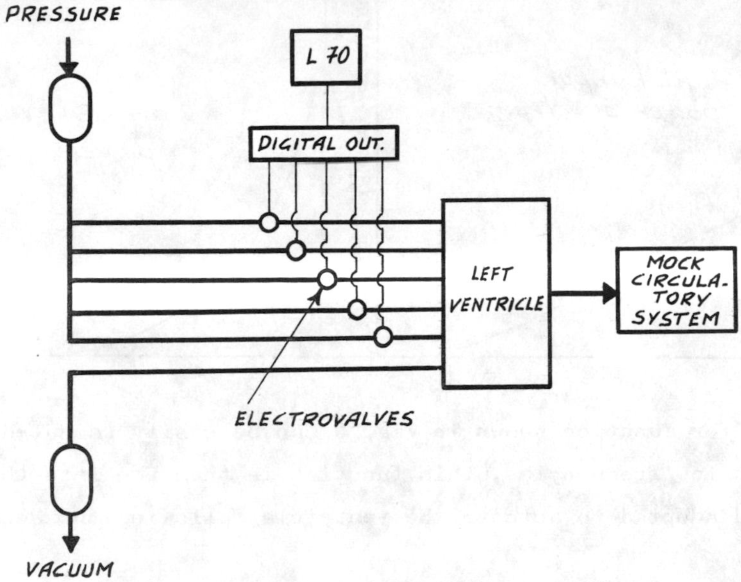
PRESSURE
L 70
DIGITAL OUT.
LEFT VENTRICLE
MOCK CIRCULATORY SYSTEM
ELECTROVALVES
VACUUM

Fig. 2

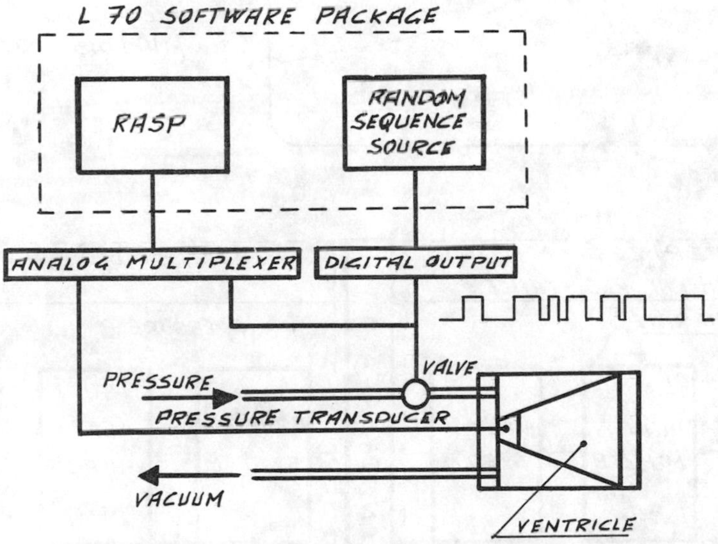
L 70 SOFTWARE PACKAGE
RASP
RANDOM SEQUENCE SOURCE
ANALOG MULTIPLEXER
DIGITAL OUTPUT
VALVE
PRESSURE
PRESSURE TRANSDUCER
VACUUM
VENTRICLE

Fig. 3

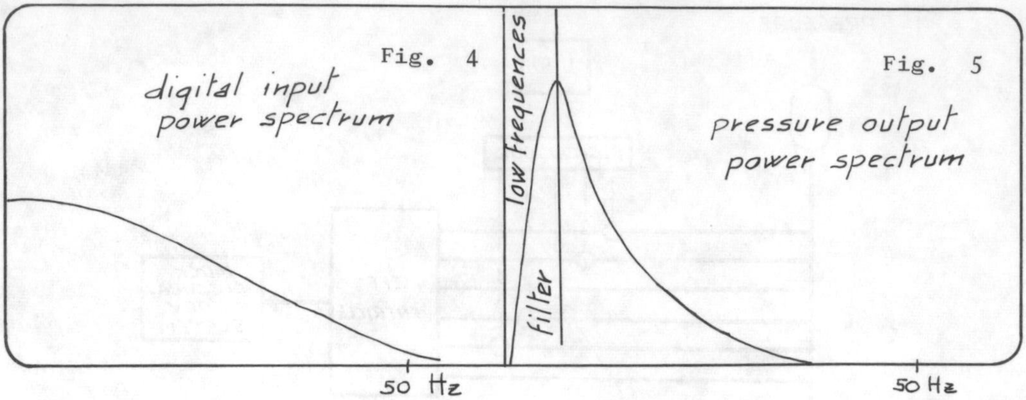

Fig. 4 Fig. 5

The transfer function shown in Fig. 6 can be easily computed and corrected at low frequencies, this function is then fed into the control model and adopted to actuate the ventricle following the scheme of Fig. 7

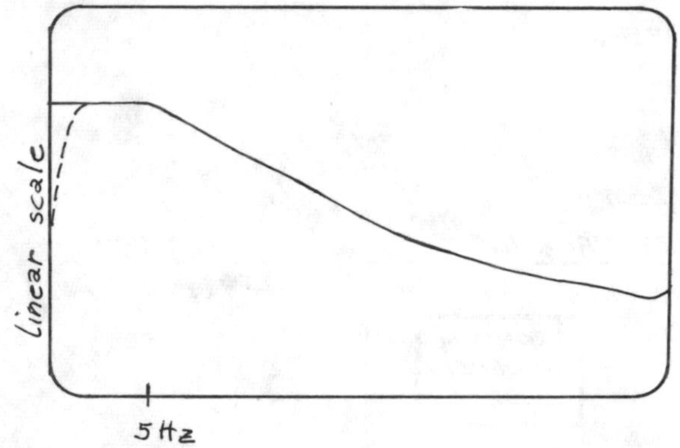

Fig. 6

actuation transfer function

– – – – – – – values to be corrected due to low frequency filtering

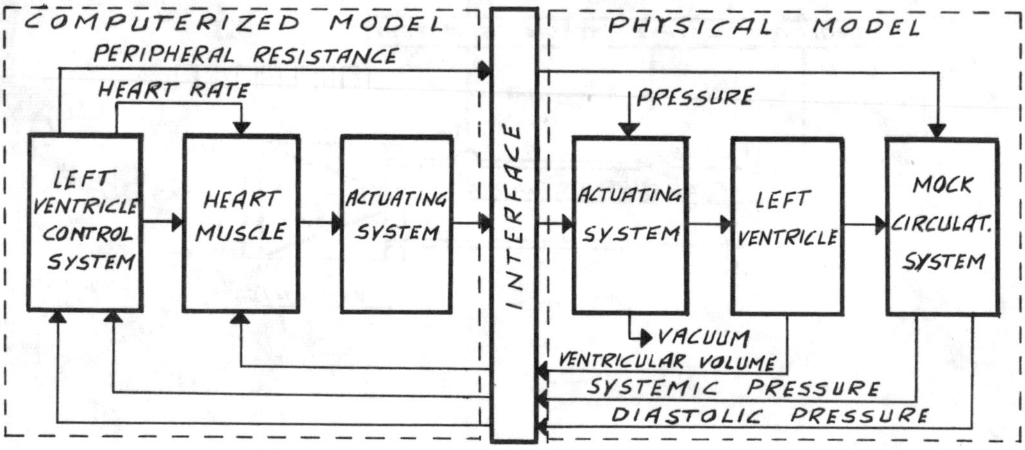

Fig. 7

Results

Results have been satisfactory, Fig. 8 shows a tipical response to sine an triangular wave at 2 Hz. Fig. 9 shows a ventricular pressure pattern and Fig. 10 the aortic one.

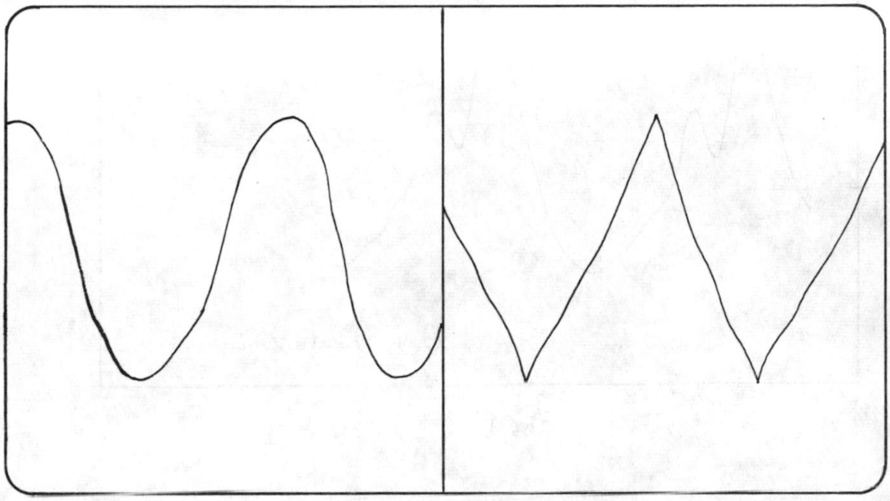

Fig. 8

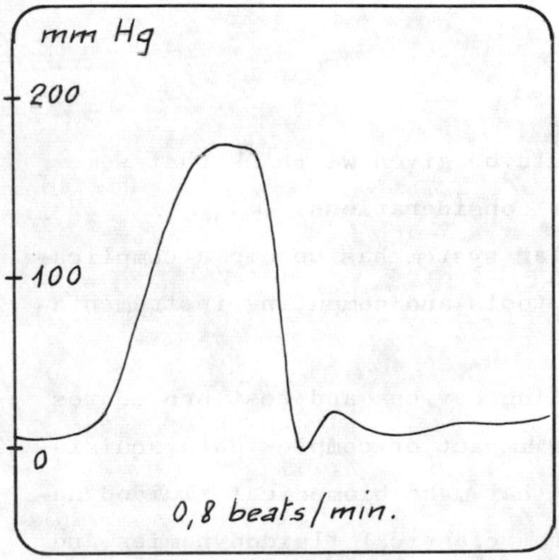

Fig. 9

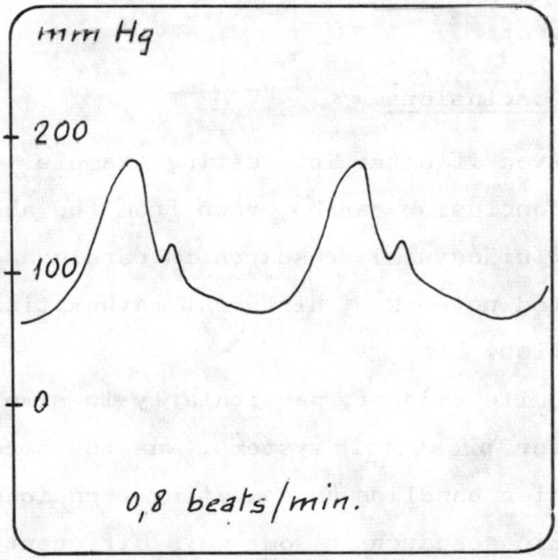

Fig. 10

The actuation control thus conceived is well capable of coping with abnormal circulatory conditions; in Fig. 11 is shown the aortic pressure after a stenosis induced in the aorta well in accordance to physiological outcomings.

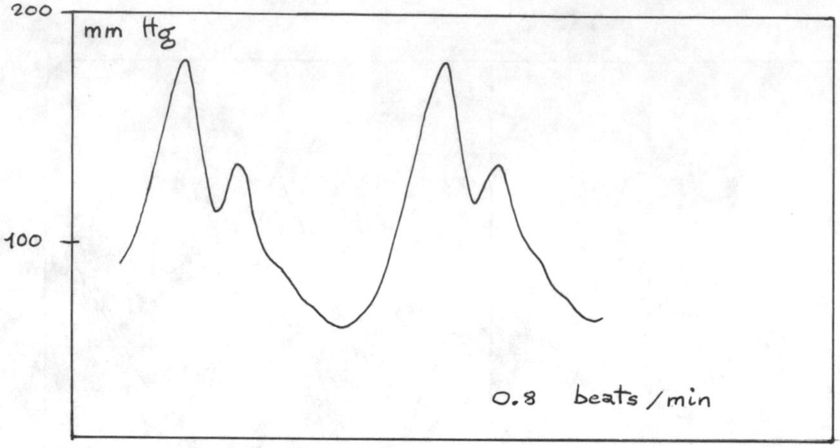

Fig 11

Conclusions

Even if other interesting examples could be given we think that some conclusion can be drawn from the above considerations.

Fluidodynamic research in cardiovascular system has become a complicated network of different mathematical tools and computing instrumentation.

Quite evident, particularly in simulating devices and test procedures for prosthetic systems, was the strong impact of complex data aquisition handling an computing techniques that make biomedical fluidodynamic research in some ways different from classical fluidodynamics and characterized by a continuos and sophisticated evolution.

This fact depend s upon many factors among which the time dependent process analized that forces a processing computer analysis and the neces-

sity of real time application that favours linearization techniques and advanced random data analysis.
The outcoming is a rich variety both of models and of the related computational techniques that build up a complex research system applied on the specific aims of the cardiovascular research.
In our work the definition and the development of the research system itself has proven to be the real point and stimulating enterprise; carefull assignement of approaches to the problems and capability and flexibility requirements have been the most difficult goals to be met.

References

Abbiati, B., Fumero, R., Montevecchi, F.M., Parrella, C. "Problemi di elaborazione di dati nella simulazione dell'apparato cardiovascolare" Convegno FAST-ANIPLA (1972).

Fumero, R., Montevecchi, F.M., Parrella, C. and Spaggiari, E. "La simulazione idromeccanica del sistema cardiocircolatorio" Convegno FAST-ANIPLA (1972).

Ghista, D.N., Patil, K.M., Woo, K.B. and Oliveres, C. Journal Biomechanics 5, 365-390 (1972).

Martin, P.J. Automatica 4, 175-191 (1970).

Pikering, W.D., Nikiforuk, P.N. and Merriman, J.E. Medical Biological Engineering, 7, 401-410 (1969).

Sonnenblick, E.H. Circulation Research 16, 441 (1965).

MODELING, SIMULATION, IDENTIFICATION AND OPTIMAL CONTROL OF LARGE BIOCHEMICAL SYSTEMS

J.P. KERNEVEZ

Faculté des Sciences
6, Boulevard Gabriel
21000 DIJON FRANCE

I. INTRODUCTION.

The aim of this paper is to describe some biochemical systems ruled by partial differential equations.

These sytems were built by a group of biochemists, led by Dr. D. Thomas ([1]). These systems are artificial membranes, made of inactive protein cocrosslinked with enzyme molecules. (For instance albumin membrane bearing glucose oxidase). Such a membrane M separates two compartments 1 and 2 in which there is some substrate which is going to diffuse into M and reacts in presence of the enzyme E (Enzymes are biological catalysts).

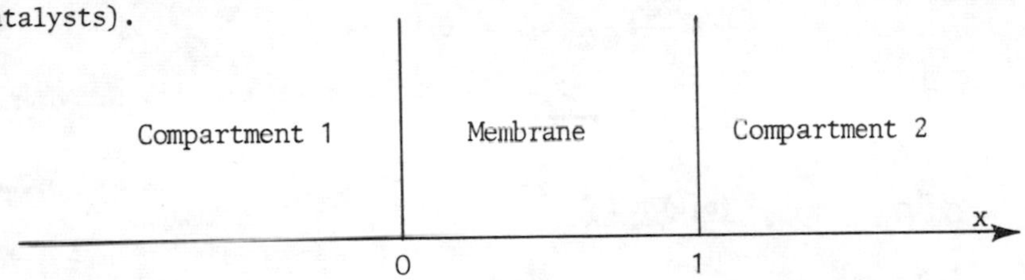

This reaction gives some product P :

$$S \xrightarrow{E} P$$

Enzyme kinetics in solution have been extensively studied since a long time. The importance of studying enzymes in insoluble phase, such as artificial membrane, is obvious when one knows that in living cells most reactions occur in such heterogeneous structures where not only reaction, but also diffusion plays a major role.

Therefore these models are useful in understanding the basic phenomena occuring in living cells, but already their interest extends to analytical, industrial and medical applications.

The plan of the paper is as follows :

Section 2 : description of the model case,
Section 3 : description of an active transport model,
Section 4 : description of a system with charge effects,
Section 5 : description of an oscillatory model,
Section 6 : description of a multienzyme monodimensional model,
Section 7 : description of multienzyme bidimensional models.

([1]) Laboratoire d'Enzymologie médicale, ERA 338 CNRS, Hôtel-Dieu (76)-ROUEN.

For modelling of other systems, we refer to Kernevez J.P., Thomas D., where are given proofs of existence and unicity for the solutions of the P.D.E., numerical methods to get the state of the system, and the study and numerical solution of control problems. For the control of these systems, see in the Proceedings, J.P. Yvon, and J.P. Kernevez, J.P. Quadrat, M.Viot. One can see also P. Penel and J.C. Brauner. For identification of kinetic parameters, one can see G. Joly.

2. THE MODEL CASE.

In the system described in Section 1 we take as unity of length the thickness of the membrane. By taking a suitable unity of concentration, the model case is then described by the following set of equations, where $s(x,t)$ (resp. $p(x,t)$) denote the substrate (resp. product) concentration $(0<x<1,\ 0<t<T)$:

(2.1) $$\frac{\partial s}{\partial t} - \frac{\partial^2 s}{\partial x^2} + F(s) = 0$$

(2.2) $$\frac{\partial p}{\partial t} - \frac{\partial^2 p}{\partial x^2} - F(s) = 0$$

(2.3) $$F(s) = \sigma \frac{1}{1+|s|}$$

(2.4) $$s(o,t) = \alpha\ ;\ s(1,t) = \beta$$

(2.5) $$p(o,t) = p(1,t) = 0$$

(2.6) $$s(x,o) = p(x,o)$$

(2.3) is the so-called Michaelis-Menten term of reaction. α , β and σ are positive, say : $\alpha = 1$, $\beta = 1$, $\sigma = 1.4$

3. THE ACTIVE TRANSPORT MODEL.

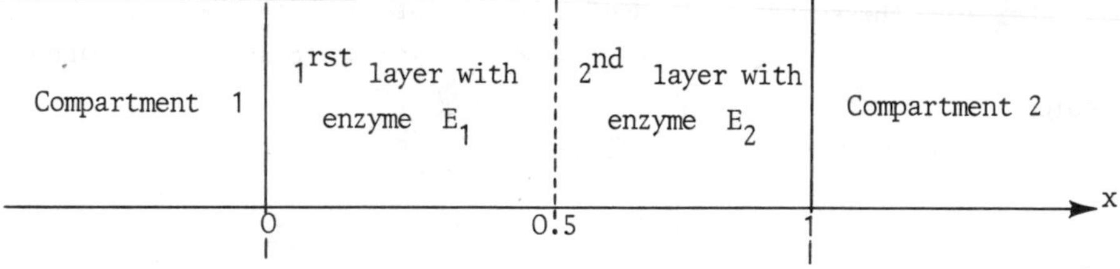

Now the membrane is made of two layers. In the first one, Enzyme E_1 is a catalyst of the reaction

$$S \xrightarrow{E_1} P$$

while in the second Enzyme E_2 is a catalyst of :

$$P \xrightarrow{E_2} S$$

so that for a given time the profile of concentration of S will be :

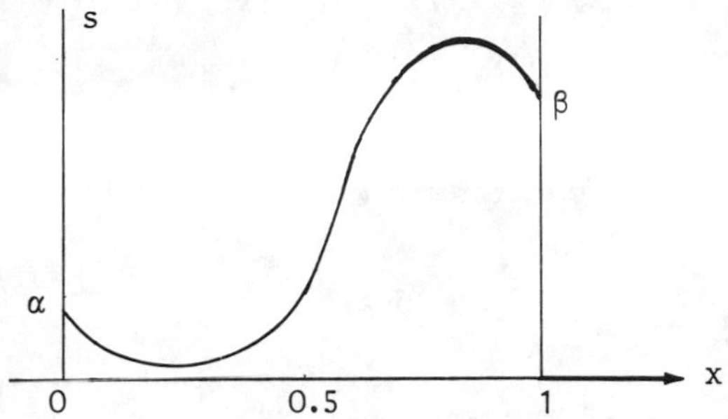

In the first layer S is consumed while in the second one it is producted. It is easy to see that substrate is entering the membrane at $x = 0$, and going out of the membrane at $x = 1$, although $\beta > \alpha$ (α and β are the substrate concentrations in compartments 1 and 2), so that there is an active transport from compartment 1 to compartment 2. Such a phenomena cannot be thought of without diffusion constraints.

Equations governing the system are the same than in Section 2, except $F(s)$ has to be replaced by :

$$(3.1) \qquad F(s,p) = \begin{cases} \sigma \dfrac{1}{1+s+p} & \text{if } 0<x<0.5 \\ -\sigma \dfrac{p}{1+p} & \text{if } 0.5<x<1 \end{cases}$$

Data are $\alpha = 1$, $\beta = 10$, $\sigma = 700.$

4. A MODEL WITH CHARGES EFFECTS.

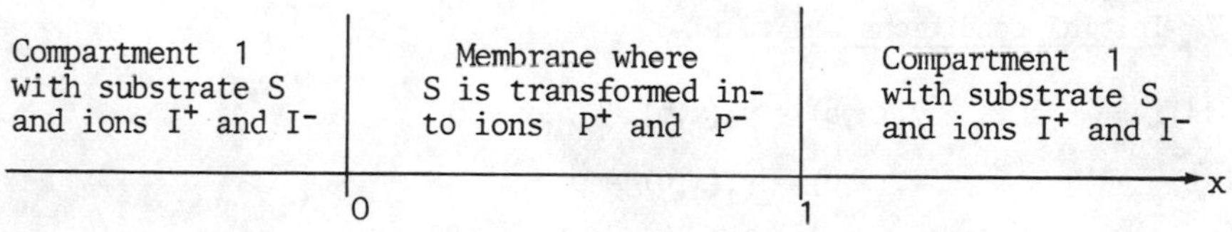

In this case, the equations, boundary and initial conditions for s are the same than in Section 2, but now one molecule of S gives one ion P^+ and one ion P^- . Moreover two ions I^+ and I^- are diffusing in the membrane, so that if we call y_1, y_2, y_3, y_4 the respective concentrations of I^+, I^-, P^+, P^-, and ψ the electric potential, we have the equations :

(4.1) $$\frac{\partial y_1}{\partial t} - \frac{\partial}{\partial x}\left(\frac{\partial y_1}{\partial x} + y_1 \frac{\partial \psi}{\partial x}\right) = 0$$

(4.2) $$\frac{\partial y_2}{\partial t} -0,1\frac{\partial}{\partial x}\left(\frac{\partial y_2}{\partial x} - y_2\frac{\partial \psi}{\partial x}\right) = 0$$

(4.3) $$\frac{\partial y_3}{\partial t} - \frac{\partial}{\partial x}\left(\frac{\partial y_3}{\partial x} + y_3 \frac{\partial \psi}{\partial x}\right) = F(s)$$

(4.4) $$\frac{\partial y_4}{\partial x} -0,1\frac{\partial}{\partial x}\left(\frac{\partial y_4}{\partial x} - y_4\frac{\partial \psi}{\partial x}\right) = F(s) \quad .$$

We make the following electromentrality hypothesis :

(4.5) $$y_1 + y_3 = y_2 + y_4$$

and

(4.6) $$J_1 + J_3 = J_2 + J_4$$

where

(4.7) $$J_1 = \frac{\partial y_1}{\partial x} + y_1 \frac{\partial \psi}{\partial x}$$

(4.8) $$J_2 = 0,1\left(\frac{\partial y_2}{\partial x} - y_2 \frac{\partial \psi}{\partial x}\right)$$

(4.9) $$J_3 = \frac{\partial y_3}{\partial x} + y_3 \frac{\partial \psi}{\partial x}$$

(4.10) $$J_4 = 0,1\left(\frac{\partial y_4}{\partial x} - y_4 \frac{\partial \psi}{\partial x}\right)$$

Boundary conditions are :

(4.11) $$s(o,t) = s(1,t) = 10$$

(4.12) $$y_1(o,t) = y_2(o,t) = 10 \; ; \; y_1(1,t) = y_2(1,t) = 1$$

(4.13) $$y_3(o,t) = y_3(1,t) = y_4(o,t) = y_4(1,t) = 0$$

Initial conditions are :

(4.14) $$y_1(x,o) = y_2(x,o) = -\,9x + 10$$

(4.15) $$y_3(x,o) = y_4(x,o) = 0 \quad .$$

5. AN OSCILLATORY MODEL.

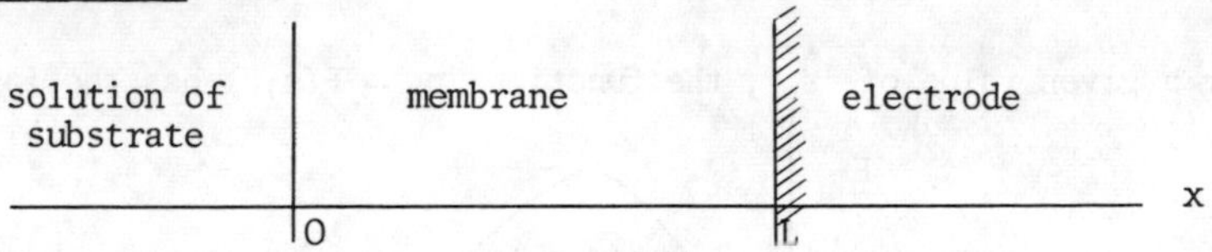

This model has been built and studied in Harvard Medical School Biophysical Laboratory in Boston by A. Naparstek , R. Caplan and D. Thomas.

Substrate and products can enter or go out of the membrane at $x = 0$ but at $x = 1$, there is an electrode, whence a boundary condition of Neumann type for $x = 1$. In this system, the product is ion H^+, whose concentration will be denoted by p. An auxiliary function $a(x,t)$ appears in the equations describing the system, which are :

(5.1) $$\frac{\partial s}{\partial t} - D_s \frac{\partial^2 s}{\partial x^2} + F(s,p) = 0 \quad , \quad o < x < L$$

(5.2) $$\frac{\partial a}{\partial t} - D_H \frac{\partial^2 a}{\partial x^2} - F(s,p) = 0$$

(5.3) $$a = p - \frac{10^{-20}}{p}$$

Boundary conditions :

(5.4) $$s(o,t) = 5.5\ 10^{-7} \quad ; \quad p(o,t) = 10^{-13}$$

(5.5) $$\frac{\partial s}{\partial x}(L,t) = \frac{\partial p}{\partial x}(L,t) = 0$$

Initial conditions :

(5.6) $$s(x,o) = 5.5\ 10^{-7} \quad ; \quad p(x,o) = 10^{-13}$$

Data $$D_s = 0.13\ 10^{-5} \quad ; \quad D_H = 0.64\ 10^{-5} \quad ; \quad L = 10^{-2} .$$

All these numerical data, as well as the following reaction term :

(5.7) $$\begin{cases} F(s,p) = Q_2 \quad Q_3 \dfrac{E}{Q_2 + Q_3 + \dfrac{Q_3\ 5{,}45\ 10^{-5}}{S}} \\[2ex] Q_2 = \dfrac{64.5}{1 + C_1 H + \dfrac{C_2}{H}} \quad ; \quad Q_3 = \dfrac{20.2}{1 + C_3 H} \\[2ex] C_1 = 10^{7.29} \quad ; \quad C_2 = 10^{-11.49} \quad ; \quad C_3 = 10^{6.92} \end{cases}$$

have been found by A. Naparstek so that oscillations appear, in the following manner:

for a given value of s , the function $p \rightarrow F(s,p)$ has the following graph :

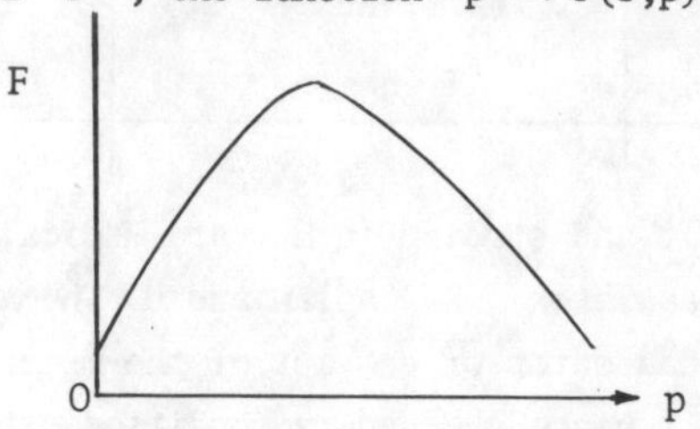

The reaction is autocatalytic but decreases when p is large enough.

for a given value of p , F is an increasing function of s

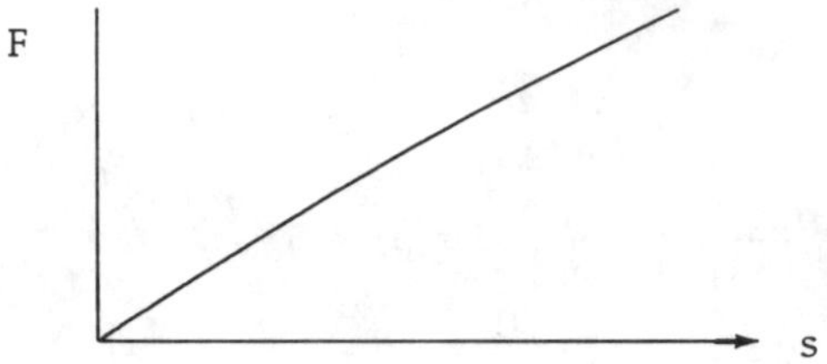

so that at first there will be a more and more important consumption of substrate.

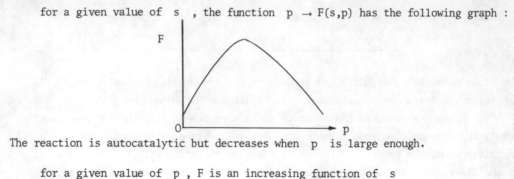

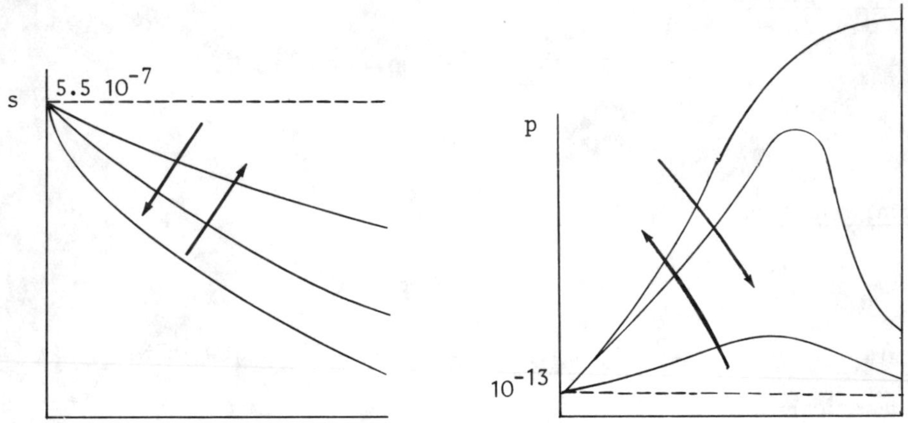

the profile of concentration of S will fall down, while the one of H^+ will go up in an explosive manner, until there is no more sufficiently substrate in the membrane to feed the reaction, and H^+ concentration is too high to have a large value of F. This, joine to the fact that the gradient of concentration of H^+ is huge (H^+ concentration has increased along the electrode from 10^{-13} to 10^{-7}, over 6 decades), constrains the profile of H^+ to go down, while S concentration is increasing by incoming flux of substrate, until S inside the membrane is large enough to feed another explosion. This cycle continues in a periodic way.

6. MULTIENZYME MONODIMENSIONAL MODEL.

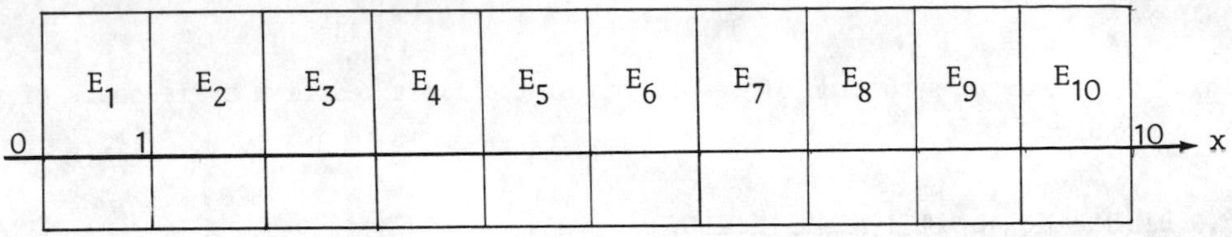

In this sequence of enzymatic membranes, each one contains a different enzyme, so that the system performs a chain of enzymatic reactions in metabolism.

In the stationary case, the equations are :

$$(6.1)\quad \begin{cases} -\dfrac{d^2y_1}{dx^2} + \chi_1 F(y_1) = 0 \\ -\dfrac{d^2y_2}{dx^2} + \chi_2 F(y_2) - \chi_1 F(y_1) = 0 \\ \cdots\cdots\cdots\cdots\cdots\cdots \\ -\dfrac{d^2y_{10}}{dx^2} + \chi_{10}F(y_{10}) - \chi_9 F(y_9) = 0 \\ -\dfrac{d^2y_{11}}{dx^2} + \chi_{10}F(y_{10}) = 0 \end{cases}$$

where

$$(6.2)\qquad \chi_i = \text{characteristic function of }]i-1,i[$$

$$(6.3)\qquad F(u_i) = \sigma \frac{y_i}{1+|y_i|} \quad , \qquad \sigma = 10$$

Boundary conditions are :

$$(6.4)\qquad y_1(o) = 1 \quad , \quad y_1(10) = 1$$

$$(6.5)\qquad y_i(o) = y_i(10) = 0 \qquad 2 \leqslant i \leqslant 11$$

A slightly different model is the one where the last product of the sequence of reactions, that is y_{11} , is an inhibitor of one of the preceding reactions, say the first ; the first reaction term has to be replaced by :

(6.6) $$\chi_1 \; F(y_1, y_{11}) = \chi_1 \; \sigma \frac{y_1}{1+|y_1|+|y_{11}|}$$

This is a frequent situation of regulation of the first reaction by feedback of the last product.

7. BIDIMENSIONAL MULTIENZYME MODELS.

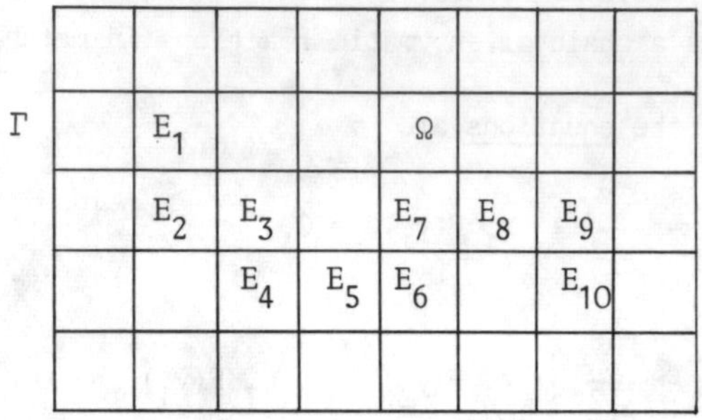

This first system is analogous to the one described in Section 6, excepted that this time, in the square containing enzyme E_J product P_J is transformed into product P_{J+1}

$$P_J \xrightarrow{E_J} P_{J+1}$$

and equations are the same. One has only to change $-\frac{d^2}{dx^2}$ into $-\Delta$ and define

(7.1) $$\chi_J = \text{characteristic function of the square containing } E_J .$$

Boundary conditions are of the same kind :

(7.2) $$\left| \begin{array}{lll} y_1 & = 1 & \text{on the boundary } \Gamma \text{ of } \Omega \\ y_i & = 0 & 2 \leqslant i \leqslant 11 \end{array} \right.$$

Another system,

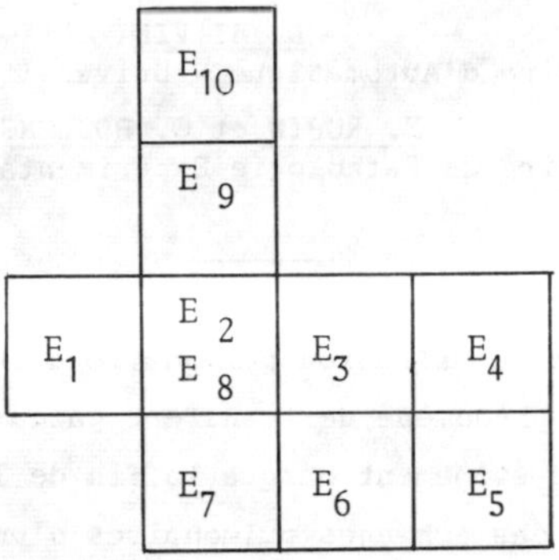

contains in the same square enzymes E_2 and E_8. Moreover y_8 is an inhibitor of E_2 so that in this square the rate of transformation of P_2 into P_3 is

(7.3) $$\chi_2 \quad \sigma \quad \frac{y_2}{1+|y_2|+|y_8|}$$

8. CONCLUSION.

A new approach of biology by biochimists has been presented. This approach consists in building simple biochemical models, well described by partial differential equations. Numerical analysis of these equations give more information about the evolution or the stationary state of the profiles of concentration inside the membranes. The numerical analyst is asked also to find optimal controls of these systems, or to identify some of their kinetic parameters. Moreover these artificial enzyme membranes offer very exciting possibilities for medical applications, biotechnology, bioengineering.

Therefore, this approach of biology seems to be a field of increasing collaboration between biochemists and applied mathematicians.

BIBLIOGRAPHY.

G. JOLY, Identification de paramètres cinétiques dans des systèmes biochimiques, Thèse de 3ème cycle, Paris, 1973.

J.P. KERNEVEZ, D. THOMAS, Book to be published, Dunod, Paris.

J.P. KERNEVEZ, J.P. QUADRAT, M. VIOT Control of a Stochastic Non Linear Boundary Value Problem, these proceedings.

P. PENEL, J.C. BRAUNER Contrôle d'un Système Biomathématique, Thèse 3ème cycle, Paris, 1972.

J.P. YVON Optimal Control of Systems Governed by Variational Inequations, these proceedings.

MODELISATION DU TRANSFERT GAZEUX PULMONAIRE ET CALCUL AUTOMATIQUE DE LA CAPACITE DE DIFFUSION*

D. SILVIE
Laboratoire d'Automatique - Université LILLE I
H. ROBIN et C. BOULENGUEZ
Laboratoire de Pathologie Expérimentale - Université LILLE II

La constante de diffusion pulmonaire a été définie par Marie KROGH en 1914 comme représentative du phénomène de transfert gazeux au travers de la "membrane alvéolo-capillaire". Cet évènement marque la fin de l'opposition de deux écoles défendant dans le cadre des échanges pulmonaires d'une part la théorie du phénomène actif de secrétion, de l'autre celle du phénomène passif de diffusion.

L'approximation qui consiste à considérer que l'alvéole pulmonaire est assimilable à une sphère permet d'utiliser la loi de diffusion de FICK à une seule dimension :

$$(1) \quad Q'_{Dx} = D_x \, (P_{\bar{A}x} - P_{\bar{c}x})$$

Q'_{Dx} = débit de diffusion pulmonaire du gaz x
D_x = capacité de diffusion pulmonaire du gaz x
$P_{\bar{A}x}$ = pression alvéolaire moyenne du gaz x
$P_{\bar{c}x}$ = pression capillaire moyenne du gaz x

De nombreuses méthodes de mesure de la capacité de diffusion ont été mises au point depuis 1914. Elles comprennent essentiellement deux grands types :

- mesures en régime stable ventilatoire (steady state)
- mesures en apnée inspiratoire (single breath)

Notre travail concerne plus particulièrement la méthode en apnée inspiratoire. La méthode en apnée que nous utilisons est une variante technique de celle proposée par KROGH (1914) élaborée par FORSTER et ROUGHTON (1954) et codifiée par COTES (1963). Elle consiste à faire inhaler au sujet au cours d'une inspiration unique succédant à une expiration forcée un mélange gazeux contenant du monoxyde de carbone à concentration connue. Après un temps d'apnée prédéterminé, un échantillon représentatif de l'air intra-alvéolaire est prélevé au cours de l'expiration et

---

* Travail bénéficiant d'une aide financière de l'I.N.S.E.R.M. - contrat personnel n°7210706 - Professeur F.GUERRIN

analysé. Les mélanges gazeux utilisés contiennent du monoxyde de carbone (maximum 0,3%) de l'oxygène (21 %), de l'hélium (14,5 %), et de l'azote. Les paramètres nécessaires au fonctionnement du modèle que nous proposons sont obtenus à partir du tracé spirographique et du résultat de l'analyse des gaz inspirés et expirés.

MODELE THEORIQUE

Pour réaliser un modèle de la diffusion pulmonaire, nous nous sommes basés sur les 3 postulats fondamentaux de KROGH :

1 - la pression intra-capillaire de monoxyde de carbone (CO) est nulle

2 - la mixique intra-pulmonaire des gaz est homogène

3 - la combinaison du monoxyde de carbone avec l'hémoglobine est immédiate

Les postulats 1 et 3 permettent donc de simplifier la loi de FICK.

(2) $$\dot{Q}_{D_{CO}} = D_{CO} \times P_{A_{CO}}$$

Le postulat 2 permet de proposer l'expression suivante de la quantité (volume) alvéolaire d'oxyde de carbone :

(3) $$Q_{A_{CO}} = Q_{I_{CO}} + Q_{E_{CO}} - Q_{D_{CO}}$$

quantité alvéol. de CO = quantité inspirée (> 0) + quantité expirée (< 0) - quantité diffusée (> 0)

La fraction d'un gaz dans un mélange est le rapport de la quantité de ce gaz en volume total, nous avons :

(4) $$F_{CO} = \frac{Q_{CO}}{V}$$

où V est le volume total du mélange.

On peut donc exprimer les quantités inspirées et expirées de CO :

(4a) $Q_{I_{CO}} = F_{I_{CO}} \times V_I$ (l'indice I signifie inspiration)

(4b) $Q_{E_{CO}} = F_{E_{CO}} \times V_E$ (l'indice E signifie expiration)

Dans le cas de l'expiration :

$F_{E_{CO}} = F_{A_{CO}}$ (l'indice A signifie alvéolaire)

$F_{I_{CO}}$ est une constante mesurée par analyse des gaz inspirés.

L'expression (4) nous permet également d'exprimer la fraction alvéolaire par rapport aux quantités alvéolaires de gaz et au volume alvéolaire :

(5) $$F_{A_{CO}} = \frac{Q_{A_{CO}}}{V_A}$$

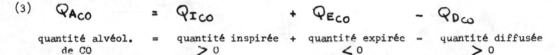

V_A est la somme de 3 volumes :

V_A	=	V_I	+	V_E	+	V_R
volume alvéolaire	=	volume inspiré >0	+	volume expiré <0	+	volume résiduel >0

Le rapport de dilution d'hélium permet la mesure du volume résiduel par l'expression :

$$V_R = (V_I - V_D)\frac{F_{IHe}}{F_{EHe}} - V_I$$

où V_D = 150 cm3 est le volume de l'espace mort anatomique qui ne participe pas aux échanges.

La quantité diffusée est l'intégrale du débit de diffusion exprimé par la loi de FICK simplifiée où la pression alvéolaire moyenne est donnée par la loi de DALTON. De ce fait l'expression (3) se transforme en :

$$F_A (V_I + V_E + V_R) = F_I V_I + F_A V_E - \int_0^t D P_A F_A \, dt.$$

Après simplification :

$$F_A (V_I + V_R) + \int_0^t D P_A F_A \, dt = F_I V_I$$

Nous allonsmaintenant proposer certaines hypothèses simplificatrices afin de résoudre cette équation.

MODELE LINEAIRE

La première simplification est de considérer la capacité de diffusion comme une constante.

Si on admet que les débits inspiratoire et expiratoire sont laminaires la pression alvéolaire correspond à :

$$P_A = P_B - \rho \dot{V}$$

où ρ est la résistance des voies aériennes pour un débit laminaire $\dot{V}$
P_B est la pression barométrique ambiante.

L'étude des courbes spirographiques du processus expérimental (796 dossiers pour 3 sujets)nous permettent sans trop grande erreur, de linéariser volumes inspirés et expirés. De ce fait :

$V_I(t) = \dot{V}_I t$ de $t = 0$ à t_I (t_I = temps de fin d'inspiration)
$\dot{V}_I$ = constante
$V_I(t) = V_I$ quant $t \longrightarrow t_I$
$V_E(t) = 0$ de $t = 0$ à t_A (t_A = temps de fin d'apnée)
$V_E(t) = \dot{V}_E (t - t_A)$ pour $t \longrightarrow t_A$

$\dot{V}_E$ = constante

La formule du modèle peut alors se résoudre en 3 temps :

a) inspiration de $t = 0$ à t_I

$\dot{V} = \dot{V}_I$ = constante.

on peut écrire:

$$F_A(\dot{V}_I + V_R) + D(P_A - \rho\dot{V}_I)\int_0^t F_A\,dt = F_I\dot{V}_I t$$

intégrale du premier ordre dont la solution est :

$$F_A(t) = F_I \times \frac{1-\left(\frac{V_R}{\dot{V}_I t + V_R}\right)^{1+D\left(\frac{P_B-\rho\dot{V}_I}{\dot{V}_I}\right)}}{1+D\left(\frac{P_B-\rho\dot{V}_I}{\dot{V}_I}\right)}$$

(V_I correspond au volume inspiré maximum)

b) apnée de $t = t_I$ à t_A

$\dot{V} = 0$; le volume reste constant et égal à $V_A = V_I + V_R$ volume alvéolaire maximum.

On écrira : $$F_A V_A + D P_B \int_0^t F_A\,dt = F_I V_I$$

dont la solution est :

$$F_A(t) = F_{AI}\, e^{-\frac{D P_B}{V_A}(t - t_I)}$$

dont la valeur en fin d'apnée sera :

$$F_A(t_A) = F_{AI}\, e^{-\frac{D P_B T_A}{V_A}}$$

où $T_A = t_A - t_I$

c) expiration de t_A à t final

$$\dot{V} = \dot{V}_E < 0$$

$$F_A V_A + D(P_B - \rho\dot{V}_E)\int_0^t F_A\,dt = F_I V_I$$

dont la solution sera :

$$F_A(t) = F_{AA}\, e^{-D\frac{(P_B - \dot{V}_E)}{V_A}(t - t_A)}$$

L'échantillon est prélevé sur l'expiration et peut s'exprimer d'après (4) par :

$$F_{ech} = \frac{(F_A \times V_E)(t_4) - (F_A \times V_E)(t_3)}{V_E(t_4) - V_E(t_3)}$$

où t_3 et t_4 sont les temps de début et de fin d'échantillon.

En posant $T_4 = t_4 - t_A$ et $T_3 = t_3 - t_A$ et en appliquant les expressions trouvées précédemment nous exprimons la valeur de la fraction échantillonnée :

$$F_{ech_{CO}} = F_{I_{CO}} \frac{1 - \left(\frac{V_R}{V_A}\right)^{1 + D \frac{P_B - \rho \dot{V}_I}{\dot{V}_I}}}{1 + D \frac{P_B - \dot{V}_I}{\dot{V}_I}} \times e^{-D \frac{P_B T_A}{V_A}} \times \ldots$$

$$\ldots \times \frac{T_4}{T_4 - T_3} e^{-D \frac{P_B - \rho \dot{V}_E}{V_A} T_4} - \frac{T_3}{T_4 - T_3} e^{-D \frac{P_B - \rho \dot{V}_E}{V_A} T_3}$$

Nous proposons sur la figure 1 un schéma bloc fonctionnel représentatif de notre modèle.

SIMULATION ANALOGIQUE

La simulation du modèle proposé peut être facilement effectuée sur calculatrice hybride. La figure 2 représente le cablage réalisé sur la calculatrice EAI 580 du laboratoire d'Automatique de l'Université de LILLE I.

Cette simulation analogique nous a permis d'étudier la sensibilité du système aux divers paramètres en fonctionnement dynamique d'une part $F_A(t)$ et sur le résultat final (Fech).

Les changements d'échelle effectués sont :

- pour les fractions de CO	: F_{CO}	1 UM = 1 %
- pour le volume	: V	1 UM = 10 l
- pour les pressions	: P	1 UM = 1000 mmHg

(1 UM = 1 unité machine)

Une valeur type permettant de tester la sensibilité des coefficients a été choisie pour les différents paramètres :

V_I = 5 l	$F_{I_{CO}}$ = 0,25 %	T_A = 10 sec.
V_R = 1,6 l	P_B = 760 mmHg	V_3 = 0,6 l
$\dot{V}_I$ = 2,5 l/sec.	ρ = 5,44 cmH_2O/l/sec.	V_4 = 1,3 l
$\dot{V}_E$ = 3 l/sec.	D = 30 ml/min/mmHg	

où V_3 est le volume de rinçage et $V_4 - V_3$ le volume de l'échantillon et T_A Le temps d'apnée.

La figure 3 concerne la courbe Fech = f (T_A) pour les valeurs types. On remarque une légère inflexion de la courbe vers 15 secondes. Elle est à rapporter au changement automatique de sensibilité introduit dans le système pour la valeur la plus faible des fractions échantillonnées. Les résultats obtenus par les divers essais corroborent l'expression analytique de la fraction échantillonnée.

La pression barométrique, la résistance des voies aériennes ainsi que le débit expiratoire n'ont que peu d'influence sur l'évolution du système.

La capacité de diffusion, le volume résiduel et le volume inspiré sont les 3 coefficients fondamentaux, car ils influencent très fortement la pente de la décroissance ainsi que la valeur initiale de la fraction échantillonnée.

La fraction inspirée et le débit inspiratoire n'ont d'influence que sur la valeur initiale du système.

La place et le volume de l'échantillon prélevé ont une importance plus grande qu'on ne pourrait s'y attendre. En effet, si leur influence sur la valeur de la fraction alvéolaire en fonction du temps est nulle (sauf par augmentation éventuelle du temps d'apnée) leur influence sur la valeur de la fraction échantillonnée est très grande aussi bien en ce qui concerne la valeur intitale que la pente de la décroissance de Fech = f (T_A). Ceci a déjà été souligné par différents auteurs et pourrait expliquer tout au moins en partie la disparité des résultats obtenus au moyen de différents modèles.

CALCUL AUTOMATIQUE DE LA CAPACITE DE DIFFUSION PULMONAIRE

L'expression analytique de notre modèle ne peut être inversée pour en extraire la capacité de diffusion ; pour cette raison nous nous sommes attachés à trouver un mode de calcul automatique de cette capacité à partir de la sortie du modèle et des paramètres réellement mesurés. Nous proposons 5 modes de calcul tous basés sur le principe de l'accumulateur analogique constitué de deux unités "track and store" montées en série et rebouclées sur l'entrée (fig.4) cet accumulateur permet d'entreprendre la recherche pas à pas du coefficient de diffusion.

La première idée est d'envoyer des incréments proportionnels à la différence entre la fraction expirée mesurée et la fraction échantillonnée calculée par le modèle. Le type de calcul (type 1) représenté sur la figure 5 ayant duré 43 périodes, nous n'avons pas recherché la proportion optimale qui aurait donné le temps de calcul le plus court.

La seconde idée est d'envoyer des incréments constants. Nous choisisons 3 types d'incréments : 0,1 UM - 0,01 UM - 0,001 UM pour qu'on obtienne des résultats avec une erreur absolue de 10^{-4} ou que l'accumulateur fournisse dix fois la valeur de la capacité de diffusion.

Les quatre calculs présentés ci-après sont des variations sur les commutations entre chaque pas de calcul :

- <u>calcul de type 2</u>

La commutation au pas inférieur se fait dès que Fech F_E et la commutation au plus petit pas si $95\ \% < F_E <$ Fech $< F_E$. Cet essai a duré 38 périodes et n'est guère plus concluant que le premier.

- <u>calcul de type 3</u>

On essaie d'améliorer le système avec commutation au plus petit pas quand $|\text{Fech} - F_E| < 5\ \%\ F_E$. L'essai dure 40 périodes. Cette contre-performance doit être liée à un mauvais réglage des seuils de commutation des divers comparateurs. De plus la précision demandée dans le cas présent est à peu près égale à la précision totale du calcul (présence de nombreux multiplieurs et diviseurs).

Nous décidons dans les deux derniers calculs de bloquer les incréments à leur plus petite valeur par mémorisation des commutations à l'aide de flip-flop.

- calcul de type 4

On passe à l'incrément 0,01 dès que $Fech > F_E$ et à l'incrément 0,001 dès que $Fech < F_E$ puis on oscille autour de la valeur finale avec ce dernier incrément. L'essai a duré 7 périodes. Essai assez satisfaisant que nous avons essayé d'améliorer.

- calcul de type 5 (fig. 6)

En posant $D_n = D_{n-1} + P$, deux solutions se présentent quand $Fech - F_E$ change de signe :

si $|Fech - F_E|_{n-1} \gg Fech - F_{E\,n} \rightarrow D_{n+1} = D_n - \frac{P}{10}$

si $|Fech - F_E|_{n-1} < Fech - F_{E\,n} \rightarrow D_{n+1} = D_{n-1} + \frac{P}{10}$

L'essai a duré 12 périodes. Deux faits peuvent expliquer ce résultat : d'une part, comme dans le cas des calculs 2 et 3, le seuil du comparateur est relativement imprécis, d'autre part la variation de Fech en fonction de D a une allure exponentielle ce qui explique l'impossibilité de traiter le problème de façon linéaire.

Le calcul de type 4 se trouve donc être le plus rapide et approxime la capacité de diffusion à 0,001 près (en unité machine). Afin d'afficher le coefficient pour une lecture directe sur un voltmètre il est multiplié par le rapport de son changement d'échelle à la tension de référence de la machine (ici 6/10) ; pour avoir la valeur réelle on multiplie par 100 la valeur lue sur le voltmètre numérique.

La nécessité pour ce type de modèle de devenir opérationnel nous obligeait à construire une maquette analogique. La difficulté de concilier qualité et prix des éléments nous avait fait opter pour l'élaboration d'une maquette de calcul incrémental qui se prête assez bien aux simulations continues et permet d'augmenter la précision des calculs surtout en ce qui concerne les multiplications. Cependant l'acquisition d'un matériel numérique assez important (calculateur MULTI 8 et chaine DIDAC 4000 Intertechnique) nous ont définitivement orientés vers la simulation numérique.

Les premiers résultats obtenus à partir du dépouillement de nos 796 dossiers expérimentaux nous conduisent à proposer un modèle non linéaire où $D = k\,V^{4/3}$ Un travail est actuellement en cours afin de vérifier la validité de ce modèle.

CONCLUSION

Le modèle que nous avons élaboré permet grâce aux méthodes modernes de calcul de se passer de modèles analytiques rigides et de décrire de façon dynamique le phénomène étudié. Les résultats obtenus nous encouragent à continuer dans cette voie afin de rendre le système opérationnel en l'introduisant dans une chaine de mesures comportant un transfert automatique des données.

Nous tentons également d'établir un modèle bi-alvéolaire qui nous permettra d'aborder le domaine pathologique.

Travail du Laboratoire d'Automatique (Professeur VIDAL) Université de LILLE I et du Laboratoire de Pathologie expérimentale (Professeur F.GUERRIN) C.H.U. - Université de LILLE II.
avec l'aide financière de l'INSERM. C.P. 7210706 Pr.F.GUERRIN

-=-=-

Bibliographie succinte

COTES J.E. : Lung Function : Assessment and application in medicine. 1st Ed. Blackwell Scientific Publications, Oxford 1965.

FORSTER R.E. : Interpretation of measurements of pulmonary diffusion capacity. Handbook of Physiology, Section 3. Respiration 2, 1453, 1964

KROGH M. : The diffusion of gases through the lungs of man. J. Physiol. (Lond) 49, 271, 1915

OGILVIE C.M., FORSTER R.E., BLAKEMORE W.S. and MORTON J.W. : A standardized breath-holding technique for the clinical measurement of the diffusing capacity of the lung for carbon monoxide. J. Clin. Invest. 36, 1, 1957

SILVIE D., ROBIN H., BOULENGUEZ C. et GUERRIN F. : Simulation du transfert pulmonaire du monoxyde de carbone et calcul automatique de la capacité de transfert. C.R. Collogque I.R.I.A. 2, 17, 1973.

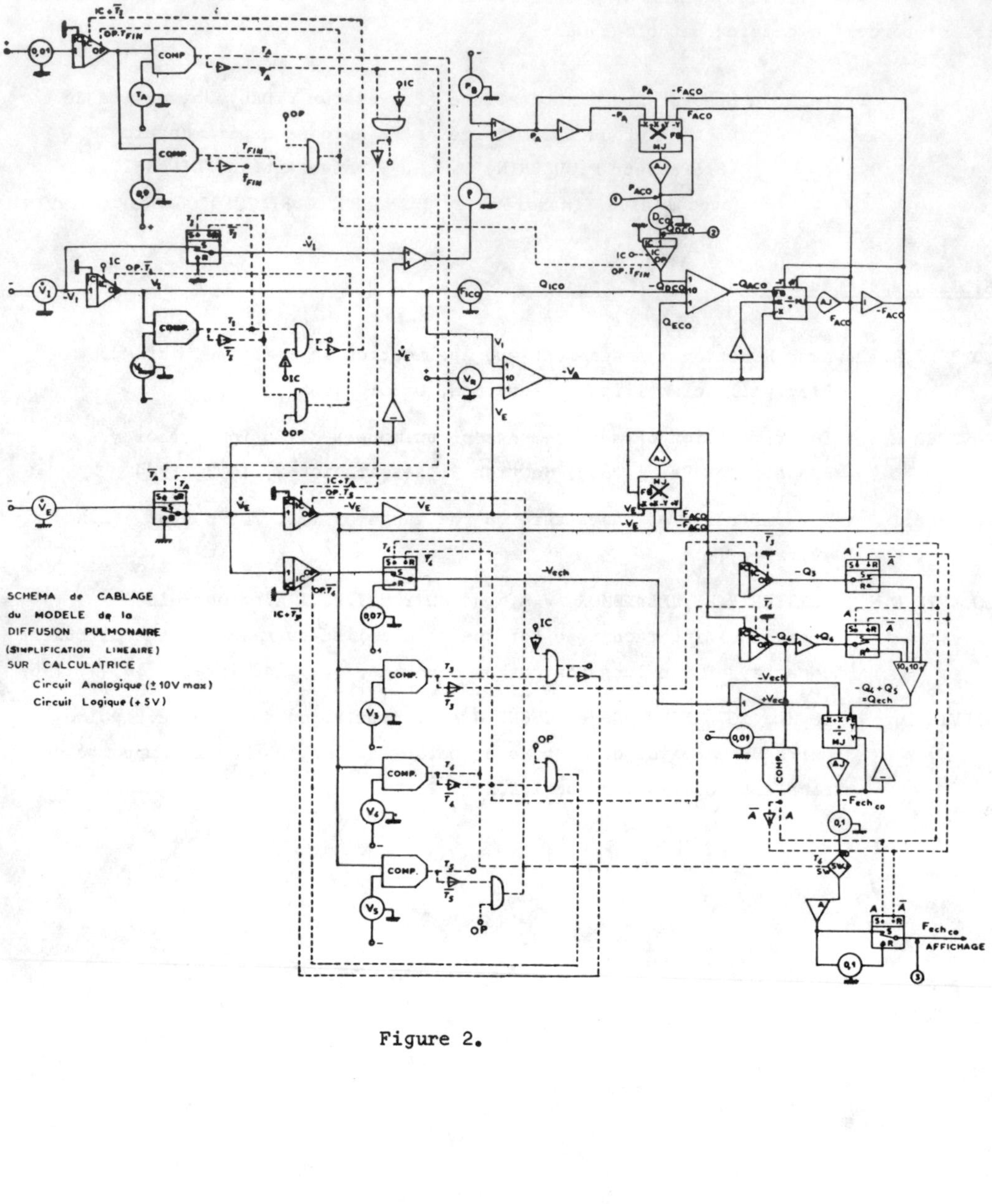
SCHEMA de CABLAGE
du MODELE de la
DIFFUSION PULMONAIRE
(SIMPLIFICATION LINEAIRE)
SUR CALCULATRICE
Circuit Analogique (± 10V max)
Circuit Logique (+ 5V)
COMP
AFFICHAGE

Figure 2.

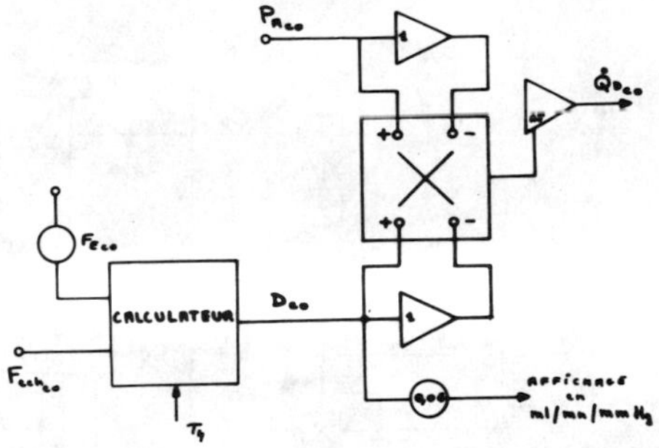

Figure 4. Schéma de principe
du calcul automatique de DpCO

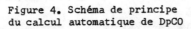

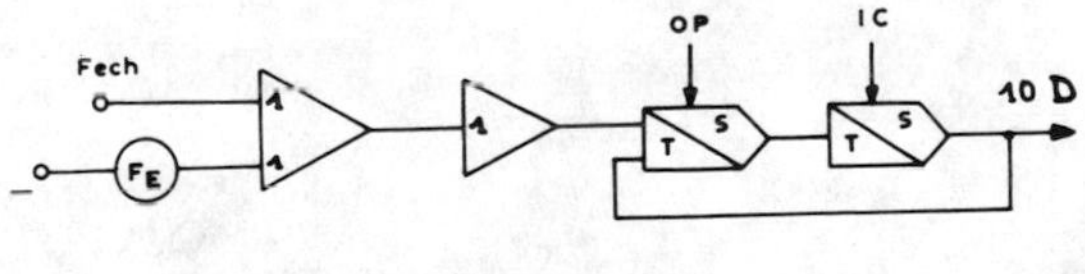

Figure 5. Calcul pas à pas de
type 1.

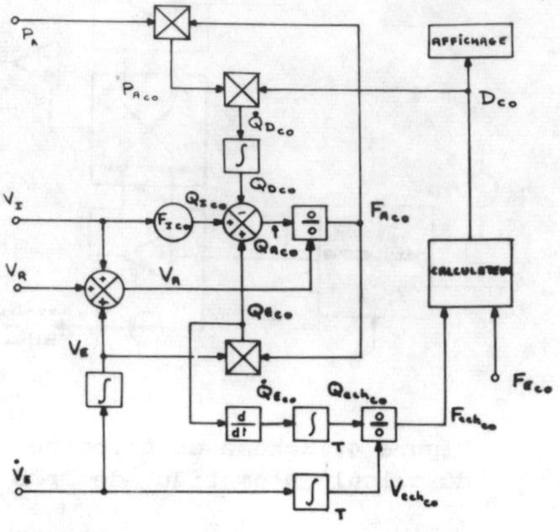

Figure 1. Modèle théorique

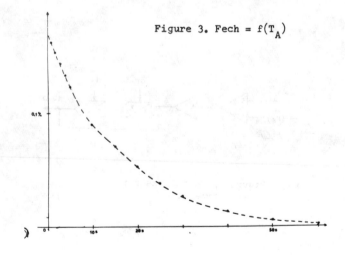

Figure 3. Fech = $f(T_A)$

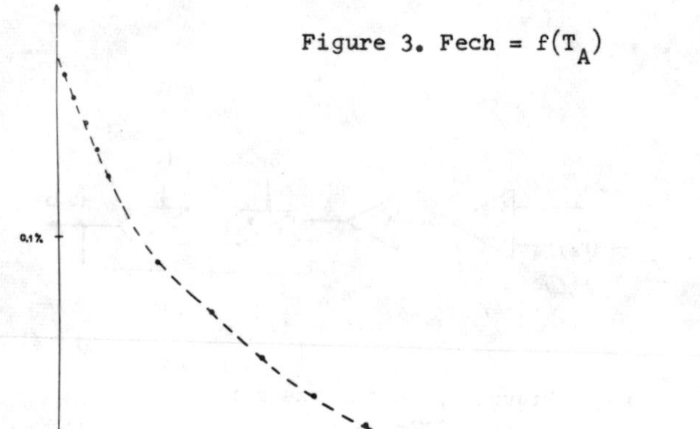

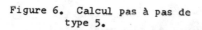

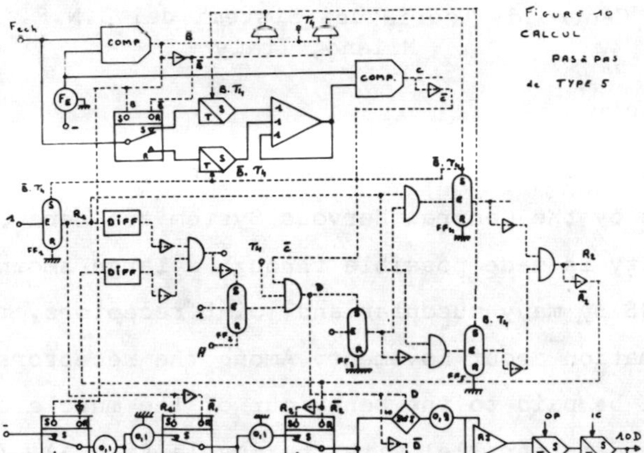

Figure 6. Calcul pas à pas de type 5.

ON SOME MODELS OF THE MUSCLE SPINDLE

C. Badi, G. Borgonovo, L. Divieti

Istituto di Elettrotecnica ed Elettronica
del Politecnico di Milano

Centro di Teoria dei Sistemi del C.N.R.
Milano, Italy

1. Introduction

The problems connected with the movement and postural control exerted on the muscles by the Central Nervous System are here considered. This control activity is made possible through a large amount of signals sent to the CNS by many muscular and joint receptors, that give a continuous information about movement. Among the receptors, particular attention will be paid to the behaviour of the muscle spindle, a receptor connected in parallel with the muscle (fig.1). This study will permit to give a first partial model of the internal structure of the CNS considered as a controller of the whole muscular system.

2. Anathomy and physiology of the system

The CNS can be considered as a controller; in so doing it will not be of interest to give a detailed anathomical and physiological description of each part. For the present purposes one can say that peripheral signals coming from receptors have a first integration level at the spinal cord level, that is the point where the innest loops close (fig.2) (see par.6); obviously many signals go to higher levels, that is brain stem, sensory motor cortex and cerebellum; they all contribute to give a conscious (motor cortex) or inconscious (cerebellum) response to peripheral signals after elaborating them, and then send down output signals to control movements. This work tries to give a model of a part of this controller; it must be noted that many interactions between input and output signals will be strongly

fig.1

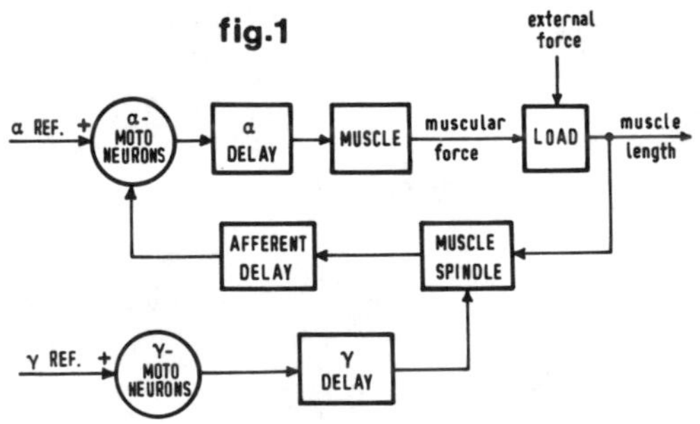

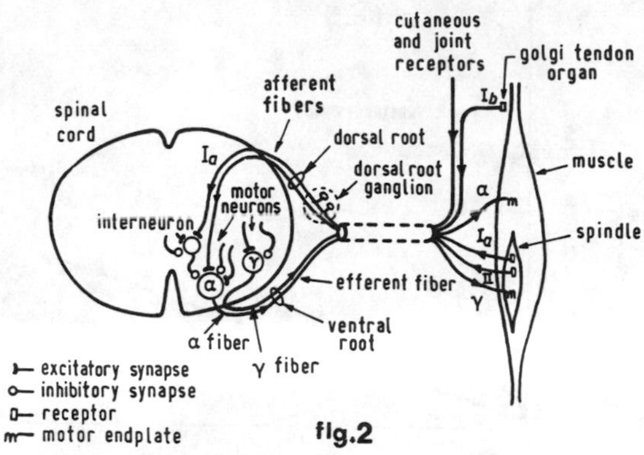

fig.2

simplified or even neglected. With regard to the muscle structure, it is enough to say that it is a viscoelastic system composed with many fibres that are contractile under α-motoneural stimulation; many mechanical models of the muscle are available; the simplest one, that is a time constant, will be used for the present modeling problem. In this model the input is the α-stimulation and the output is the muscular force acting on the system dynamics (limb weight etc.). The muscle spindle is a receptor connected in parallel with the contractile fibres of voluntary muscles; its function consists in giving information about applied stretches and their velocity (1) (fig.3). Spindles are composed with many fibres that can be subdivided into two groups with regard to their anathomical and physiological properties (1) (14) :

1) Nuclear bag fibres 2) nuclear chain fibres

The first ones have a central "bag", few myofibrils and a large number of nuclei; their length and thickness are larger that those of the second ones, which also have an uniform distribution of nuclei and

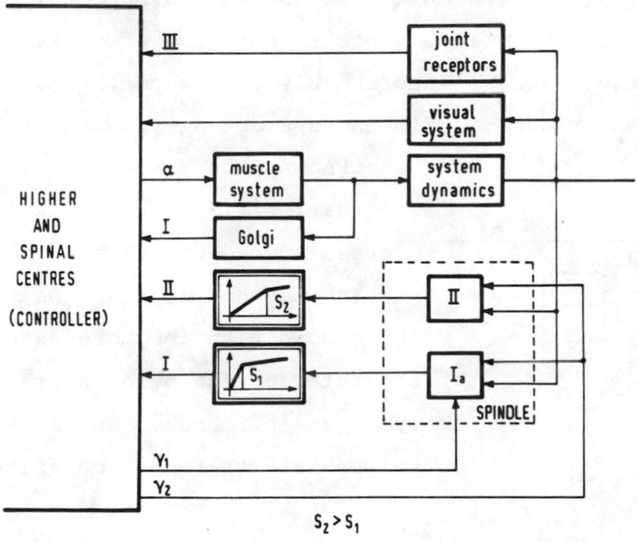

fig.3

fig.4

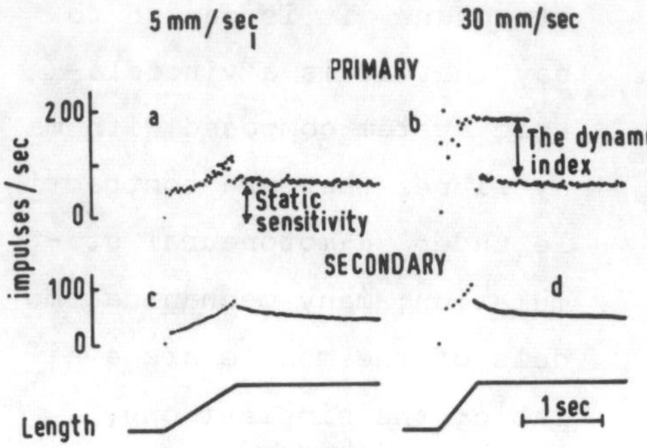

fig.5

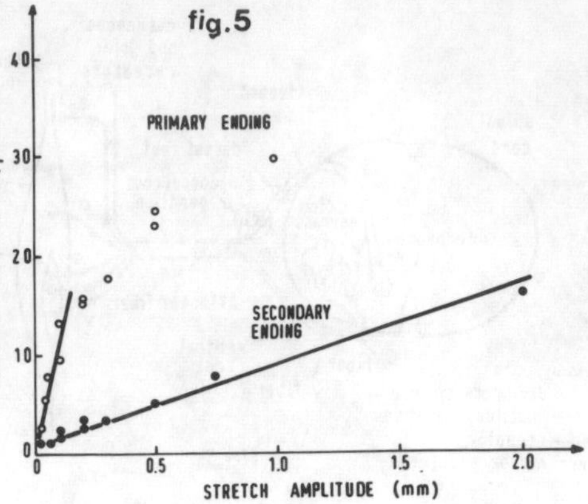

myofibrils.

The efferences to the muscle spindle consist on γ-motor fibres; with reference to Boyd (15) can be classified as γ_1 and γ_2 ; the first ones have plate endings particularly on the polar region of nuclear bag fibres; the second ones have trail endings distributed along the whole length of nuclear chain fibres. With regard to the γ-fibres, it is known that many of them can converge on the same spindle, though it has not yet been clarified either the ratio between the number of γ_1 and γ_2 motor fibres acting on a single spindle, or their highest total number. The first group has been called "dynamic", since upon stimulation of these fibers the spindle sensory discharge increase markedly, particularly when the muscle is suddenly stretched. The other group has been called "static", since stimulation of these fibers increases the spindle discharge appreciably only when the muscle is at rest. Two different kinds of afferences starting from the spindle are also known: the primaries (Ia) and the secondaries (II). The primaries are wrapped around the equatorial region of nuclear bag and nuclear chain fibres. The secondaries are more extensively distributed on the other parts ot the nuclear chain fibres. Generally only one primary starts from each spindle, while

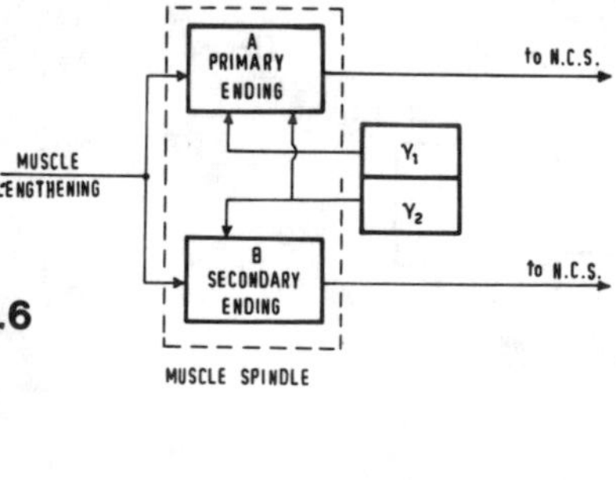

fig.6

fig.7

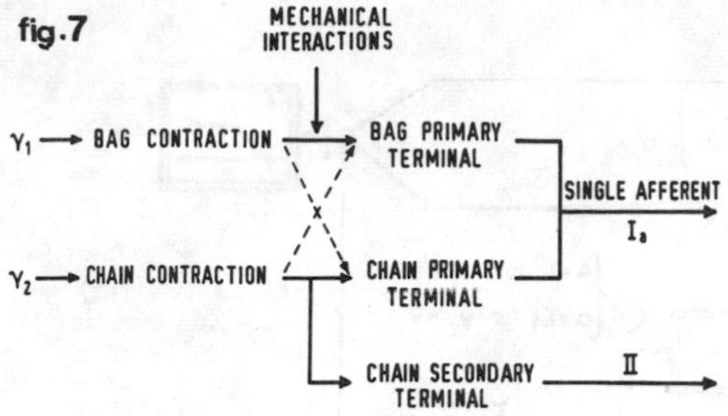

fig.8

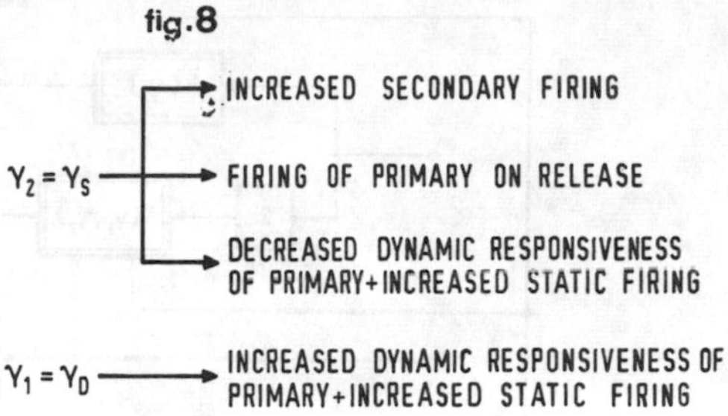

the secondaries do not undergo this limitation in number. Another anathomical difference between primary and secondary endings consists in a greater conduction velocity and section of the first ones than the second ones.

As previously said, the spindle measures lengthenings and rate of lengthenings; that is it can be considered a sort of physiological derivative gauge. More precisely, primary endings mostly show derivative effects while the secondary endings are mostly proportional to lengthenings. On their ramp response some characteristic spindle parameters can be defined (fig.4); moreover, the first ones have a smallest linearity zone of stretch amplitudes than the second ones (fig.5). A first block diagram of the spindle is shown in fig. 6 : so doing it is possible to build two separate models for primary and secondary endings (fig.6) (9) . The linear models of the first ones are based on the transfer function: $F(s) = H+K\,s$; (8) this is the simplest model that was furtherly improved through many other mechanical models (4) (10) (5) . But it is possible to show that all these models have a bad performance at the lower frequencies and within the physiological amplitudes zone. For this reasons a nonlinear model of primary ending was proposed (18) (24). Γ-stimulation induces contractions in spindle fibers; this modifies the parameters previously defined, so that also the the spindle response shows some changes (figg. 7-8). Yet considering γ_1 and γ_2 stimulation, it can be observed that they both act mostly contemporarily on the spindle, so that they could even cancel their effects each other.

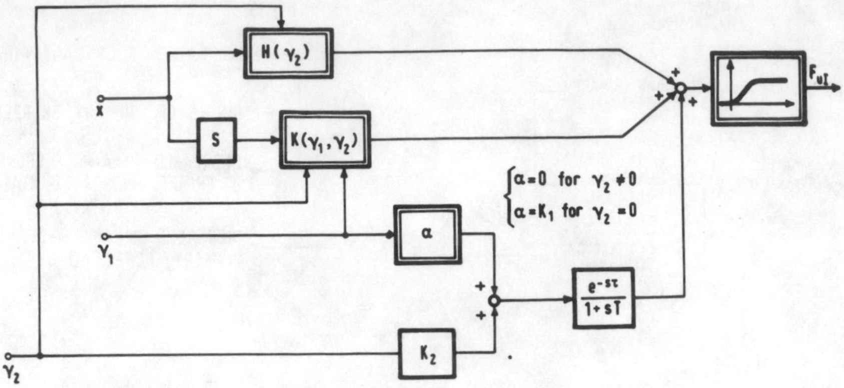

$\tau = 10$ m sec
$T = 25\text{-}30$ m sec
$K_1 = 70$
$K_2 = 120$
} adult cat intercostal muscles

fig.9

This makes difficult measuring "in vivo" contemporary γ_1 and γ_2 action. However it is possible to establish a particular type of $\gamma_1 - \gamma_2$ interaction, that is "occlusion" (12). With this word is defined the effect that when both γ_1 and γ_2 are present, the resultant effect on the spindle response and parameters is that one could observe in presence of γ_2 stimulation only. With only γ action on the spindle and no changes in lenght, the spindle shows a linear response, described with the transfer function $\frac{e^{-\tau s}}{1+sT}$ suitable for both γ_1 and γ_2 stimulation. (fig.9) (13). This response is still small if compared with that obtained under external imposed lengthenings.

3. The tendon jerk and Ia fibres role

The muscle receptors clearly exist in order to feed back information to the CNS about the mechanical state of muscle. For several reasons the various computations required to exert effective motor control may be expected to be carried out at a hierarchy of levels in the nervous system and, correspondingly, the central projection of the information carried by the muscle afferents may be expected to be widespread; in particular, suitable spinal reflex connections might enable the muscle afferents to be used to produce rapid automatic corrections to the motor outflow, leaving it to the higher centres to take slower action on the basis of information received from a wider field.

At a first glance, the initial central evaluation of, and response to, all this information would seen to be prerogative of the spinal cord for this would avoid any further delay in simple conduction; particular interest therefore attaches to the study of the direct reflex actions of the three main outputs of the muscle receptors: that is Ia, II and Ib fibres. In other terms it is of interest testing the most simple closed loop in the whole muscle control mechanism; for this purpose, the study of the tendon jerk is comparable to get the impulse response of a linear control system. The tendon jerk consist in the fact that, when the muscle is stretched with a sudden tap, there is a consequent muscle contraction due to a feedback excitation of the α-motoneurones.

For many years there were some uncertainties about what afferent pathways are interested in the tendon jerk, but now it can be said that the Ia fibres can produce the tendon jerk unaided by other afferent inputs. In fact no other receptor is significantly excited by a brief tap of a previously non contracting muscle, and even if it were, the resulting discharge would arrive too late to be able to influence the motoneurones before they fired under Ia excitation; it was also shown that the whole tendon jerk happens without γ stimulation, that could be, at a first glance, considered to contribute to this reflex. This demonstration does not of itself exclude the possibility of the same Ia further exciting, in other phenomena, the α-motoneurones via polysynaptic patways.

fig.10

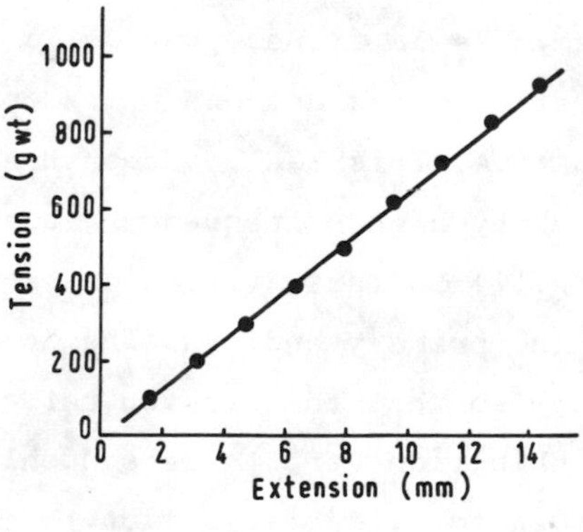

fig.11

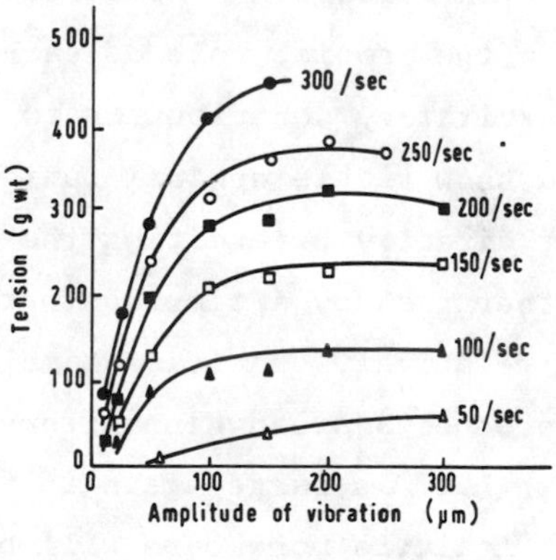

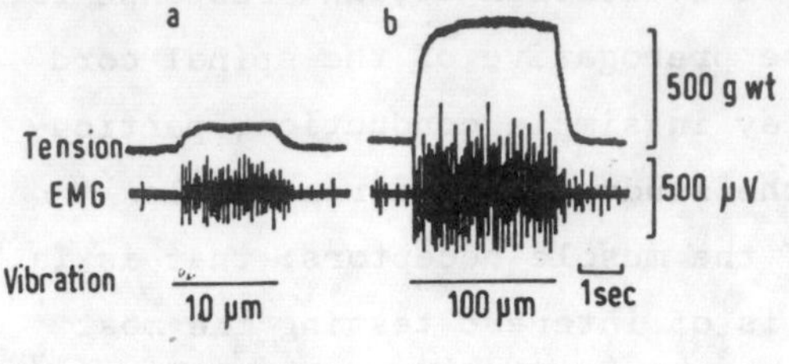

fig.12

4. The stretch reflex

The stretch or myotatic reflex was demonstrated by showing that an extensor muscle in the decerebrate produced a greater tension when it was stretched with the reflex arc intact than it did when its nerve was cut (6) (17). Stretch reflexes seem to be largerly the prerogative of extensor muscles, though weaker ones may also be seen in flexors. The stretch reflex is a localised reflex, for on stretching one head of a muscle it is rarely overtly present in another head of the same muscle. It is rather closely related to the magnitude of the stimulus. On applying a series of stretches of increasing amplitude of several millimetres extent the reflexly elicited tension is often found to be almost linearly related to the extension (fig.10). However, the reflex response evoked by small alternating stretches and releases is often larger than might have been expected on the basis of the response to large stretch reflexes (fig.11). (20) (7). In this respect the behaviour of the stretch reflex seems to have the valuable property of being at its best in resisting disturbance from a set point, but without being driven into the saturation which would be shown if the same high sensitivity were maintained for stretches of all sizes. Steady stretch of a muscle excites a steady afferent discharge in each of the three main types of muscular receptors fibres (that is Ia, II, Ib); each of them may be expected to play some part in modulating the stretch reflex: the present problem is to determine the precise role of each of them. The Ia fibres provide a steady excitatory contribution to the stretch reflex and one which appears to show little central fatigue with time. This can be demonstrated most directly by exciting the Ia fibres by high - frequency vibration rather than by a simple stretch; (fig.12) in fact, vibration provides a relatively specific stimulus for the primary endings. The Ib fibres produce a steady inhibitory feedback so that the stretch reflex is abouth half as large again if the Ib inhibition were to be suddenly removed. A little more care will be devolved to II fibres action.

fig.13

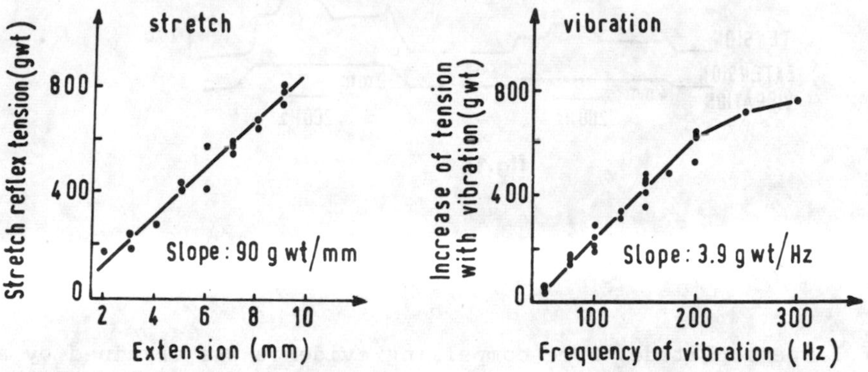

5. <u>II fibres and their excitatory feedback action</u>

The experiments using electrical stimulation of nerve trunks led to the belief that the secondary fibres of the spindle merely produced a non-specific flexor reflex. Laporte and Bessou (21) found that when the Ia fibres were blocked electrically, moderate stretching of soleus produced autogenetic inhibition of its motoneurones. This effect seemed to be caused by II afferents on α-motoneurones. In the latest years was found that in the decerebrate cat the spindle group II fibres of extensor muscles produced autogenetic excitation rather than inhibition. The starting point for the suggestion was the new finding that in the decerebrate cat the reflex response to simple stretch was often rather stronger than the reflex response of the same muscle to high frequency vibration (22). Stretch excites both Ia and II fibres whereas vibration powerfully excites the Ia fibres without significantly affecting the group II fibres.

Thus, if the II fibres should indeed be excitatory, stretch would be a more potent stimulus than vibration. Figure 13 compares the relative strengths of the two kinds of reflex and thereby furthers the arguments; one can note that the stretch reflex tension is approximately linearly related to the extension and the vibration reflex response is also linearly related to the frequency of vibration. On this basis one can estimate the increase of Ia firing required to produce, a required stretch reflex: one can say that the experimentally observed values in comparable experiments are often very much lower for stret

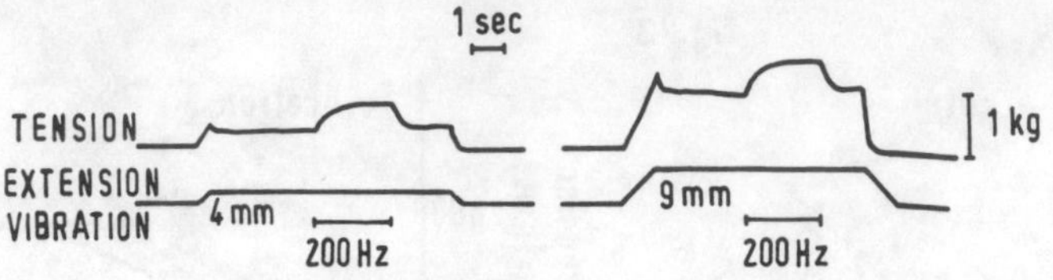

fig.14

ches of large amplitude. More compelling evidence was obtained by studying the interaction between the reflex response to stretch and vibration. If they both depended only upon the Ia pathway any increase in the stretch reflex with increasing stretch should occlude the response to vibration by an equivalent amount. This is because any increase in Ia firing with stretch will leave so much less additional excitatory effect available for the action of vibration on the Ia fibres. However, as shown in fig. 14 no occlusion occurs on combining stretch and vibration. There seems no explanation for this if the Ia fibres provides the only excitatory component of the stretch reflex, but any problem is removed if the spindle II fibres also produce excitation. The attribution of the missing excitatory action to the II fibres rests on argument by exclusion, for no other kinds of afferent fibres are suitably stretch sensitive.

Further support for an excitatory action of the II fibres has been obtained by studying the effect of the application of procaine (fig.15) to the nerve to the soleus to the combined stimuli of stretch and vibration (23). Progressive procaine paralysis diminishes the reflex response to simple stretch at a time whenthe α-motor fibres appear to be still conducting quite normally. Experiments have shown that the γ-efferents are paralysed long before α-motoneurones and Ia afferents, and that these two groups of large fibres are about equally sensitive to the procaine. Thus the initial early decline in the stretch reflex appears at least partly due to an equivalently early paralysis of the efferents. This would abolish the pre-existing γ bias of the spindle and would thus lead to a decrease in the spindle afferent firing elicited by a standard streetch. The early decline could well be entirely attributed to a reduction of Ia firing, but such a suggestion is not acceptable beacause procaine causes a similar early diminution in the combined reflex response to stretch plus vibration. The amplitude of vibration is presumed sufficiently large to ensure that all Ia fibres are driven to discharge at the vibration frequency. But vibration has no action on secondary endings so that during the combined stimuli of stretch plus vibration, efferent paralysis will, as usual, reduce the response of II fibers to the stretch component of the combined stimuli. Thus experimental evidence supports the belief that II fibres may make an appreciable contribution to the stretch reflex.

6. <u>The model</u>

Basing on the previous physiological data on Ia and II fibres reflex action, the proposed model is reported in fig. 16; Ia, II signals, α reference coming from the CNS and possibly Ib from Golgi receptor, converge upon a summing point to give the resulting α stimulation to the muscle. The Ia monosynaptic loop is straightforward after past works on tendon jerk and the stretch reflex; one must observe that just because of the derivative high sensitivity of Ia output from the spindle, this feedback carries signals that take into account vibrations where there are small variations in lengthening but high frequencies. When, as previously described, Ia fibers give a maximal reflex response this does not mean that α motoneurones cannot increase their firing; in fact II fibres, merely sensitive to stretch in the physiological conditions, send to the summing point their signal that contributes to further increasing of α discharge. Thus a correspondence with physiological data is reached. In this

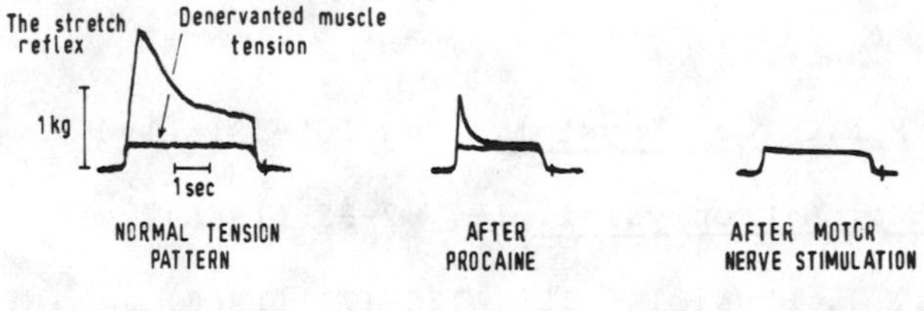

fig.15

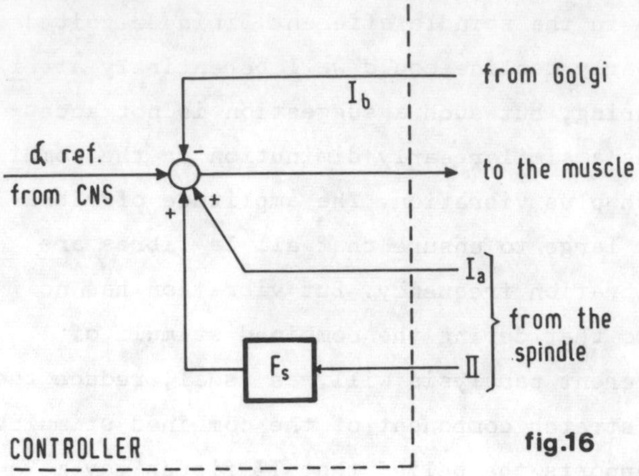

fig.16

model one must take into account the different saturation levels of Ia and II fibres that contribute to explain the greater aptitude of II fibers to measure large lengthenings. The summing point does not necessarily represent a single monosynaptic junction; in fact it is possible that II fibres act through a polisynaptic pathway. A digital computer simulation of this model has been performed; as a first step, the simplest linear models for the muscle and spindle have been adopted, that is, for the first one a time constant, and for the spindle a derivative block in parallel with a gain representing Ia transfer function: while a simple gain F_s for the II response signal has been used. First results about tendon jerk simulation and the stretch reflex response are in fair agreement with physiological data; nevertheless one must observe that nonlinearities of the various blocks of the system play an important role in the real behaviour of posture and movement control. In the future checks of this model we shall take into account that:

1) The spindle itself, particularly for Ia output, has a nonlinear behaviour.
2) There is a continuous interaction between α and γ signals (α - γ_2 linkage) and a continuous parameter adjustment via γ_1 and γ_2 on Ia and II outputs.

References

(1) Crowe, A., Matthews, P.P.C., J. Physiol. 174, 109-131 (1964)

(2) Stuart, D., et Al., Exp. Neurophysiol. 13, 82-95 (1965)

(3) Appelberg, B., et Al., J. Physiol. 185, 160-171 (1966)

(4) Gottlieb, G.L., et Al., IEEE Trans. MMS-10, 17-27 (1969)

(5) Angers, D., Delisle, G.Y., IEEE Trans.BME-18 175-180 (1971)

(6) Lippold, O.C.J., et Al., J. Physiol. 144, 373-386 (1958)

(7) Bianconi, R., Van Der Meulen, J.P., J. Neurophysiol. 26, 177--190 (1963)

(8) Poppele, R.E., Terzuolo, C.A., Science 159, 743-745 (1968)

(9) Matthews, P.B.C., Stein, R.B., J. Physiol. 200, 723-743 (1969)

(10) Poppele, R.E. Bowman, R.J., J. Neurophysiol. 33, 59-72 (1970)

(11) Lennerstrand, G., Thoden, U., Acta Physiol Scand. 74, 30-49(1968)

(12) Lennerstrand, G., Thoden, U.,Acta Physiol.Scand.74, 153-165(1968)

(13) Andersson, B.F., et Al., Acta Physiol. Scand. 74, 301-316 (1968)

(14) Matthews, P.B.C. Mammalian Muscle Receptors and their Central Actions, 67-95 (1972)

(15) Boyd, I.A., J. Physiol. 153, 23-24P (1960)

(16) Boyd, I.A., et Al., The Role of the Gamma System in Movement and Posture, 29-48 (1968)

(17) Boyd, I.A., et Al., The Role of the Gamma System in Movement and Posture, 40-70 (1962)

(18) Badi, C., Borgonovo, G., Divieti, L. Atti del II° Congresso Nazionale di Cibernetica - Casciana Terme, 287-296 (1972)

(19) Matthews, P.B.C., J. Physiol. 147, 521-546 (1959)

(20) Matthews, P.B.C., J. Physiol. 184, 450-472 (1966)

(21) Laporte, Y., Bessou, P., J. Physiol. Paris 51, 897-908 (1959)

(22) Grillner, S. Acta Physiol. Scand. 78, 431-432 (1970)

(23) McGrath, G.J., Matthews, P.B.C., J. Physiol. 210, 176-177P (1970)

(24) Badi, C., Borgonovo, G. Divieti, L. Il fuso muscolare Internal report LCA 73-2 Ist. Elettrotecnica ed Elettronica - Politecnico di Milano (1973).